PLASTICITY: Theory and Application

PLASTICITY: Theory and Application

ALEXANDER MENDELSON
Professor of Civil Engineering
Case Western Reserve University

ROBERT E. KRIEGER PUBLISHING COMPANY
MALABAR, FLORIDA

Original Edition 1968
Reprint Edition 1983 w/new preface, 1986

Printed and Published by
ROBERT E. KRIEGER PUBLISHING COMPANY, INC.
KRIEGER DRIVE
MALABAR, FLORIDA 32950

Copyright © 1968 by Alexander Mendelson
Reprinted by Arrangement

All rights reserved. No part of this book may be reproduced in any form or by any electronic or mechanical means including information storage and retrieval systems without permission in writing from the publisher.

Printed in the United States of America

Library of Congress Cataloging in Publication Data

Mendelson, Alexander.
   Plasticity : theory and application.

   Reprint. Originally published: New York : Macmillan, 1968. (Macmillan series in applied mechanics) With new pref.
   Bibliography: p.
   Includes index.
   1. Plasticity. I. Title. II. Series: Macmillan series in applied mechanics.
TA418.14.M4 1983   620.1'1233   82-21231
ISBN 0-89874-582-9

10 9 8 7 6 5

*To my wife Charlotte
for her patience, understanding,
and encouragement*

PREFACE

With the advent of the jet age followed closely by the space age, the theory of plasticity has been brought forcibly into the forefront of engineering application and design. Modern aircraft, missiles, and space vehicles must be designed on the basis of minimum weight, which invariably means designing into the plastic range to obtain maximum load to weight ratios. Moreover, the facts of economic life have made the saving of material and more efficient design a necessity for even the more earthbound industrial applications.

This book is the result of the author's teaching for several years of a graduate course in plasticity for engineers at Case Western Reserve University. It was soon realized that although a number of excellent books on plasticity were available, none of them adequately met the requirements of the course. The available books were either too theoretical and mathematical for the average engineer and designer, or the main emphasis was placed on problems of large plastic deformations such as are encountered in metal-forming processes. Very little has been published in textbook form on the most important class of elastoplastic problems, in which the plastic strains are of the same order of magnitude as the elastic strains, which are of such prime concern to today's engineer. Furthermore, where such problems are treated, the usual assumptions of perfect plasticity are used, no attempt being made to take into account the strainhardening properties of real materials.

The present book includes the basic theory of plasticity but places primary emphasis on the solution of elastoplastic problems for materials with strain hardening. In particular, it should be emphasized that with the present availability of high-speed computing facilities, many of the simplifying assumptions hitherto commonplace in plasticity calculations are no longer necessary.

Although the book first appeared in print more than a decade ago, it would appear to be as relevant or even more so today than at the time. This is due to the fact that basic plasticity theory hasn't changed very much in the past decade. Furthermore, the new generations of computers have made the methods and techniques, described herein, even more appealing and viable than heretofore for engineers and designers. Thus, for example, elastoplastic solutions for three-dimensional bodies with crack have recently been obtained with very low computing cost.

Following a brief introduction, Chapter 2 discusses some of the basic experiments concerning the elastoplastic behavior of metals. Chapters 3 and 4 describe the basic properties of the stress and strain tensors. Tensor notation is introduced and is frequently used together with the longhand notation, but a knowledge of tensor properties is not needed. Chapter 5 describes briefly the elastic stress-strain relations. Chapter 6 discusses the various yield criteria and their experimental verification. In Chapter 7 the plasticity flow rules, or stress-strain relations, are derived and discussed, including a new set of equations which relate plastic strain increments to total strains rather than stresses.

A series of practical problems for both ideally plastic and strain-hardening materials is presented in Chapters 8 through 11. Chapter 8 deals with problems of spheres and cylinders. Chapter 9 is devoted entirely to the powerful method of successive elastic solutions, by means of which a large class of otherwise intractable problems can be solved. First introduced by Ilyushin some thirty years ago, this method has not yet gained wide acceptance in this country and, to the author's knowledge, is not even mentioned in any other current book in the English language.

Chapter 10 discusses plate problems, both for the plane stress and plane strain cases. Chapter 11 gives the general solution to the elastoplastic torsion problem. The theory of the slip-line field as applied to the plane strain problem of plastic-rigid materials is then presented in Chapter 12, and limit analysis of framed structures in Chapter 13. Chapter 14 discusses problems of creep at elevated temperatures and shows how the previously discussed plasticity methods can be applied to creep problems. It is realized that to treat these last three subjects adequately would require a book for each of them. It is hoped, however, that sufficient information is furnished herein to provide the reader with a worthwhile introduction to, and basic understanding of these subjects.

Preface

In the author's experience the material included can be covered adequately in a one-semester graduate course. Chapers 3 through 5 may be omitted by those familiar with basic elasticity theory. Sections 6.3, 6.5, 7.6, 7.8, 12.6, 12.8, and 13.6 may also be omitted on a first reading or if time is short. It is hoped that this book will be found useful as a graduate text and as an aid to engineers and designers faced with the problem of designing into the plastic range.

CONTENTS

Chapter 1. Introduction   1

Chapter 2. Basic Experiments   4
    2-1 Tensile Test   4
    2-2 True Stress–Strain Curve   7
    2-3 Compression Test and the Bauschinger Effect. Anisotropy   13
    2-4 Effects of Strain Rate and Temperature   15
    2-5 Influence of Hydrostatic Pressure. Incompressibility   16
    2-6 Idealization of the Stress–Strain Curve. Dynamic and Kinematic Models   16
    2-7 Empirical Equations for Stress–Strain Curves   20

Chapter 3. The Stress Tensor   24
    3-1 Tensor Notation   25
    3-2 Stress at a Point   27
    3-3 Principal Stresses. Stress Invariants   30
    3-4 Maximum and Octahedral Shear Stresses   34
    3-5 Mohr's Diagram   37
    3-6 Stress Deviator Tensor   39
    3-7 Pure Shear   41

Chapter 4. The Strain Tensor — 44
- 4–1 Strain at a Point — 44
- 4–2 Physical Interpretation of Strain Components — 48
- 4–3 Finite Deformations — 51
- 4–4 Principal Strains. Strain Invariants — 53
- 4–5 Maximum and Octahedral Shear Strains — 55
- 4–6 Strain Deviator Tensor — 58
- 4–7 Compatibility of Strains — 59

Chapter 5. Elastic Stress–Strain Relations — 64
- 5–1 Equations of Elasticity — 64
- 5–2 Elastic Strain Energy Functions — 67
- 5–3 Solution of Elastic Problems — 68

Chapter 6. Criteria for Yielding — 70
- 6–1 Examples of Multiaxial Stress — 70
- 6–2 Examples of Yield Criteria — 71
- 6–3 Yield Surface. Haigh–Westergaard Stress Space — 79
- 6–4 Lode's Stress Parameter. Experimental Verification of Yield Criteria — 88
- 6–5 Subsequent Yield Surfaces. Loading and Unloading — 92

Chapter 7. Plastic Stress–Strain Relations — 98
- 7–1 Distinction between Elastic and Plastic Stress–Strain Relations — 98
- 7–2 Prandtl–Reuss Equations — 100
- 7–3 Plastic Work. Two Measures of Work Hardening — 104
- 7–4 Stress–Strain Relations Based on Tresca Criterion — 108
- 7–5 Experimental Verification of Prandtl–Reuss Equations — 109
- 7–6 General Derivation of Plastic Stress–Strain Relations — 110
- 7–7 Incremental and Deformation Theories — 119
- 7–8 Convexity of Yield Surface. Singular Points — 121
- 7–9 Plastic Strain–Total Strain Plasticity Relations — 123
- 7–10 Complete Stress–Strain Relations. Summary — 127

Chapter 8. Elastoplastic Problems of Spheres and Cylinders — 135
- 8–1 General Relations — 135
- 8–2 Thick Hollow Sphere with Internal Pressure and Thermal Loading — 138

Contents   xiii

    8–3   Hollow Sphere. Spread of Plastic Zone.
           Pressure Loading Only   141
    8–4   Hollow Sphere. Residual Stresses. Pressure Loading   145
    8–5   Hollow Sphere. Thermal Loading Only   148
    8–6   Hollow Sphere of Strain-Hardening Material   150
    8–7   Plastic Flow in Thick-Walled Tubes   156

Chapter 9. The Method of Successive Elastic Solutions   164
    9–1   General Description of the Method   164
    9–2   Thin Flat Plate   172
    9–3   Thin Circular Shell   183
    9–4   Long Solid Cylinder   193
    9–5   Rotating Disk with Temperature Gradient   197
    9–6   Circular Hole in Uniformly Stressed Infinite Plate   208

Chapter 10. The Plane Elastoplastic Problem   213
    10–1   General Relations   213
    10–2   Elastoplastic Thermal Problem for a Finite Plate   218
    10–3   Elastoplastic Problem of the Infinite Plate with a Crack   223
    10–4   Strain-Invariance Principle   230

Chapter 11. The Torsion Problem   234
    11–1   Torsion of Prismatic Bar. General Relations   234
    11–2   Elasticity Solutions   240
    11–3   Membrane Analogy   245
    11–4   Elastoplastic Torsion. Perfect Plasticity   246
    11–5   Elastoplastic Torsion with Strain Hardening   248
    11–6   Bar with Rectangular Cross Section   250
    11–7   Bar with Circular Cross Section   253

Chapter 12. The Slip-Line Field   260
    12–1   Plane Strain Problem of a Rigid Perfectly Plastic Material   260
    12–2   Velocity Equations   266
    12–3   Geometry of the Slip-Line Field   268
    12–4   Some Simple Examples   272
    12–5   Numerical Solutions of Boundary-Value Problems   276
    12–6   Geometric Construction of Slip-Line Fields   279
    12–7   Complete Solutions. Upper and Lower Bounds   284
    12–8   Slip Lines as Characteristics   285

Chapter 13. Limit Analysis   300

    13–1   DESIGN OF STRUCTURES   300
    13–2   SIMPLE TRUSS   301
    13–3   PURE BENDING OF BEAMS   305
    13–4   BEAMS AND FRAMES WITH CONCENTRATED LOADS   307
    13–5   THEOREMS OF LIMIT ANALYSIS   312
    13–6   METHOD OF SUPERPOSITION OF MECHANISMS   318
    13–7   LIMIT DESIGN   323

Chapter 14. Creep   327

    14–1   BASIC CONCEPTS   327
    14–2   MULTIDIMENSIONAL PROBLEMS   331
    14–3   UNIAXIAL CREEP IN INFINITE STRIP   333
    14–4   CREEP IN ROTATING DISKS   335

Index   347

CHAPTER **1**

INTRODUCTION

The history of plasticity as a science began in 1864 when Tresca [1] published his results on punching and extrusion experiments and formulated his famous yield criterion. A few years later, using Tresca's results, Saint-Venant [2] and Lévy [3] laid some of the foundations of the modern theory of plasticity. For the next 75 years progress was slow and spotty, although important contributions were made by von Mises [4], Hencky [5], Prandtl [6], and others. It is only since approximately 1945 that a unified theory began to emerge. Since that time, concentrated efforts by many researchers have produced a voluminous literature which is growing at a rapid rate. Brief but excellent historical sketches are furnished by Hill [7] and Westergaard [8].

The theories of plasticity fall into two categories: *physical theories* and *mathematical theories*. The physical theories seek to explain why metals flow plastically. Looking at materials from a microscopic viewpoint, an attempt is made to determine what happens to the atoms, crystals, and grains of a material when plastic flow occurs. The mathematical theories, on the other hand, are phenomenological in nature and attempt to formalize and put into useful form the results of macroscopic experiments, without probing very deeply into their physical basis. The eventual hope, of course, is for a merger of these two approaches into one unified theory of plasticity which will both explain the material behavior and provide the engineer and scientist with the necessary tools for practical application. The present treatise is concerned with the second of these categories, i.e., the mathematical theories of plasticity and their application, as distinct from the physical theories. The latter belong to the realm of the metal physicist or solid-state physicist.

We start by defining roughly and intuitively what is meant by a metal flowing plastically. If one takes a thin strip of a metal such as aluminum and clamps one end and applies a bending force to the other end, the end of the strip will deflect. Upon removal of this force, if this force is not too large, the end of the strip will spring back to its original position, and there will be no apparent permanent deformation. If a sufficiently large load is applied to the end, the end will not spring back all the way upon the removal of the load but will remain permanently deformed, and we say that *plastic flow* has occurred. Our objective in this case will not be to determine why the permanent deformation took place but to describe what has happened in terms of stresses, strains, and loads. Solutions of this particular problem can be found, for example, in references [9] and [10].

In short, plasticity is the behavior of solid bodies in which they deform permanently under the action of external loads, whereas elasticity is the behavior of solid bodies in which they return to their original shape when the external forces are removed. Actually, however, the elastic body is an idealization, because all bodies exhibit more or less plastic behavior even at the smallest loads. For the so-called elastic body, however, this permanent deformation is so small as to be practically not measurable, if the loads are sufficiently small. Plasticity theory thus concerns itself with situations in which the loads are sufficiently large so that measurable amounts of permanent deformation occur. It should further be noted that plastic deformation is independent of the time under load. Time-dependent deformations are discussed briefly in Section 2.4 and in Chapter 14.

The theory of plasticity can conveniently be divided into two ranges. At one end are metal-forming processes such as forging, extrusion, drawing, rolling, etc., which involve very large plastic strains and deformations. For these types of problems the elastic strains can usually be neglected and the material can be assumed to be *perfectly plastic*. At the other end of the scale are a host of problems involving small plastic strains on the order of the elastic strains. These types of problems are of prime importance to the structural and machine designer. With the great premium currently placed on the saving of weight in aircraft, missile, and space applications, the designer can no longer use large factors of safety and "beef up" his design. He must design for maximum load to weight ratio, and this inevitably means designing into the plastic range. Even in more prosaic industrial applications the competitive market is forcing the application of more efficient design.

In this book emphasis will be placed primarily on the second type of problem, i.e., the *elastoplastic problems*, where the plastic strains are of the same order of magnitude as the elastic strains. Problems of large plastic deformations will be treated only briefly, as will problems of creep and limit design.

In Chapter 2 some simple experiments to determine several basic facts about the elastoplastic behavior of metals will be discussed.

References

1. H. Tresca, Sur l'ecoulement des corps solids soumis à de fortes pression, *Compt. Rend.*, **59**, 1864, p. 754.
2. B. de Saint-Venant, Memoire sur l'établissement des équations différentielles des mouvements intérieurs opérés dans les corps solides ductiles au delà des limites où l'élasticité pourrait les ramener à leur premier état, *Compt. Rend.*, **70**, 1870, pp. 473–480.
3. M. Lévy, Memoire sur les équations générales des mouvements intérieurs des corps solides ductiles au delà des limites où l'élasticité pourrait les ramener à leur premier état, *Compt. Rend.*, **70**, 1870, pp. 1323–1325.
4. R. von Mises, Mechanik der festen Koerper im plastisch deformablen Zustant, *Goettinger Nachr., Math.-Phys. Kl.*, 1913, pp. 582–592.
5. H. Hencky, Zur Theorie plastischer Deformationen und der hierdurch im Material hervorgerufenen Nebenspannungen, *Proceedings of the 1st International Congress on Applied Mechanics, Delft*, Technische Boekhandel en Druckerij, J. Waltman, Jr., 1925, pp. 312–317.
6. L. Prandtl, Spannungsverteilung in plastischen Koerpern, *Proceedings of the 1st International Congress on Applied Mechanics, Delft*, 1924, pp. 43–54.
7. R. Hill, *The Mathematical Theory of Plasticity*, Oxford University Press, London, 1950.
8. H. M. Westergaard, *Theory of Elasticity and Plasticity*, Harvard University Press, Cambridge, 1952.
9. B. W. Shaffer and R. N. House, The Elastic–Plastic Stress Distribution Within a Wide Curved Bar Subjected to Pure Bending. *J. Appl. Mech.*, **22**, No. 3, 1955, pp. 305–310.
10. B. W. Shaffer and R. N. House, Displacements in a Wide Curved Bar, *J. Appl. Mech.*, **24**, No. 3, 1957, pp. 447–452.

CHAPTER **2**

BASIC EXPERIMENTS

In this chapter the results of some basic experiments on the behavior of metals is presented. The stress–strain curve in tension, one of the basic ingredients necessary in applying plasticity theory, is described in some detail. The effects of reverse loading, strain rate, temperature, and hydrostatic pressure are briefly discussed. Idealizations of the stress–strain curve and various models of material behavior are described.

2–1 TENSILE TEST

The simplest and most common experiment, as well as the most important, is the standard *tensile test*. A cylindrical test specimen such as shown in Figure 2.1.1 is inserted into the tensile machine, the load is increased, and the readings of the load, the extension of the gage length inscribed on the specimen, and/or the decrease in diameter are recorded. A typical load extension diagram is shown in Figure 2.1.2.

The *nominal stress*, defined as the load divided by the original cross-sectional area, is plotted against the *conventional* or *engineering strain*, defined as the increase in length per unit original length. Nominal stress is represented by

$$\sigma_n = \frac{P}{A_0} \tag{2.1.1}$$

Sec. 2-1] Tensile Test

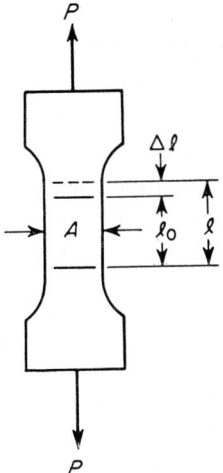

FIGURE 2.1.1 Tensile specimen.

and conventional strain by

$$\varepsilon = \frac{l - l_0}{l_0} \tag{2.1.2}$$

Initially the relation between stress and strain is essentially linear. This linear part of the curve extends up to the point A, which is called the *proportional limit*. It is in this range that the linear theory of elasticity, using Hook's law, is valid. Upon further increase of the load, the strain no longer increases linearly with stress, but the material still remains elastic; i.e., upon removal of the load the specimen returns to its original length. This condition will

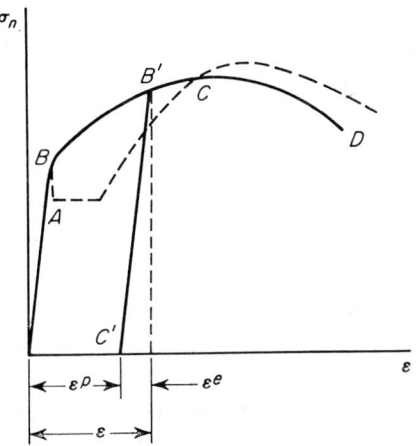

FIGURE 2.1.2 Conventional stress–strain curve.

prevail until some point *B*, called the *elastic limit*, or *yield point*, is reached. In most materials there is very little difference between the proportional limit *A* and the elastic limit *B*. For our purpose, we shall consider them to be the same. Furthermore, the values of these points depend on the sensitivity of the measuring instruments. For some materials the yield point is so poorly defined that it is arbitrarily taken to be at some fixed value of permanent strain, such as 0.2 per cent. The stress at this point is usually called the *offset yield strength*, or the *proof strength*. Beyond the elastic limit, permanent deformation, called *plastic deformation*, takes place. The strain at the elastic limit is of the order of magnitude of 0.001, or 0.1 per cent. As the load is increased beyond the elastic limit, the strain increases at a greater rate. However, the specimen will not deform further unless the load is increased. This condition is called *work hardening*, or *strain hardening*. The stress required for further plastic flow is called *flow stress*. Finally a point is reached, *C*, where the load is a maximum. Beyond this point, called the *point of maximum load*, or *point of instability*, the specimen "necks down" rapidly and fractures at *D*. Beyond *C* a complicated triaxial state of stress exists. The point *C* therefore represents the limit of the useful part of the tensile test as far as plasticity theory is concerned. The stress at the maximum load point *C* is called the *tensile strength*, or *ultimate stress*.

If at any point between the elastic limit *B* and the maximum load point *C* the load is removed, unloading will take place along a line parallel to the elastic line, as shown in the figure by $B'C'$. Part of the strain is thus recovered and part remains permanently. The total strain can therefore be considered as being made up of two parts, ε^e, the elastic component, and ε^P, the plastic component:

$$\varepsilon = \varepsilon^e + \varepsilon^P \qquad (2.1.3)$$

Upon reloading, the unloading line, $B'C'$, is retraced with very minor deviations. Actually a very thin hysteresis loop is formed, which is usually neglected. Plastic flow does not start again until the point B' is reached. With further loading, the stress–strain curve is continued along $B'C$ as if no unloading had occurred. Point B' can thus be considered as a new yield point for the *strain-hardened* material.

A few materials, such as annealed mild steel, exhibit a sharp drop in yield after the upper yield point *B* is reached, as shown by the dashed line. The specimen will then extend at approximately constant load to a strain of about 10 times the initial yield before the load will start increasing again as the material begins to work harden. The flat portion of the curve, called the *lower yield*, actually represents an average of a series of unstable jumps between the upper and lower yields caused by the propagation of *Luder bands* across the specimen. The upper yield point is very sensitive to small bending stresses or to irregularities in the specimen as well as to the rate of loading. Furthermore, very little plastic flow takes place at the upper yield point. The lower yield point should therefore always be used for design purposes and for plastic flow calculations.

2-2 TRUE STRESS-STRAIN CURVE

We have discussed the plot of the nominal stress versus the conventional strain. It is evident, however, that this nominal stress is not the true stress acting in the specimen, since the cross-sectional area of the specimen is decreasing with load. At stresses up to and near the yield, this distinction is of no importance. At higher stresses and strains this difference becomes important. The *true stress* can readily be obtained from the nominal stress as follows. If small changes in volume are neglected, i.e., the material is assumed to be incompressible, then

$$A_0 l_0 = A l$$

where A_0 and l_0 are the original cross-sectional area and gage length and A and l are the current values. If P is the load, then the true stress σ is

$$\sigma = \frac{P}{A} = \frac{Pl}{A_0 l_0}$$

The nominal stress σ_n is $\sigma_n = P/A_0$ and the conventional strain is $\varepsilon = (l/l_0) - 1$. Therefore,

$$\sigma = \sigma_n(1 + \varepsilon) \tag{2.2.1}$$

In a somewhat similar fashion, one recognizes that the conventional or engineering strain cannot be completely correct, since it is based on initial length, whereas the length is continuously changing. A different definition was therefore introduced by Ludwik [1] based on the changing length. Thus the increment of strain for a given length is defined as

$$d\bar{\varepsilon} = \frac{dl}{l} \tag{2.2.2}$$

and the total strain in going from some initial length l_0 to the length l is

$$\bar{\varepsilon} = \int_{l_0}^{l} \frac{dl}{l} = \ln \frac{l}{l_0} \tag{2.2.3}$$

$\bar{\varepsilon}$ is called the *natural, logarithmic,* or *true strain* and it represents a sort of average strain in going from the length l_0 to the length l. Its relation to the conventional strain is readily found, since $l/l_0 = 1 + \varepsilon$:

$$\bar{\varepsilon} = \ln(1 + \varepsilon) \qquad (2.2.4)$$

For small strains the two are practically identical, and for most problems considered the conventional strain will be used. The natural strain, however, has several advantages. Natural strains are additive, but conventional strains are not. Second, if a ductile material is tested in compression and in tension, the true-stress versus true-strain curves are almost identical, whereas they are quite different if conventional strain is used. Finally, the *incompressibility condition* to be used later becomes simply

$$\bar{\varepsilon}_1 + \bar{\varepsilon}_2 + \bar{\varepsilon}_3 = 0 \qquad (2.2.5)$$

whereas in terms of conventional strains it is

$$(1 + \varepsilon_1)(1 + \varepsilon_2)(1 + \varepsilon_3) - 1 = 0 \qquad (2.2.6)$$

which reduces to

$$\varepsilon_1 + \varepsilon_2 + \varepsilon_3 = 0$$

only in the case of small strains.

If a plot is now made of true stress versus true strain for the tensile test previously described, the curve will be essentially the same up to and slightly above the yield point. Beyond this point the two types of plots will diverge. The true stress will always increase until the rupture point and does not have a maximum at the point where the load starts dropping. The true stress at the point of maximum load can be found as follows. Since

$$P = \sigma A$$

at the point of maximum load

$$dP = \sigma\, dA + A\, d\sigma = 0$$

or

$$\frac{d\sigma}{\sigma} = -\frac{dA}{A}$$

Also

$$A_0 l_0 = Al$$

$$A\, dl + l\, dA = 0$$

Sec. 2-2] True Stress-Strain Curve

or

$$\frac{dA}{A} = -\frac{dl}{l}$$

Hence

$$\frac{d\sigma}{\sigma} = \frac{dl}{l} = d\bar{\varepsilon}$$

or

$$\frac{d\sigma}{d\bar{\varepsilon}} = \sigma$$

$$\frac{d\sigma}{d\varepsilon} = \frac{\sigma}{1+\varepsilon} \tag{2.2.7}$$

On a plot of σ versus $\bar{\varepsilon}$, the value of σ at which the load is a maximum occurs where the slope is equal to the stress; i.e., one must draw a tangent to that point of the curve for which the subtangent is equal to 1, as shown in Figure 2.2.1. Discussions of the stress-strain curve and the strain distributions

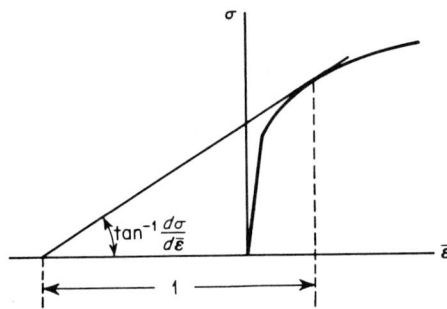

FIGURE 2.2.1 True stress-strain curve.

in the neck of a tensile specimen after necking has started can be found in references [2] and [3].

Alternatively, the true stress-strain curve can be obtained by measuring the *diametral strain* rather than the longitudinal strain, provided the tensile specimen has a circular section. Thus, if ε_D is the strain in the diametral direction, then

$$\varepsilon_D = \frac{D - D_0}{D_0} \tag{2.2.8}$$

where D_0 is the initial diameter and D is the diameter at the true stress, σ. The true diametral strain is

$$\bar{\varepsilon}_D = \ln(1 + \varepsilon_D) = \ln\frac{D}{D_0} \tag{2.2.9}$$

and from equation (2.2.5) the longitudinal true strain is

$$\bar{\varepsilon} = -2\bar{\varepsilon}_D = 2 \ln \frac{D_0}{D} \qquad (2.2.10)$$

The true strain at any load can therefore be determined by measuring the change in diameter of the specimen.

From equation (2.2.10) it is seen that the true strain can also be written

$$\bar{\varepsilon} = \ln \frac{A_0}{A} \qquad (2.2.11)$$

The quantity on the right of equation (2.2.11) is called the *true reduction in area*. Equation (2.2.11) states that the true strain is equal to the true reduction in area.

Figure 2.2.2 (from reference [4]) shows the true stress–strain curves for a

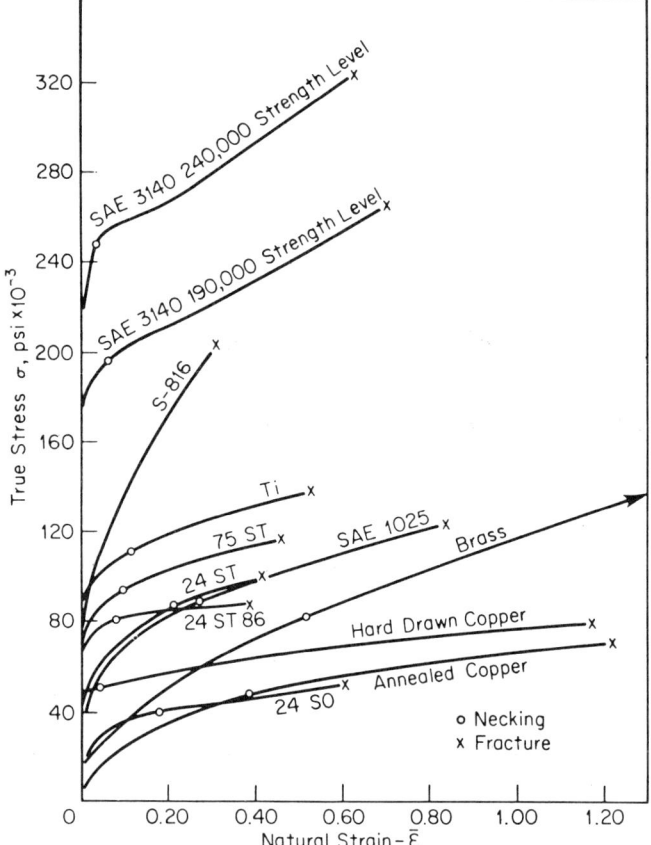

FIGURE 2.2.2 True stress–strain curves for several materials.

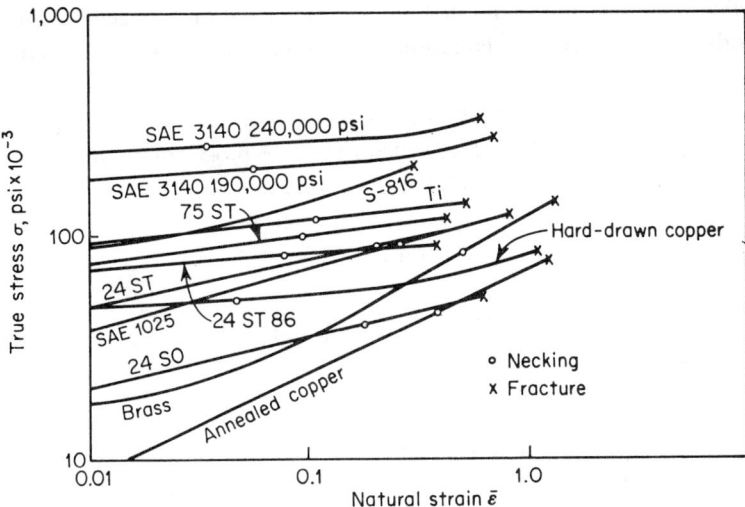

FIGURE 2.2.3 True stress–strain curves on log–log coordinates.

variety of materials. The ends of the curves represent the points of fracture and the circle on each curve represents the maximum load point or instability point for that curve. To show the complete curves to fracture, the abscissa is such that the elastic parts of the curves are too small to be seen. These curves are also shown replotted on log–log coordinates in Figure 2.2.3. Note that most of the curves appear as straight lines on this log–log plot. This indicates that they can be represented by an equation of the form

$$\sigma = A\bar{\varepsilon}^n \qquad (2.2.12)$$

where A and n are material constants with n the slope of the curve when plotted on log–log coordinates. A is called the *strength coefficient* and n is called the *strain-hardening exponent*.

It follows from equation (2.2.7) that for a material which behaves according to equation (2.2.12), the true strain at the point of maximum load is given by

$$\bar{\varepsilon} = n \qquad (2.2.13)$$

The simple relation (2.2.13) has been found useful in fracture studies. It also provides a simple method for determining the instability point on the true stress–strain curve.

Equation (2.2.12) will, of course, not fit all materials, nor will it be valid for small strains or very large strains. However, Marin [5] has studied 31 different materials and found that the average deviation between the theoretical values of $\bar{\varepsilon}$ as given by equation (2.2.13) and the actual values was 2 per cent.

A single quantity which represents the ability of a material to deform plastically is the *ductility* of the material. The most common measure of ductility is the *per cent elongation* in the tensile test, i.e., the per cent strain at fracture. Thus, if l_f is the gage length at fracture and l_0 is the initial gage length, then the per cent elongation is

$$\varepsilon_f = \frac{l_f - l_0}{l_0} \times 100 \qquad (2.2.14)$$

Together with the per cent elongation as given by equation (2.2.14), one must also specify the initial gage length l_0, since the per cent elongation will depend on the gage length used. This is due to the fact that once the instability point is reached and *necking* starts, most of the deformation occurs in the smallest cross section, with only a relatively small amount of additional deformation occurring throughout the rest of the gage length. The longer the gage length used, the smaller the per cent elongation will be. The ductility is therefore reported as the per cent elongation for a given gage length.

A better measure for ductility, however, is the *true strain at fracture*:

$$\bar{\varepsilon}_f = \ln \frac{l_f}{l_0} \qquad (2.2.15)$$

Alternatively, equation (2.2.15) can be written in terms of the *reduction in area* at fracture. From equation (2.2.11) it follows that

$$\bar{\varepsilon}_f = \ln \frac{A_0}{A_f} \qquad (2.2.16)$$

where A_0 is the initial area and A_f is the area at fracture. As mentioned previously, ductility is an important property which measures the ability of a material to deform. A material with low ductility cannot deform very much under load and will behave in a brittle fashion. Unexpectedly high loads may cause such a material to fracture, whereas a material with high ductility will deform under similar loads without fracturing. A cyclic load above the yield will cause a low-ductility material to fail in relatively few cycles, whereas a high-ductility material will fail after a much larger number of cycles (at least for very low cycle fatigue). In metal-forming processes such as rolling, drawing, forging, etc., a sufficient amount of ductility is needed to prevent fracture during the forming process.

2-3 COMPRESSION TEST AND THE BAUSCHINGER EFFECT. ANISOTROPY

If instead of a tensile test one runs a compression test and plots nominal stress against conventional strain, a different curve will be obtained than for the tensile test. However, if the true stress is plotted against the true strain, practically identical curves are usually obtained. The yield points in tension and compression will, for example, generally be the same. If, however, a metal is first deformed by uniform tension and the load is removed and the specimen is reloaded in compression, the yield point obtained in compression will be considerably less than the initial yield in tension. This has been explained as being the result of the residual stresses left in the material due to the tensile deformations [6]. A perhaps better explanation is based on the anisotropy of the dislocation field produced by loading [7]. This effect is called the *Bauschinger effect*, and is present whenever there is a reversal of the stress field. The Bauschinger effect is very important in cyclic plasticity studies. Unfortunately, however, it enormously complicates the problem and is therefore usually ignored.

There are several simplified models used to describe the Bauschinger effect. These are illustrated in Figure 2.3.1 (from reference [8]). At one extreme it is assumed that the elastic unloading range will be double the initial yield stress. If the initial yield stress in tension is σ_0, then the specimen will yield in compression after being stressed in tension to $\sigma = \sigma_1$ when

$$\sigma = \sigma_1 - 2\sigma_0$$

This is shown as path *ABCDE* in Figure 2.3.1. According to this theory, then, the total elastic range of the material remains constant, the initial compressive yield being reduced by the same amount as the tensile yield is raised.

At the other extreme there is *isotropic hardening*. This theory assumes that the mechanism that produces hardening acts equally in tension and compression. Thus compressive yielding will occur when

$$\sigma = -\sigma_1$$

as shown by the path *ABCFG*. This is the simplest of the theories to apply and is consequently the one most frequently used.

Between these theories there is a theory which assumes that the tensile and compressive yields are independent of each other. The compressive yield stress is independent of the amount of tensile hardening and remains at

$$\sigma = -\sigma_0$$

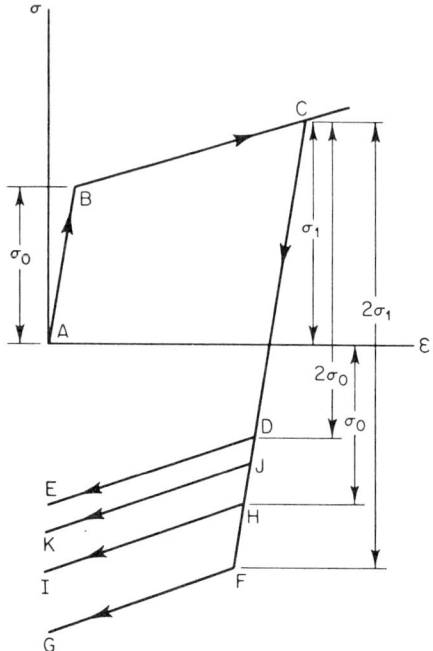

FIGURE 2.3.1 Theories for Bauschinger effect.

as shown by *ABCHI*. Actually experiments indicate that the compressive yield stress usually lies between points H and D of Figure 2.3.1, such as at J. It should be noted that in this figure the curves after yielding are shown as a set of parallel lines for simplicity. Actually a real stress–strain curve will show continuous curvature and varying slope after yielding when the load is reversed.

As an allied effect to the Bauschinger effect, any initial isotropy which is present is usually destroyed upon loading into the plastic range; i.e., if originally the tensile yield point was the same in all directions, it will no longer be so. Both the compressive and tensile yield values are changed in all directions by plastic yielding in one direction. Thus plastic deformation is *anisotropic*. For example, cold-rolled sheet has markedly different properties in the thickness direction than in the plane of the sheet, and usually somewhat different yield points in the rolling than in the transverse direction.

We see that the material may have initial anisotropy due to the manufacturing process, and it may also develop anisotropy due to plastic yielding. For small plastic strains the second effect is probably not too important. As for the first effect, the material being used can be tested for anisotropy. If a large amount of anisotropy is found, a much more complicated anisotropic theory of plasticity may have to be used.

2-4 EFFECTS OF STRAIN RATE AND TEMPERATURE

Tests on the effect of the rate of straining and of temperature on the properties of mild steel were carried out by Manjoine [9], among others. The effect of increasing the strain rate is generally to increase the tensile yield, as shown in Figure 2.4.1. For materials with a lower yield, such as mild steel, the stress–

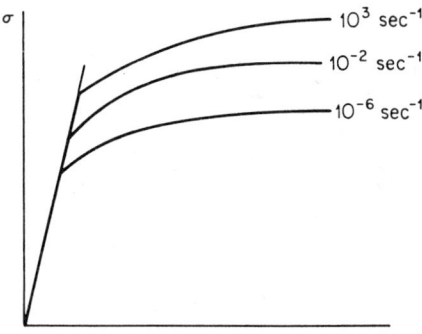

FIGURE 2.4.1 Effect of strain rate.

strain curve may approach that of a perfectly plastic material. For other materials the reverse will be true, and the strain hardening will increase with strain rate [16]. These effects are important in some metal-forming processes which are performed at very high strain rates. These types of processes will not be discussed in this text.

Temperature has a very important effect on metal properties. At very low temperatures metals which are very ductile can become very brittle. This is illustrated in Figure 2.4.2 (from reference [10]). The temperature at which the ductility changes so rapidly is called the *transition temperature*. Such strong temperature and strain-rate effects occur more generally in metals with body-centered-cubic structures.

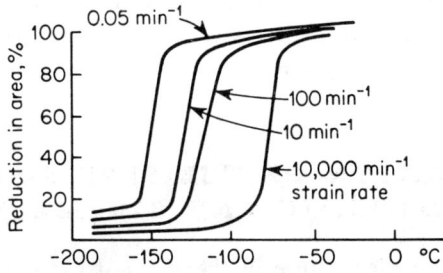

FIGURE 2.4.2 Effect of temperature.

At the other end of the time and temperature scales is the phenomenon of *creep*. Creep is a continuous deformation with time under constant load and occurs primarily at high temperatures, although some metals, e.g., lead, will creep at room temperature. Although it is questionable whether plasticity theories can be applied to the creep phenomenon, it is the usual practice to do so and, in Chapter 14, we shall describe how this is done. A typical set of creep curves is shown in Figure 2.4.3.

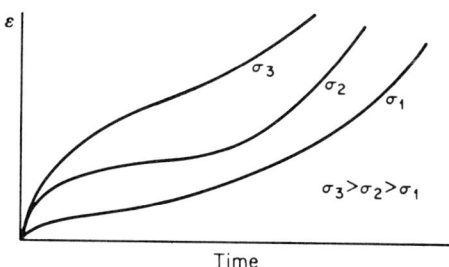

FIGURE 2.4.3 Creep curves.

2-5 INFLUENCE OF HYDROSTATIC PRESSURE. INCOMPRESSIBILITY

Bridgman, in a series of classical experiments [11, 12] in which he carried out tensile tests under conditions of hydrostatic pressures up to 25,000 atm, showed that hydrostatic pressure has negligible effect on the yield point until extremely high pressures are reached. Furthermore, the shape of the stress–strain curve remains unaltered in the small-strain range. The major effect of hydrostatic pressure is to increase the ductility of the material and to permit much larger deformations prior to fracture.

It has also been shown that the density, and consequently the volume, does not change even for very large plastic deformations. Thus, in the plastic range, a material can be considered as incompressible. These two experimental facts, i.e., the lack of influence of hydrostatic pressure and incompressibility, are very important in the development of plastic flow theories.

2-6 IDEALIZATION OF THE STRESS–STRAIN CURVE. DYNAMIC AND KINEMATIC MODELS

Because of the complex nature of the stress–strain curve, it has become customary to idealize this curve in various ways [13]. Figure 2.6.1 shows

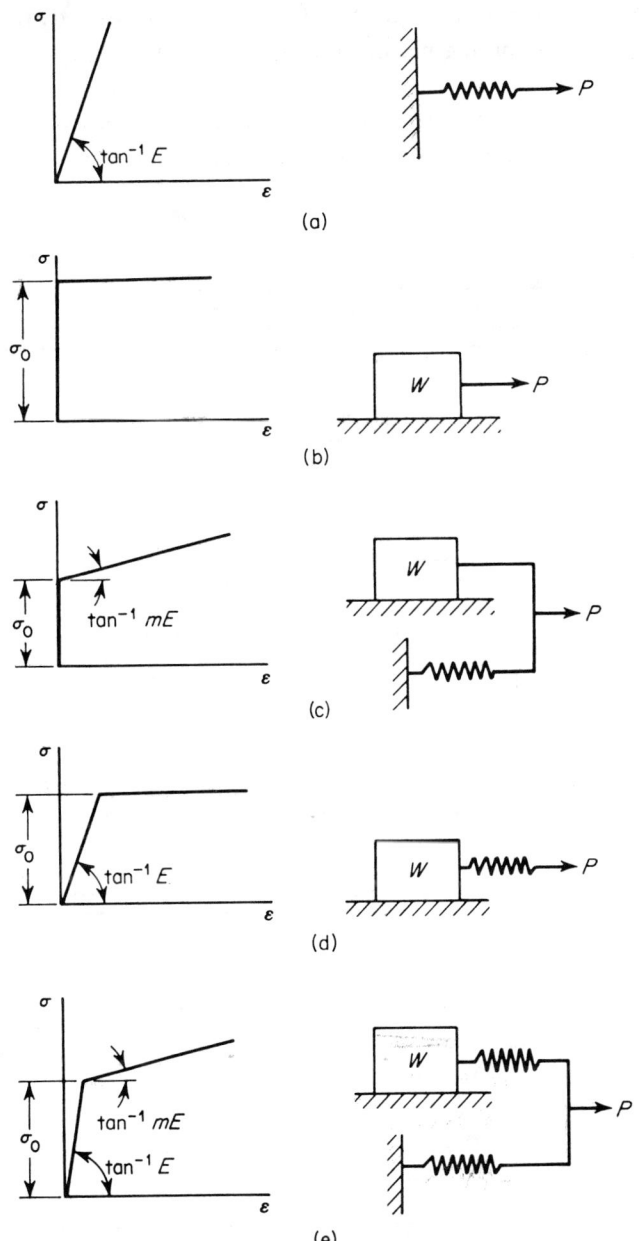

FIGURE 2.6.1 Idealized stress–strain curves: (a) perfectly elastic, brittle; (b) rigid, perfectly plastic; (c) rigid, linear strain hardening; (d) elastic, perfectly plastic; (e) elastic, linear strain hardening.

idealized curves as well as corresponding dynamic models which can be used to describe the material behavior. It will be shown subsequently that with the use of modern computing machinery, these idealizations are in many cases not necessary.

The models shown in Figure 2.6.1 are designated as *dynamic models*. They replace stresses by forces and strains by displacements. To devise models of

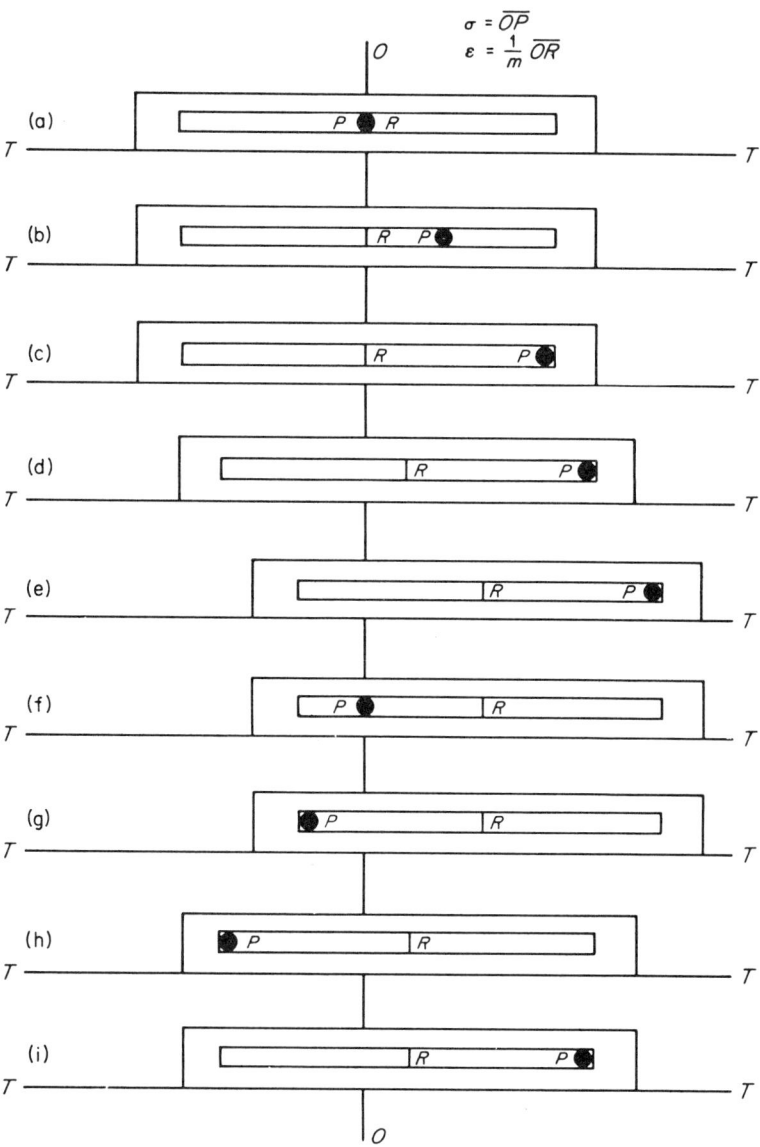

FIGURE 2.6.2 Kinematic model.

Sec. 2–6] Idealization of the Stress–Strain Curve

this type to represent combined stresses acting in several directions would be extremely difficult. For this reason Prager [14] introduced ingenious *kinematic models* in which both the stresses and strains are represented by displacements.

Figure 2.6.2 illustrates this type of model for the case of the rigid linear-hardening plastic material [8]. The model is taken to be a slotted bar, as shown. The bar is free to move along its length on the frictionless table T. But for the bar to move, the pin P must engage the end of the bar. Initially pin P is at the center R of the slot and this point is marked as point O on the table. The distance from P to either end of the slot is taken equal to the yield stress σ_0 of the rigid linear-hardening material.

The distance OP of the pin P from the fixed point O is taken equal to the stress. Then the distance OR from the center of the slot R to the fixed point O is proportional to the strain, i.e., $\varepsilon = OR/m$, where $\tan^{-1} m$ is the slope of the plastic stress–strain curve. Thus plastic flow will take place when the pin is engaged at one end or the other of the slot. Figures 2.6.2 and 2.6.3 illustrate the different positions of the kinematic model and the corresponding stress–strain diagram. Note that for this particular model it has been assumed that the elastic unloading range EG is equal to twice the initial yield, so that the yield point in compression G is less than the initial yield C.

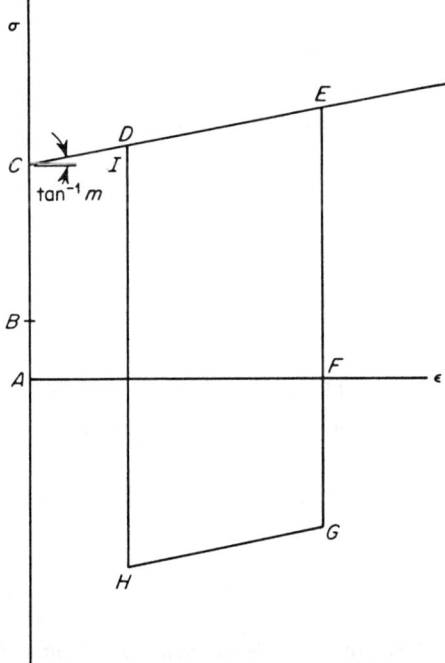

FIGURE 2.6.3 Stress–strain curve for model of Figure 2.6.2.

In a two-dimensional state of stress, two stress variables, σ_1 and σ_2, are specified instead of the single variable σ. Therefore, if the position of the pin is to indicate the state of stress, it must be free to move in two directions. In this case the rigid region must be a two-dimensional frame rather than a slot. This will be discussed in greater detail in Chapter 6, after the discussion of yield surfaces.

2-7 EMPIRICAL EQUATIONS FOR STRESS–STRAIN CURVES

It is sometimes useful to represent the stress–strain curve of a given material by an equation obtained empirically by fitting the experimental data. Equation (2.2.12) is such an equation which will frequently fit most of a given stress–strain curve, but, as was previously mentioned, will not usually fit at the low-strain and high-strain ends of the stress–strain curve. One of the first such empirical equations was proposed by Ludwik [1]. It has the form

$$\sigma = \sigma_0 + m\varepsilon^n \qquad (2.7.1)$$

A frequently used form, due to Ramberg and Osgood [15], is

$$\varepsilon = \frac{\sigma}{E} + k\left(\frac{\sigma}{E}\right)^n \qquad (2.7.2)$$

Some other forms that have been proposed are

$$\sigma = a + (b - a)(1 - e^{-n\varepsilon}) \qquad (2.7.3)$$

$$\sigma = c(a + \varepsilon)^n \qquad 0 \leq n \leq 1 \qquad (2.7.4)$$

$$\sigma = \sigma_0 \tanh\left(\frac{\varepsilon}{\varepsilon_0}\right) \qquad (2.7.5)$$

where e is the base of natural logarithms, ε_0 the yield strain, σ_0 the yield stress, E the elastic modulus, and m, n, k, a, b, and c constants.

It is also possible to fit the plastic part of the stress–strain curve as accurately as desired by a polynomial of arbitrary degree, i.e.,

$$\sigma = \begin{cases} E\varepsilon & \varepsilon \leq \varepsilon_0 \\ a_0 + a_1\varepsilon + a_2\varepsilon^2 + \cdots + a_m\varepsilon^m & \varepsilon \geq \varepsilon_0 \end{cases} \qquad (2.7.6)$$

where ε_0 is the yield strain. For *linear strain hardening* all the coefficients beginning with a_2 are zero.

Problems

1. Show that natural strains are additive whereas conventional strains are not.
2. Assume that a material behaves elastically up to the point of instability. Show that the natural strain at this point is unity.
3. Derive equation (2.2.13).
4. Let the stress–strain curve of a material be given by $\sigma = A\varepsilon^n$, where ε is the conventional strain. Show that at the point of instability

$$\varepsilon = \frac{n}{1-n}$$

5. In a standard tensile test using a $\frac{1}{2}$-in.-diameter specimen with a 1-in. gage length, the following data were obtained. At a load of 10,000 lb, the conventional strain was 0.10, and at a load of 12,000 lb, the conventional strain was 0.60. Find the true stresses and strains for these two conditions. Determine the strength coefficient A, the strain-hardening exponent n, the change in gage length at the maximum load, and the maximum load assuming equation (2.2.12) to hold.
6. A tensile load is applied to a thin-walled hollow circular cylinder. Determine the change in wall thickness and in mean radius at the point of maximum load, if the stress–strain curve is given by $\sigma = A\varepsilon^n$, where ε is the conventional strain and σ is the true stress.
7. Derive the incompressibility conditions (2.2.5) and (2.2.6).
8. The following data were obtained in a tensile test on a 0.505-in.-diameter specimen:

Diameter, in.	Load, lb	Diameter, in.	Load, lb
0.487	6,750	0.419	11,000
0.481	9,250	0.402	10,800
0.472	10,400	0.375	10,200
0.463	10,900	0.361	9,700
0.450	11,100	0.354	9,500
0.438	11,200	0.326	8,950
		Fracture	

   (a) Plot the true stress–strain curve.
   (b) Determine the strength coefficient A and the strain-hardening exponent n.
   (c) Determine the maximum load from the stress–strain curve and compare it with that obtained using equation (2.2.13).
9. Consider a material whose stress–strain curve is given by $\sigma = 30{,}000 + 1.5 \times 10^6 \varepsilon$, $\sigma > 30{,}000$ psi. If a tensile specimen of this material is stretched to a strain of 0.004 in./in., at what stress will it yield in compression when the load is reversed, for each of the assumptions in Figure 2.3.1?
10. For the dynamic models of Figure 2.6.1, show the relations between the model constants and the parameters of the stress–strain curve. Denote the spring constants by k (k_1 and k_2 for the last model), the weight of the block

by W, the friction coefficient by μ, and the force by P. For example, for the first model, the equation of the stress–strain curve is $\sigma = E\varepsilon$ and the corresponding model equation is $P = kx$. Thus

$$P \leftrightarrow \sigma$$
$$k \leftrightarrow E$$
$$x \leftrightarrow \varepsilon$$

11. For the kinematic model of Figure 2.6.2, show that $\varepsilon = OR/m$.
12. Describe a kinematic model similar to that shown in Figure 2.6.2 for isotropic hardening.
13. Sketch typical stress–strain curves that would be obtained using Ludwik's expression for the following cases:
    (a) $n = 1$.
    (b) $0 \leq n < 1$.
    (c) $\sigma_0 = 0$, $n = 0, \frac{1}{2}, 1$.

References

1. P. Ludwik, *Elemente der technologischen Mechanik*, Springer, Berlin, 1909.
2. J. D. Lubahn and R. P. Felgar, *Plasticity and Creep of Metals*, Wiley, New York, 1961.
3. G. E. Dieter, Jr., *Mechanical Metallurgy*, McGraw-Hill, New York, 1961.
4. H. Schwartzbart and W. F. Brown, Jr., Notch-Bar Tensile Properties of Various Materials and their Relation to the Unnotch Flow Curve and Notch Sharpness, *Trans. ASM*, **46**, 998, 1954.
5. J. Marin, *Mechanical Behavior of Engineering Materials*, Prentice-Hall, Englewood Cliffs, N.J., 1962.
6. R. Hill, *The Mathematical Theory of Plasticity*, Oxford Univ. Press, London, 1950.
7. D. Mclean, *Mechanical Properties of Metals*, Wiley, New York, 1962.
8. J. N. Goodier and P. G. Hodge, Jr., *Elasticity and Plasticity*, Wiley, New York, 1958.
9. M. J. Manjoine, Influence of Rate of Strain and Temperature on Yield Stresses of Mild Steel, *J. Appl. Mech.*, **11**, A-211, 1944.
10. A. W. Magnusson and W. M. Baldwin, Low Temperature Brittleness, *J. Mech. Phys. Solids*, **5**, 172, 1957.
11. P. W. Bridgman, The Effect of Hydrostatic Pressure on the Fracture of Brittle Substances, *J. Appl. Phys.*, **18**, 246, 1947.
12. P. W. Bridgman, *Studies in Large Plastic Flow and Fracture with Special Emphasis on the Effects of Hydrostatic Pressure*, McGraw-Hill, New York, 1952.
13. W. Johnson and P. B. Mellor, *Plasticity for Mechanical Engineers*, Van Nostrand, Princeton, N.J., 1962.
14. W. Prager, The Theory of Plasticity—A Survey of Recent Achievements, *Proc. Inst. Mech. Engrs.*, London, **169**, 41, 1955.

15. W. Ramberg and W. R. Osgood, Description of Stress–Strain Curves by Three Parameters, *NACA Technical Note No. 902*, July 1943.
16. T. A. Trozera, O. D. Sherby, and J. E. Dorn, Effect of Strain Rate and Temperature on the Plastic Deformation of High Purity Aluminum, *Univ. Calif. (Berkeley) Tech. Rept., Ser. 22, Issue 44, Contract N7-ONR-295*, Dec. 1955.

General References

Drucker, D. C., Stress–Strain Relations in the Plastic Range—A Survey of Theory and Experiment, *ONR Rept. NR–041–032*, 1950.

Goodier, J. N., and P. J. Hodge, Jr., *Elasticity and Plasticity*, Wiley, New York, 1958.

Hill, R., *The Mathematical Theory of Plasticity*, Oxford Univ. Press, London, 1950.

Johnson, W., and P. B. Mellor, *Plasticity for Mechanical Engineers*, Van Nostrand, London, 1962.

CHAPTER **3**

THE STRESS TENSOR

It is assumed that the reader is familiar with the basic concepts of the theory of elasticity, including the definitions of stress and strain. However, to avoid having the student refer to other texts and to refresh the memory of those readers who have not recently done any work in elasticity, we shall briefly review in the next three chapters some of these basic concepts, with particular emphasis on those properties of the stress and strain tensors which are particularly important in the development of plasticity theory. The reader thoroughly familiar with elasticity theory may skip directly to Chapter 6.

Stress and strain are second-order *tensors*, and although we shall not concern ourselves with tensors and their properties as such, it is important that the student be familiar with the subscript notation known as *tensor notation*. This notation is not only a time saver in writing out long formulas or expressions, but it is also extremely useful in derivations and in the proof of theorems. Furthermore, a major part of the past and present literature on the subject utilizes tensor notation, and a knowledge of this notation is essential in following the literature. In solving specific problems the usual longhand notation must, however, always be used. We shall therefore start with a brief description of tensor notation.

3-1 TENSOR NOTATION

A tensor is a system of numbers or functions which transform according to a certain law, when the independent variables undergo a linear transformation. We shall not concern ourselves here with the transformation laws for tensors but will merely record a few elementary properties, including the tensor notation. We shall accept the fact that stress and strain are tensors.

A subscript notation is used which is really very simple. The coordinate axes are designated by the letter x, with a latin subscript. Thus x_i does not mean just one quantity, but three quantities, x_1, x_2, and x_3, where x_1, x_2, and x_3 are used instead of x, y, and z (or r, θ, z, etc.). Any other subscript, such as j, k, l, m, etc., can be used equally well. For two-dimensional problems the subscript is understood to have a range of only two rather than three.

A double subscript indicates a system of nine components if the range of each of the subscripts is three, or a system of four components if the range is two. For example, the stress tensor is designated by σ_{ij} and stands for nine components:

$$\sigma_{ij} \equiv \begin{bmatrix} \sigma_{11} & \sigma_{12} & \sigma_{13} \\ \sigma_{21} & \sigma_{22} & \sigma_{23} \\ \sigma_{31} & \sigma_{32} & \sigma_{33} \end{bmatrix} = \begin{bmatrix} \sigma_x & \tau_{xy} & \tau_{xz} \\ \tau_{yx} & \sigma_y & \tau_{yz} \\ \tau_{zx} & \tau_{zy} & \sigma_z \end{bmatrix} \qquad (3.1.1)$$

Similarly, the nine components of the strain tensor are designated by ε_{ij}.

Two subscripted quantities are said to be equal if their corresponding components are equal. Thus if $A_{ij} = B_{ij}$, then $A_{11} = B_{11}$, $A_{12} = B_{12}$, etc.

If two subscripted quantities are added, their corresponding components are added. Thus

$$A_{ij} + B_{ij} = C_{ij}$$

means

$$A_{11} + B_{11} = C_{11}$$
$$A_{12} + B_{12} = C_{12}$$
$$A_{21} + B_{21} = C_{21} \qquad \text{etc.}$$

A one-subscript system is called a system of first order, a double subscript system one of second order, etc. It is evident from the definitions of equality and addition above, that these can apply only to systems of equal order.

We now come to the only "tricky" part of tensor notation—the summation convention. Whenever a subscript is repeated, this indicates summation over the range of the subscript. Thus

$$\tau_{ii} = \tau_{11} + \tau_{22} + \tau_{33} = \tau_{jj} = \tau_{kk}$$

Such a subscript is called a *dummy subscript* and it must be a letter not a number; i.e., σ_{11} does not mean summation.

A more complicated example is given by the increment of work per unit volume:

$$dW = \tau_{ij}\, d\varepsilon_{ij} \tag{3.1.2}$$

The advantage here becomes apparent, since we have written down one term instead of nine. Also the work increment stands out as the scalar product of the strain increment and the stress. Furthermore, $\tau_{ij}\, d\varepsilon_{ij}$ represents the work increment without being tied to any particular system of axes. It would thus include $\tau_{11}\, d\varepsilon_{11} + \tau_{22}\, d\varepsilon_{22} + \tau_{33}\, d\varepsilon_{33}$, where these are the principal stresses and strains.

A system having any number of subscripts is said to be *symmetric* in two of these subscripts if the components of the system are unaltered when the two subscripts are interchanged. Thus a second-order system is called symmetric if

$$A_{ij} = A_{ji}$$

The stress and strain tensors are usually symmetric.

A system is said to be *skew-symmetric* or *antisymmetric* if the interchange of the indices changes the signs of the components. Thus for a second-order skew-symmetric system

$$A_{ij} = -A_{ji}$$

This shows immediately that

$$A_{11} = -A_{11} = 0$$
$$A_{22} = -A_{22} = 0$$
$$A_{33} = -A_{33} = 0$$

Therefore a skew-symmetric tensor of second order is characterized by just three quantities:

$$p_1 = A_{32} = -A_{23}$$
$$p_2 = A_{13} = -A_{31}$$
$$p_3 = A_{12} = -A_{21}$$

It can readily be shown that every second-order tensor A_{ij} may be decomposed into the sum of a symmetric tensor e_{ij} and a skew-symmetric tensor p_{ij}.

Sec. 3–2] Stress at a Point

The simplest second-order symmetric tensor is the *Kronecker delta* or *substitution tensor*, defined by

$$\delta_{ij} = 0 \quad i \neq j$$
$$\delta_{ij} = 1 \quad i = j \tag{3.1.3}$$
$$= \begin{bmatrix} 1 & 0 \\ 0 & 1 \end{bmatrix}$$

It is called the substitution tensor because

$$\delta_{ij} A_j = A_i \tag{3.1.4}$$
$$\delta_{ij} A_{ik} = A_{jk}$$

Finally, the convention is used to designate partial differentiation by a comma. Thus

$$A_{,i} = \frac{\partial A}{\partial x_i}$$
$$\sigma_{ij,i} = \frac{\partial \sigma_{1j}}{\partial x_1} + \frac{\partial \sigma_{2j}}{\partial x_2} + \frac{\partial \sigma_{3j}}{\partial x_3} \tag{3.1.5}$$

3–2 STRESS AT A POINT

Consider a body as shown in Figure 3.2.1 subjected to a system of external forces $P_1 P_2 \ldots P_8$. Now consider a plane AB passing through the body dividing it into parts I and II, as shown. If we consider part I, it is seen that

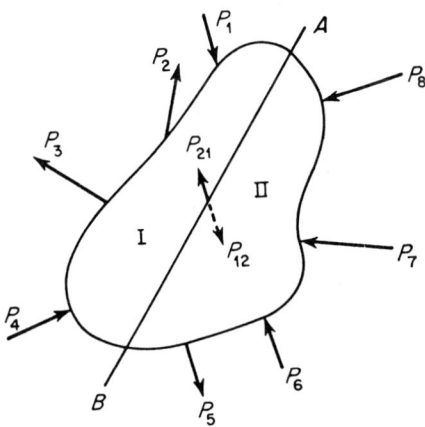

FIGURE 3.2.1 Loaded body.

it is in equilibrium under the action of forces P_1, P_2, P_3, P_4, and the force P_{12} that part II exerts on part I, P_{12} being the resultant of the continuous distribution of forces on the plane AB that part II exerts on part I. If a small area ΔA is taken in this plane with a force ΔP acting on it, then the *unit stress* acting at this point is defined as

$$p = \lim_{\Delta A \to 0} \frac{\Delta P}{\Delta A} \qquad (3.2.1)$$

The important thing to note here is that the unit stress, p, must be referred to a particular plane. For any other plane passing through the same point, it is obvious from consideration of Figure 3.2.1 that the force distribution on this plane, and hence the unit stress, will be different.

The unit stress, p, of course, need not be perpendicular to the plane AB. In practice, therefore, the stress, p, is decomposed into two components, one normal to the plane of reference, called the *normal stress*, and one parallel to this plane, called the *shearing stress*. The normal stress is taken as positive when it is tensile in nature and negative when it is compressive.

To completely specify the stress at a point it is necessary to specify the stresses at that point on three mutually perpendicular planes passing through the point. The stress on any arbitrary plane through the point can then be determined in terms of the stresses on the three perpendicular planes, as will shortly be shown. Let the three mutually perpendicular planes be the planes perpendicular to the x, y, and z coordinate axes. Then the stresses acting on these planes at their point of intersection are as designated in Figure 3.2.2. The stresses as shown are all positive. The subscripts denote the direction of the stress. The first subscript designates the normal to the plane under consideration, and the second subscript designates the direction of the stress. Thus τ_{xy} denotes a shearing stress acting on the face of the element that is perpendicular to the x axis, the stress acting in the direction of the y axis. As mentioned previously, the normal stress is taken positive when it produces tension and negative when it produces compression. The positive directions of the components of shearing stress on any side of the cubic element are taken as the positive directions of the coordinate axes, if a tensile stress on the same side would have the positive direction of the corresponding axis.

It is seen from the figure that the complete specification of the stress at a point is given by the nine quantities

$$\begin{matrix} \sigma_x & \tau_{xy} & \tau_{xz} \\ \tau_{yx} & \sigma_y & \tau_{yz} \\ \tau_{zx} & \tau_{zy} & \sigma_z \end{matrix}$$

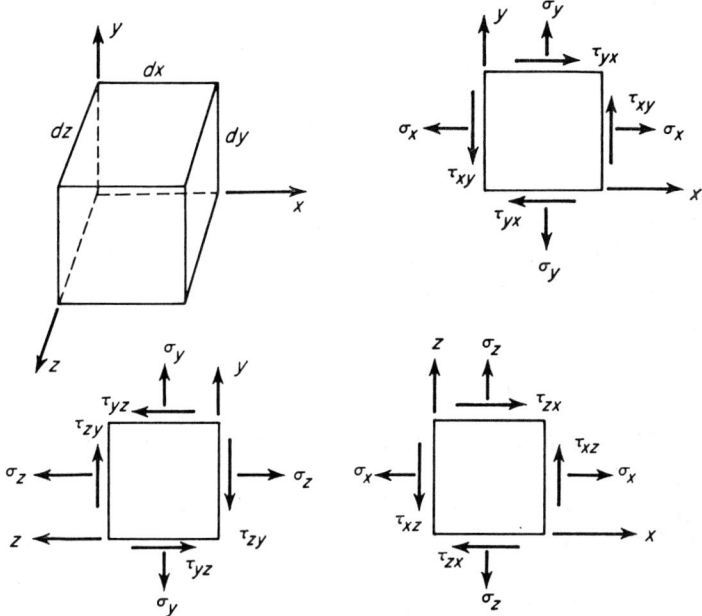

FIGURE 3.2.2 Convention for stresses.

It is customary in engineering practice to delete the second subscript on the normal stress and write σ_x instead of σ_{xx}, etc. In tensor notation the nine quantities are designated simply by σ_{ij} or, by some authors, τ_{ij}.

If one considers an infinitesimal rectangular parallelepiped surrounding a given point in a body, then it readily follows, as is shown in standard texts, that static equilibrium of forces and moments requires that the stresses at this point satisfy the following equations:

$$\frac{\partial \sigma_x}{\partial x} + \frac{\partial \tau_{yx}}{\partial y} + \frac{\partial \tau_{zx}}{\partial z} = -F_x$$

$$\frac{\partial \tau_{xy}}{\partial x} + \frac{\partial \sigma_y}{\partial y} + \frac{\partial \tau_{zy}}{\partial z} = -F_y \qquad (3.2.2)$$

$$\frac{\partial \tau_{xz}}{\partial x} + \frac{\partial \tau_{yz}}{\partial y} + \frac{\partial \sigma_z}{\partial z} = -F_z$$

where F_j are the components of the body forces per unit volume. Also

$$\tau_{yx} = \tau_{xy} \qquad \tau_{yz} = \tau_{zy} \qquad \tau_{xz} = \tau_{zx} \qquad (3.2.3)$$

In tensor notation these become simply

$$\sigma_{ij,i} = -F_j$$
$$\sigma_{ij} = \sigma_{ji}$$
(3.2.4)

The second line of (3.2.4) expresses the fact that the stress tensor is symmetric. There are therefore in general only six independent components of stress at a point rather than nine. (*Note*: There are some peculiar conditions for which the stress tensor will not be symmetric, as in the case when body moments act [1].)

3-3 PRINCIPAL STRESSES. STRESS INVARIANTS

If we are given the six components of stress at a point with respect to some coordinate system (x, y, z), we can determine the stresses acting on any plane through this point. This can be done by consideration of the static equilibrium of an infinitesimal tetrahedron formed by this plane and the coordinate planes, as shown in Figure 3.3.1. In this figure we have shown the stresses acting on the three coordinate planes. These stresses are assumed to be known. We wish to find the stresses acting on the plane ABC whose normal ON has direction cosines l, m, and n. Let the area of the infinitesimal triangle ABC be designated by ΔA. Then the areas of the faces AOB, COB, and AOC are equal to $m\,\Delta A$, $l\,\Delta A$, and $n\,\Delta A$, respectively. Now let the stress vector acting on the face ABC be designated by S and its x, y, and z components by S_x, S_y, and S_z as shown in Figure 3.3.1(b). Then for equilibrium of forces in the x direction,

$$S_x \Delta A = \sigma_x l\,\Delta A + \tau_{yx} m\,\Delta A + \tau_{zx} n\,\Delta A$$

or

$$S_x = l\sigma_x + m\tau_{yx} + n\tau_{zx}$$

Similarly,

$$S_y = l\tau_{xy} + m\sigma_y + n\tau_{zy}$$
$$S_z = l\tau_{xz} + m\tau_{yz} + n\sigma_z$$
(3.3.1)

To obtain the normal stress S_n acting on this plane we project the stresses S_x, S_y, and S_z onto the normal ON, to get

$$S_n = lS_x + mS_y + nS_z$$
(3.3.2)

Sec. 3-3] Principal Stresses. Stress Invariants

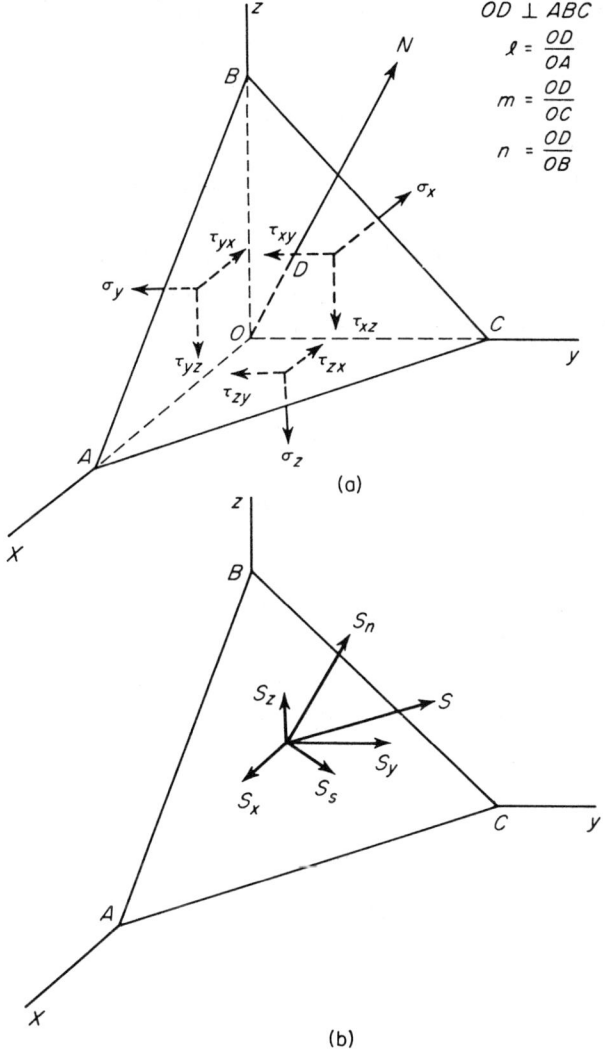

FIGURE 3.3.1 Forces on infinitesimal tetrahedron.

and substituting from (3.3.1) results in

$$S_n = l^2\sigma_x + m^2\sigma_y + n^2\sigma_z + 2(lm\tau_{xy} + mn\tau_{yz} + nl\tau_{zx}) \quad (3.3.3)$$

Furthermore, to obtain the resultant shear stress S_s acting on this plane,

$$\begin{aligned}S_s^2 &= S^2 - S_n^2 \\ &= S_x^2 + S_y^2 + S_z^2 - S_n^2\end{aligned} \quad (3.3.4)$$

Equations (3.3.1) give the x, y, and z components of the stress acting on this plane, and equations (3.3.3) and (3.3.4) give the normal and shear stresses.

Equations (3.3.1) can also be considered as the boundary conditions that have to be satisfied by the stress components σ_{ij} at any point on the boundary of the body. Thus if the element of area ABC is considered to be an element of the boundary whose normal has the direction cosines l, m, and n, and S_x, S_y, and S_z are the components of the applied boundary forces at the point O, then equations (3.3.1) are precisely the boundary conditions that must be satisfied by the stress tensor. In tensor notation, if we replace l, m, and n by l_1, l_2, and l_3, we can write (3.3.1) as

$$S_j = l_i \sigma_{ij} \tag{3.3.5}$$

Suppose the plane element ABC of Figure 3.3.1 is so oriented that the resultant stress S on this plane element is normal to the plane; i.e., $S = S_n$ and $S_s = 0$. The plane is then called a *principal plane* at the point, its normal direction is called a *principal direction*, and the stress $S = S_n$ is called a *principal stress*. At every point in a body there are at least three principal directions. These principal stresses and principal directions can readily be found as follows. Assume the element ABC to lie in a principal plane at point O so that $S = S_n$. Then S has the same direction cosines l, m, and n as the normal. The components of S in the x, y, and z directions are then

$$S_x = lS$$
$$S_y = mS$$
$$S_z = nS$$

and equations (3.3.1) give immediately

$$l(\sigma_x - S) + m\tau_{yx} + n\tau_{zx} = 0$$
$$l\tau_{xy} + m(\sigma_y - S) + n\tau_{zy} = 0 \tag{3.3.6}$$
$$l\tau_{xz} + m\tau_{yz} + n(\sigma_z - S) = 0$$

or in tensor notation (replacing l, m, and n by l_1, l_2, and l_3)

$$l_i(\sigma_{ij} - \delta_{ij}S) = 0 \tag{3.3.7}$$

For equation (3.3.6) to have a nontrivial solution for l, m, and n, the determinant of the coefficients must vanish, resulting in

$$|\sigma_{ij} - \delta_{ij}S| = \begin{vmatrix} \sigma_x - S & \tau_{yx} & \tau_{zx} \\ \tau_{xy} & \sigma_y - S & \tau_{zy} \\ \tau_{xz} & \tau_{yz} & \sigma_z - S \end{vmatrix} = 0 \tag{3.3.8}$$

Sec. 3-3] Principal Stresses. Stress Invariants

Expanding the determinant gives a cubic equation for S:

$$S^3 - I_1 S^2 - I_2 S - I_3 = 0 \tag{3.3.9}$$

where

$$\begin{aligned}
I_1 &= \sigma_x + \sigma_y + \sigma_z \\
I_2 &= \tau_{xy}^2 + \tau_{yz}^2 + \tau_{zx}^2 - (\sigma_x \sigma_y + \sigma_y \sigma_z + \sigma_z \sigma_x) \\
I_3 &= \sigma_x \sigma_y \sigma_z + 2\tau_{xy}\tau_{yz}\tau_{zx} - (\sigma_x \tau_{yz}^2 + \sigma_y \tau_{zx}^2 + \sigma_z \tau_{xy}^2)
\end{aligned} \tag{3.3.10}$$

It can be proved [2] that the cubic equation (3.3.9) has three real roots and consequently there are (at least) three principal stresses, which will be designated by σ_1, σ_2, and σ_3. Substituting any of these solutions back into equations (3.3.6) enables one to solve for the corresponding direction cosines l, m, and n, if in addition the identity $l^2 + m^2 + n^2 = 1$ is used. If the three roots σ_1, σ_2, and σ_3 are distinct, the three corresponding principal directions will be unique and orthogonal. If two of these roots are equal, one direction will be unique but the other two directions can be any two directions, orthogonal to the first. If all three roots are equal, there are no unique principal directions and any three directions can be chosen. This corresponds to a state of hydrostatic stress.

Suppose instead of the axes x, y, and z, a different set of axes, x', y', and z', were chosen at the point O. Then the equation for determining the principal stresses, (3.3.9), would be the same, except that I_1, I_2, and I_3 would be defined in terms of the stresses σ'_x, σ'_y, σ'_z, etc., with respect to the new coordinate axes, i.e.,

$$\begin{aligned}
I_1 &= \sigma'_x + \sigma'_y + \sigma'_z \\
I_2 &= \tau'^2_{xy} + \tau'^2_{yz} + \cdots \quad \text{etc.}
\end{aligned}$$

But the principal stresses are physical quantities and obviously do not depend on the coordinate axes chosen. Hence the numbers I_1, I_2, and I_3 appearing in (3.3.9) must be the same, no matter what coordinate axes are used, in order that they give the same values for σ_1, σ_2, and σ_3. Thus

$$I_1 = \sigma_x + \sigma_y + \sigma_z = \sigma'_x + \sigma'_y + \sigma'_z$$

and similarly for I_2 and I_3. I_1, I_2, and I_3 are therefore called the first, second, and third *invariants* of the stress tensor, respectively. We will show later that the first and second invariants are particularly important in plasticity theory.

If we choose the principal directions as the directions of the coordinate axes, then the stress invariants take on the simple form

$$I_1 = \sigma_1 + \sigma_2 + \sigma_3$$
$$I_2 = -(\sigma_1\sigma_2 + \sigma_2\sigma_3 + \sigma_3\sigma_1) \qquad (3.3.11)$$
$$I_3 = \sigma_1\sigma_2\sigma_3$$

It is to be noted that the invariants I_1, I_2, and I_3 appearing in (3.3.11) are three independent quantities which specify the state of stress just as well as σ_1, σ_2, and σ_3. That is, given σ_1, σ_2, and σ_3, we can calculate I_1, I_2, and I_3, and given I_1, I_2, and I_3, we can calculate σ_1, σ_2, and σ_3, one set of these numbers uniquely determining the other set. Furthermore, any three independent combinations of these invariants will obviously also be invariants and can specify the state of stress just as σ_1, σ_2, and σ_3 do.

3-4 MAXIMUM AND OCTAHEDRAL SHEAR STRESSES

Let us take the coordinate axes in the principal directions. Then the shear stresses referred to these axes are zero and the normal and shear stresses on some oblique plane, with direction cosines with respect to these axes of l, m, and n, are, from equations (3.3.3) and (3.3.4),

$$S_n = l^2\sigma_1 + m^2\sigma_2 + n^2\sigma_3$$
$$S_s^2 = S_1^2 + S_2^2 + S_3^2 - S_n^2 \qquad (3.4.1)$$

But, from (3.3.1), with the shear stresses zero,

$$S_1^2 = l^2\sigma_1^2 \qquad S_2^2 = m^2\sigma_2^2 \qquad S_3^2 = n^2\sigma_3^2$$

Therefore,

$$S_s^2 = l^2\sigma_1^2 + m^2\sigma_2^2 + n^2\sigma_3^2 - (l^2\sigma_1 + m^2\sigma_2 + n^2\sigma_3)^2 \qquad (3.4.2)$$

We already know that on the principal planes the shear stress is a minimum—zero. Let us now find the planes for which it is a maximum; i.e., we seek the values of l, m, and n, such that S_s as given by equation (3.4.2) is a maximum. In addition to equation (3.4.2), there is a restriction on the direction cosines, $l^2 + m^2 + n^2 = 1$; i.e., only two of them can be independent. Substituting into (3.4.2) $n^2 = 1 - m^2 - l^2$, differentiating the resultant equation with

Sec. 3-4] Maximum and Octahedral Shear Stresses

respect to l and m, and equating these derivatives to zero, the following equations are obtained for l and m:

$$l[(\sigma_1 - \sigma_3)l^2 + (\sigma_2 - \sigma_3)m^2 - \tfrac{1}{2}(\sigma_1 - \sigma_3)] = 0$$
$$m[(\sigma_1 - \sigma_3)l^2 + (\sigma_2 - \sigma_3)m^2 - \tfrac{1}{2}(\sigma_2 - \sigma_3)] = 0 \qquad (3.4.3)$$

One solution is obviously $l = m = 0$ and $n = \pm 1$. Another solution is obtained by taking $l = 0$ but not m. Then from the second equation, $m = \pm\sqrt{1/2}$. Also, taking $m = 0$ gives, from the first equation, $l = \pm\sqrt{1/2}$. There are in general no solutions for l and m both nonzero except in the special case $\sigma_1 = \sigma_2$.

If the above calculations are carried out by eliminating from equation (3.4.2) first m and then l, the following table is obtained:

l	0	0	± 1	0	$\pm\sqrt{1/2}$	$\pm\sqrt{1/2}$
m	0	± 1	0	$\pm\sqrt{1/2}$	0	$\pm\sqrt{1/2}$
n	± 1	0	0	$\pm\sqrt{1/2}$	$\pm\sqrt{1/2}$	0

The first three columns obviously give the direction cosines of the coordinate planes which are principal planes, and therefore the shearing stresses on these planes are zero; i.e., they are a minimum. The three last columns give direction cosines of 45° angles. These planes, therefore, bisect the angles between the coordinate axes. On these planes the shearing stresses are maximum. Designating these stresses by τ_i and substituting these direction cosines into equation (3.4.2), the values of the shearing stresses are obtained as

$$\tau_1 = \pm\tfrac{1}{2}(\sigma_2 - \sigma_3)$$
$$\tau_2 = \pm\tfrac{1}{2}(\sigma_1 - \sigma_3) \qquad (3.4.4)$$
$$\tau_3 = \pm\tfrac{1}{2}(\sigma_1 - \sigma_2)$$

Thus the maximum shearing stress acts on the plane bisecting the angle between the largest and smallest principal stresses and is equal to half the difference between these principal stresses.

If we compute the normal stresses on these planes and designate them by N_i we get, from (3.4.1),

$$N_1 = \tfrac{1}{2}(\sigma_2 + \sigma_3)$$
$$N_2 = \tfrac{1}{2}(\sigma_1 + \sigma_3) \qquad (3.4.5)$$
$$N_3 = \tfrac{1}{2}(\sigma_1 + \sigma_2)$$

so that the normal stress on each of these planes is equal to the average of the principal stresses on the two planes whose angle it bisects.

If the plane ABC of Figure 3.3.1 is so oriented that $OA = OB = OC$, then the normal ON will make equal angles with all the axes, and

$$l = m = n = \pm \frac{1}{\sqrt{3}} \tag{3.4.6}$$

The normal stress acting on this plane is

$$S_n = \tfrac{1}{3}(\sigma_1 + \sigma_2 + \sigma_3) \equiv \sigma_m \tag{3.4.7}$$

That is, the normal stress on this plane is equal to the mean stress, and the shear stress is

$$\begin{aligned}S_s^2 &= \tfrac{1}{3}(\sigma_1^2 + \sigma_2^2 + \sigma_3^2) - \tfrac{1}{9}(\sigma_1 + \sigma_2 + \sigma_3)^2 \\ &= \tfrac{1}{9}[(\sigma_1 - \sigma_2)^2 + (\sigma_2 - \sigma_3)^2 + (\sigma_3 - \sigma_1)^2]\end{aligned} \tag{3.4.8}$$

or, making use of (3.4.7),

$$S_s^2 = \tfrac{1}{3}[(\sigma_1 - \sigma_m)^2 + (\sigma_2 - \sigma_m)^2 + (\sigma_3 - \sigma_m)^2]$$

It is apparent that a tetrahedron similar to this one can be constructed in each of the four quadrants above the xy plane and in each of the four quadrants below the xy plane. On the oblique face of each of these eight tetrahedra the condition $1^2 = m^2 = n^2 = \tfrac{1}{3}$ will apply. The difference between the tetrahedra will be in the signs attached to l, m, and n. The eight tetrahedra together will form an octahedron, and on each of the eight planes forming the faces of this octahedron the normal stress and the shear stress will be as given by equations (3.4.7) and (3.4.8). These planes are called the *octahedral planes*, and the shear stresses acting on them are called the *octahedral shear stresses*. We shall designate this stress by τ_{oct}. Thus

$$\begin{aligned}\tau_{\text{oct}} &= \tfrac{1}{3}[(\sigma_1 - \sigma_2)^2 + (\sigma_2 - \sigma_3)^2 + (\sigma_3 - \sigma_1)^2]^{1/2} \\ &= \frac{1}{\sqrt{3}}[(\sigma_1 - \sigma_m)^2 + (\sigma_2 - \sigma_m)^2 + (\sigma_3 - \sigma_m)^2]^{1/2}\end{aligned} \tag{3.4.9}$$

In terms of the stress invariants the octahedral shear stress can be written

$$\tau_{\text{oct}} = \frac{\sqrt{2}}{3}(I_1^2 + 3I_2)^{1/2} \tag{3.4.10}$$

Sec. 3-5] Mohr's Diagram

and in terms of general nonprincipal stresses it becomes

$$\tau_{oct}^2 = \tfrac{2}{9}[(\sigma_x + \sigma_y + \sigma_z)^2 + 3(\tau_{xy}^2 + \tau_{yz}^2 + \tau_{zx}^2) - (\sigma_x\sigma_y + \sigma_y\sigma_z + \sigma_z\sigma_x)]$$
$$= \tfrac{1}{9}[(\sigma_x - \sigma_y)^2 + (\sigma_y - \sigma_z)^2 + (\sigma_z - \sigma_x)^2 + 6(\tau_{xy}^2 + \tau_{yz}^2 + \tau_{zx}^2)] \quad (3.4.11)$$

which gives the octahedral shear stress in terms of the stress components referred to an arbitrary set of axes.

3-5 MOHR'S DIAGRAM

The result of the previous section can be obtained graphically by the use of a diagram proposed by Mohr [3]. This approach is particularly useful in discussing a pair of parameters called Lode's variables which are frequently encountered in plasticity theory. A full discussion of Lode's variables will be postponed until later. Mohr's diagram can be obtained in the following way. Let us take the principal directions for the coordinate axes and consider the normal and shear stresses acting on an arbitrary plane oblique to these axes. Instead of S_n and S_s we will for convenience use σ and τ for the normal and shear stresses on this plane. Then in terms of the principal stresses equations (3.4.1) and (3.4.2) become

$$\sigma = l^2\sigma_1 + m^2\sigma_2 + n^2\sigma_3$$
$$\tau^2 + \sigma^2 = l^2\sigma_1^2 + m^2\sigma_2^2 + n^2\sigma_3^2 \quad (3.5.1)$$

With the identity $l^2 + m^2 + n^2 = 1$, we have three equations to solve for l^2, m^2, and n^2. Solving these results in

$$l^2 = \frac{\tau^2 + (\sigma - \sigma_2)(\sigma - \sigma_3)}{(\sigma_1 - \sigma_2)(\sigma_1 - \sigma_3)}$$
$$m^2 = \frac{\tau^2 + (\sigma - \sigma_3)(\sigma - \sigma_1)}{(\sigma_2 - \sigma_3)(\sigma_2 - \sigma_1)} \quad (3.5.2)$$
$$n^2 = \frac{\tau^2 + (\sigma - \sigma_1)(\sigma - \sigma_2)}{(\sigma_3 - \sigma_1)(\sigma_3 - \sigma_2)}$$

Without loss of generality we can order the axes so that

$$\sigma_1 > \sigma_2 > \sigma_3$$

Then since $l^2 \geq 0$ and $(\sigma_1 - \sigma_2)(\sigma_1 - \sigma_3) \geq 0$, from the first of equations (3.5.2),

$$\tau^2 + (\sigma - \sigma_2)(\sigma - \sigma_3) \geq 0 \tag{3.5.3}$$

Equation (3.5.3) can also be written

$$\tau^2 + \left(\sigma - \frac{\sigma_2 + \sigma_3}{2}\right)^2 \geq \left(\frac{\sigma_2 - \sigma_3}{2}\right)^2 \tag{3.5.4}$$

If the equality sign is used in equation (3.5.4), this equation represents a circle in the $\tau - \sigma$ plane, as shown by C_1 in Figure 3.5.1. The center is

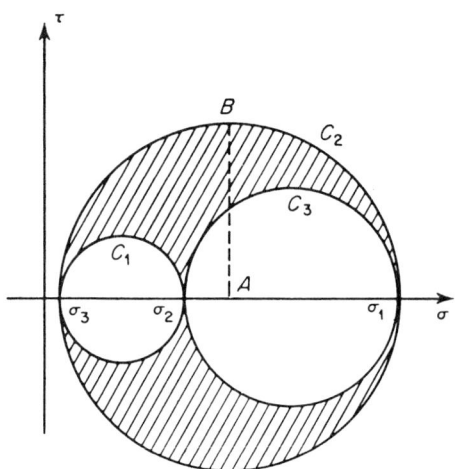

FIGURE 3.5.1 Mohr's diagram.

obviously at 0, $(\sigma_2 + \sigma_3)/2$, and it crosses the σ axis at $\sigma = \sigma_2$ and $\sigma = \sigma_3$. Therefore the region defined by equation (3.5.4) lies outside this circle and includes it as a boundary, since it represents circles with radii equal to or greater than $(\sigma_2 - \sigma_3)/2$.

Now considering the second equation of (3.5.2), since $\sigma_2 - \sigma_3 > 0$ and $\sigma_2 - \sigma_1 < 0$, the numerator must be less than or equal to zero:

$$\tau^2 + (\sigma - \sigma_3)(\sigma - \sigma_1) \leq 0$$

or
$$\tau^2 + \left(\sigma - \frac{\sigma_1 + \sigma_3}{2}\right)^2 \leq \left(\frac{\sigma_1 - \sigma_3}{2}\right)^2 \tag{3.5.5}$$

which represents a family of circles with centers at $(\sigma_1 + \sigma_3)/2$ and radii equal to or less than $(\sigma_1 - \sigma_3)/2$, as shown by the interior of circle C_2 in Figure 3.5.1. Finally, the third of equations (3.5.3) gives

$$\tau^2 + (\sigma - \sigma_1)(\sigma - \sigma_2) \geq 0$$

or
$$\tau^2 + \left(\sigma - \frac{\sigma_1 + \sigma_2}{2}\right) \geq \left(\frac{\sigma_1 - \sigma_2}{2}\right)^2 \quad (3.5.6)$$

which represents the region exterior to the circle C_3 with center at $(\sigma_1 + \sigma_2)/2$ and crossing the σ axis at σ_1 and σ_2.

It follows therefore from (3.5.4), (3.5.5), and (3.5.6) that the admissible values of τ and σ lie in the shaded region shown in Figure 3.5.1 bounded by the circles C_1, C_2, and C_3. The maximum shearing stress, as is clear from the figure, is represented by the largest ordinate, AB, which is the radius of the circle C_2 and is therefore equal to $(\sigma_1 - \sigma_3)/2$. To determine the orientation of the plane that has this shearing stress, we make use of equation (3.5.2). The value of σ corresponding to the maximum shearing stress is equal to $(\sigma_1 + \sigma_3)/2$ (the center of the circle C_2). Substituting these values of τ and σ into equations (3.5.2), we get

$$l^2 = n^2 = \tfrac{1}{2} \qquad m^2 = 0$$

We have thus obtained the same values for the maximum shearing stress as in the previous section.

3-6 STRESS DEVIATOR TENSOR

It is convenient in plasticity theory to split the stress tensor into two parts, one called the *spherical stress tensor* and the other the *stress deviator tensor*. The spherical stress tensor is the tensor whose elements are $\sigma_m \delta_{ij}$, where σ_m is the mean stress, i.e.,

$$p_{ij} \equiv \sigma_m \delta_{ij} = \begin{bmatrix} \sigma_m & 0 & 0 \\ 0 & \sigma_m & 0 \\ 0 & 0 & \sigma_m \end{bmatrix} \quad (3.6.1)$$

where
$$\sigma_m = \tfrac{1}{3}(\sigma_1 + \sigma_2 + \sigma_3) = \tfrac{1}{3}(\sigma_x + \sigma_y + \sigma_z) = \tfrac{1}{3}I_1 \quad (3.6.2)$$

From (3.6.2) it is apparent that σ_m is the same for all possible orientations of the axes; hence the name spherical stress. Also, since σ_m is the same in all directions, it can be considered to act as a hydrostatic stress. Now, it was shown in Section 2.5 that even very large hydrostatic pressure has a negligible effect on yielding and plastic flow. Therefore, in plastic flow considerations one can consider, if desired, the stress system obtained by subtracting the spherical state of stress from the actual state, rather than working with the actual state of stress. We therefore define a stress tensor called the stress deviator tensor S_{ij} as follows:

$$S_{ij} = \sigma_{ij} - p_{ij} = \sigma_{ij} - \sigma_m \delta_{ij}$$

$$= \begin{bmatrix} \sigma_x - \sigma_m & \tau_{xy} & \tau_{xz} \\ \tau_{yx} & \sigma_y - \sigma_m & \tau_{yz} \\ \tau_{zx} & \tau_{zy} & \sigma_z - \sigma_m \end{bmatrix}$$

$$= \begin{bmatrix} \dfrac{2\sigma_x - \sigma_y - \sigma_z}{3} & \tau_{xy} & \tau_{xz} \\ \tau_{xy} & \dfrac{2\sigma_y - \sigma_x - \sigma_z}{3} & \tau_{yz} \\ \tau_{xz} & \tau_{yz} & \dfrac{2\sigma_z - \sigma_x - \sigma_y}{3} \end{bmatrix} \quad (3.6.3)$$

It is apparent that subtracting a constant normal stress in all directions will not change the principal directions. The principal directions are therefore the same for the deviator tensor as for the original stress tensor. In terms of the principal stresses the deviator tensor is

$$S_{ij} = \begin{bmatrix} \sigma_1 - \sigma_m & 0 & 0 \\ 0 & \sigma_2 - \sigma_m & 0 \\ 0 & 0 & \sigma_3 - \sigma_m \end{bmatrix}$$

$$= \begin{bmatrix} \dfrac{2\sigma_1 - \sigma_2 - \sigma_3}{2} & 0 & 0 \\ 0 & \dfrac{2\sigma_2 - \sigma_3 - \sigma_1}{3} & 0 \\ 0 & 0 & \dfrac{2\sigma_3 - \sigma_1 - \sigma_2}{3} \end{bmatrix} \quad (3.6.4)$$

To obtain the invariants of the stress deviator tensor, replace S by $S' + \frac{1}{3}I_1$ in equation (3.3.9). This results in

$$S'^3 - J_1 S'^2 - J_2 S' - J_3 = 0 \quad (3.6.5)$$

where
$$J_1 = 0$$
$$J_2 = \tfrac{1}{3}(I_1^2 + 3I_2) \tag{3.6.6}$$
$$J_3 = \tfrac{1}{27}(2I_1^3 + 9I_1 I_2 + 27 I_3)$$

One advantage of using the stress deviator tensor is now apparent. The first invariant of this tensor is always zero. This can also be seen by taking the sum of the diagonal elements in equation (3.6.3) or (3.6.4).

The invariants J_2 and J_3 can, of course, be written in terms of the stress components σ_{ij}. For example,

$$J_2 = \tfrac{1}{6}[(\sigma_x - \sigma_y)^2 + (\sigma_y - \sigma_z)^2 + (\sigma_z - \sigma_x)^2 + 6(\tau_{xy}^2 + \tau_{yz}^2 + \tau_{zx}^2)] \tag{3.6.7}$$

Or, in terms of principal stresses,

$$\begin{aligned} J_2 &= \tfrac{1}{6}[(\sigma_1 - \sigma_2)^2 + (\sigma_2 - \sigma_3)^2 + (\sigma_3 - \sigma_1)^2] \\ J_3 &= (\sigma_1 - \sigma_m)(\sigma_2 - \sigma_m)(\sigma_3 - \sigma_m) \end{aligned} \tag{3.6.8}$$

In terms of the principal stress deviators S_1, S_2, and S_3 defined by $S_i = \sigma_i - \sigma_m$,

$$\begin{aligned} J_2 &= -(S_1 S_2 + S_2 S_3 + S_3 S_1) \\ J_3 &= S_1 S_2 S_3 \end{aligned} \tag{3.6.9}$$

Finally, comparing (3.6.7) to (3.4.11) we see that

$$J_2 = \tfrac{3}{2}\tau_{oct}^2 \tag{3.6.10}$$

This relationship between J_2 and the octahedral shear stress is sometimes used to lend credence on physical grounds to some plasticity theories, as will be discussed later.

3-7 PURE SHEAR

An important state of stress is the one designated as *pure shear*, or simple shear. If at a point in a body a set of coordinate axes can be found such that $\sigma_x = \sigma_y = \sigma_z = 0$, then the point is said to be in a state of pure shear. The stress state in a cylindrical bar in torsion is an example of pure shear. Another example is the stress state

$$\sigma_1 = -\sigma_2 \qquad \sigma_3 = 0 \tag{3.7.1}$$

It can readily be shown [4] that a necessary and sufficient condition for a state of pure shear to exist is

$$\sigma_{ii} = \sigma_x + \sigma_y + \sigma_z = 0 \qquad (3.7.2)$$

In conclusion it should be emphasized that all the equations developed and presented in this chapter are independent of the material properties and are therefore equally valid for bodies which behave elastically or plastically.

Problems

1. Write out the expression for the work increment

$$dW = \tau_{ij}\, d\varepsilon_{ij}$$

2. Given the tensor $\tau_{ik}\tau_{kj}$.
    (a) What is the order of this tensor?
    (b) If i, j, and k range from 1 to 3, how many components does this tensor have?
    (c) Write out the components.
3. The general Hooke's law is given by

$$\varepsilon_{ij} = c_{ijkl}\tau_{kl}$$

   (a) Write out the equations this represents.
   (b) What is the order of this tensor?
4. Prove that every second-order tensor a_{ij} may in a *unique* way be decomposed into the sum of a symmetric tensor e_{ij} and a skew-symmetric tensor p_{ij}.
5. Verify relations (3.4.4).
6. Verify equations (3.4.6), (3.4.7), (3.4.8), (3.4.10), and (3.4.11).
7. Discuss the Mohr diagram for the case $\sigma_2 = \sigma_3$ and determine the orientation of surface elements experiencing extreme shearing stresses. Consider also the case where $\sigma_1 = \sigma_2 = \sigma_3$.
8. Derive equations (3.6.7), (3.6.8), and (3.6.9).
9. Let $\sigma_x = \sigma_y = \sigma_z = 0$, $\tau_{xy} = \tau_{yz} = \tau_{zx} = a$. Calculate the octahedral shear stress, J_1, J_2, and J_3, the principal stresses, the greatest shearing stress, and the direction cosines for the greatest principal stress.
10. Show that subtracting a hydrostatic stress from a given stress state will not change the principal directions.
11. Show that the second and third invariants of the stress deviator tensor can also be written

$$J_2 = \tfrac{1}{2}(S_1^2 + S_2^2 + S_3^2)$$
$$J_3 = \tfrac{1}{3}(S_1^3 + S_2^3 + S_3^3)$$

12. Determine the invariants of the stress tensor and the octahedral stresses for the case of uniaxial state of stress.
13. The case of plane stress is partially characterized by $\sigma_z = \tau_{xz} = \tau_{yz} = 0$. Determine the invariants of the stress tensor, the principal stresses, the maximum shear stresses, and the octahedral stresses for this case.

14. Given the following stress tensor at a point:

$$\begin{pmatrix} 10{,}000 & 1{,}000 & -8{,}000 \\ 1{,}000 & -6{,}000 & 6{,}000 \\ -8{,}000 & 6{,}000 & 20{,}000 \end{pmatrix}$$

Determine the stress invariants, the stress deviator tensor, the principal stresses, the maximum shear stress, and the plane upon which it acts and the normal and shear stresses acting on a plane with direction cosines

$$l = \frac{1}{\sqrt{3}} \qquad m = \frac{1}{\sqrt{2}} \qquad n = \frac{1}{\sqrt{6}}$$

15. A circular cylinder 2 in. in diameter is subjected to a tensile force of 30,000 lb., a bending moment of 35,000 in.-lb, and a twisting moment of 50,000 in.-lb. Determine the stress invariants, the principal stresses, the maximum shear stress, and the stress deviator tensor.
16. Show that if a uniform tension σ_1 is applied to the two opposite sides of a square and a uniform compression $-\sigma_1$ is applied to the other two sides, a state of *pure shear* is obtained.

References

1. I. S. Sokolnikoff, *Mathematical Theory of Elasticity*, 2nd ed., McGraw-Hill, New York, 1956, p. 42.
2. *Ibid.*, p. 17.
3. O. Mohr, *Abhandlungen aus dem Gebiet der technischen Mechanik*, 2nd ed., Wilhelm Ernst und Sohn, Berlin, 1914, p. 192.
4. C. E. Pearson, *Theoretical Elasticity*, Harvard University Press, Cambridge, 1959, p. 57.

General References

Sokolnikoff, I. S., *Mathematical Theory of Elasticity*, 2nd ed., McGraw-Hill, New York, 1956.
Timoshenko, S., and J. N. Goodier, *Theory of Elasticity*, McGraw-Hill, New York, 1951.

CHAPTER **4**

THE STRAIN TENSOR

When the relative position of any two points in a continuous body is changed the body is said to be *deformed* or *strained*. If the distance between every pair of points in a body remains constant during motions of the body, the body is called *rigid*. The displacements of a rigid body consist of *translations* and *rotations*; translations and rotations are therefore called *rigid body displacements*. The analysis of strain consists of the study of the deformations of bodies which is essentially a geometric problem and is unrelated to the material properties. The specification of strain at a point is therefore the same for elastic and for plastically deforming bodies.

4–1 STRAIN AT A POINT

We start the analysis of the strain at a point in a body by considering the body referred to an orthogonal set of axes as shown in Figure 4.1.1. Consider two arbitrary neighboring points P_0 and P in the unstrained body. After straining, the points move to P_0' and P', respectively. The coordinates of P_0 are x_0, y_0, and z_0 and of P, x, y, and z. The coordinates of P_0' and P' are x_0', y_0', and z_0' and x', y', and z', respectively. The vector $\overline{A}$ deforms into the vector $\overline{A}'$. The components of $\overline{A}$ are A_x, A_y, and A_z and the components of $\overline{A}'$ are

$$A_x' = A_x + \delta A_x \quad A_y' = A_y + \delta A_y \quad A_z' = A_z + \delta A_z$$

Sec. 4-1] Strain at a Point

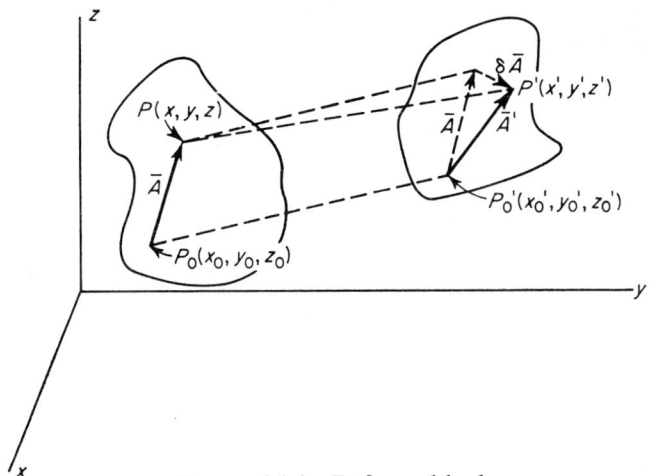

FIGURE 4.1.1 Deformed body.

The components of displacement of the point P_0 are

$$u_0 = x_0' - x_0$$
$$v_0 = y_0' - y_0 \quad (4.1.1)$$
$$w_0 = z_0' - z_0$$

The components of displacement of the point P are

$$u = x' - x$$
$$v = y' - y \quad (4.1.2)$$
$$w = z' - z$$

We assume that the displacements are single-valued continuous functions of the coordinates x, y, and z. (Actually, as will be apparent later, we have to assume that all their derivatives through the third are also continuous.) We can therefore expand the displacements at P in a Taylor series about P_0 as follows:

$$u = u_0 + \frac{\partial u}{\partial x} A_x + \frac{\partial u}{\partial y} A_y + \frac{\partial u}{\partial z} A_z$$

$$v = v_0 + \frac{\partial v}{\partial x} A_x + \frac{\partial v}{\partial y} A_y + \frac{\partial v}{\partial z} A_z \quad (4.1.3)$$

$$w = w_0 + \frac{\partial w}{\partial x} A_x + \frac{\partial w}{\partial y} A_y + \frac{\partial w}{\partial z} A_z$$

where higher-order terms are neglected, since P is taken in the neighborhood of P_0. Making use of (4.1.1) and (4.1.2),

$$(x' - x) - (x'_0 - x_0) = \frac{\partial u}{\partial x} A_x + \frac{\partial u}{\partial y} A_y + \frac{\partial u}{\partial z} A_z$$

$$(y' - y) - (y'_0 - y_0) = \frac{\partial v}{\partial x} A_x + \frac{\partial v}{\partial y} A_y + \frac{\partial v}{\partial z} A_z \qquad (4.1.4)$$

$$(z' - z) - (z'_0 - z_0) = \frac{\partial w}{\partial x} A_x + \frac{\partial w}{\partial y} A_y + \frac{\partial w}{\partial z} A_z$$

Now the change in the components of the vectors $\bar{A}$ and $\bar{A}'$ can be written

$$\begin{aligned}
\delta A_x &= A'_x - A_x = (x' - x'_0) - (x - x_0) = (x' - x) - (x'_0 - x_0) \\
\delta A_y &= A'_y - A_y = (y' - y'_0) - (y - y_0) = (y' - y) - (y'_0 - y_0) \qquad (4.1.5) \\
\delta A_z &= A'_z - A_z = (z' - z'_0) - (z - z_0) = (z' - z) - (z'_0 - z_0)
\end{aligned}$$

Substituting into (4.1.4) gives

$$\delta A_x = \frac{\partial u}{\partial x} A_x + \frac{\partial u}{\partial y} A_y + \frac{\partial u}{\partial z} A_z$$

$$\delta A_y = \frac{\partial v}{\partial x} A_x + \frac{\partial v}{\partial y} A_y + \frac{\partial v}{\partial z} A_z \qquad (4.1.6)$$

$$\delta A_z = \frac{\partial w}{\partial x} A_x + \frac{\partial w}{\partial y} A_y + \frac{\partial w}{\partial z} A_z$$

or, in tensor notation, simply

$$\delta A_i = u_{i,j} A_j$$

The transformation (4.1.6) is called an *affine transformation*, and if it is further assumed that the u_i and their derivatives are small, then it becomes an *infinitesimal affine transformation* [1]. The tensor

$$u_{i,j} = \begin{bmatrix} \dfrac{\partial u_1}{\partial x_1} & \dfrac{\partial u_1}{\partial x_2} & \dfrac{\partial u_1}{\partial x_3} \\ \dfrac{\partial u_2}{\partial x_1} & \dfrac{\partial u_2}{\partial x_2} & \dfrac{\partial u_2}{\partial x_3} \\ \dfrac{\partial u_3}{\partial x_1} & \dfrac{\partial u_3}{\partial x_2} & \dfrac{\partial u_3}{\partial x_3} \end{bmatrix} = \begin{bmatrix} \dfrac{\partial u}{\partial x} & \dfrac{\partial u}{\partial y} & \dfrac{\partial u}{\partial z} \\ \dfrac{\partial v}{\partial x} & \dfrac{\partial v}{\partial y} & \dfrac{\partial v}{\partial z} \\ \dfrac{\partial w}{\partial x} & \dfrac{\partial w}{\partial y} & \dfrac{\partial w}{\partial z} \end{bmatrix} \qquad (4.1.7)$$

Sec. 4-1] Strain at a Point

is called the *relative displacement tensor*. In general, as can be seen, it is not symmetric.

Nothing has been said so far as to whether the displacements of P_0 and P represent rigid body motions or not, or what part of them does. Since rigid body motions play no part in the analysis of strain, it is best to eliminate them from consideration at the start. This can be done as follows. A rigid body motion, as mentioned earlier, is characterized by the fact that the length of a line joining any two points remains unchanged. Consider the vector $\bar{A}$ shown in Figure 4.1.1 and assume the vector $\bar{A}'$ to be obtained from $\bar{A}$ by rigid body displacements. Then

$$A^2 = A'^2 = (A_x + \delta A_x)^2 + (A_y + \delta A_y)^2 + (A_z + \delta A_z)^2$$
$$= A_x^2 + A_y^2 + A_z^2 + 2(A_x \delta A_x + A_y \delta A_y + A_z \delta A_z)$$
$$= A^2 + 2(A_x \delta A_x + A_y \delta A_y + A_z \delta A_z)$$

where we have neglected higher-order terms in δA_i since we are considering only infinitesimal transformations. Therefore,

$$A_x \delta A_x + A_y \delta A_y + A_z \delta A_z = 0$$

or
$$A_i \delta A_i = 0 \tag{4.1.8}$$

But, from (4.1.6),

$$A_i \delta A_i = u_{i,j} A_i A_j = 0 \tag{4.1.9}$$

or

$$\frac{\partial u}{\partial x} A_x^2 + \left(\frac{\partial u}{\partial y} + \frac{\partial v}{\partial x}\right) A_x A_y + \left(\frac{\partial u}{\partial z} + \frac{\partial w}{\partial x}\right) A_x A_z$$
$$+ \frac{\partial v}{\partial y} A_y^2 + \left(\frac{\partial v}{\partial z} + \frac{\partial w}{\partial y}\right) A_y A_z + \frac{\partial w}{\partial z} A_z^2 = 0$$

Since equation (4.1.9) must be true for all values of A_x, A_y, and A_z, a necessary and sufficient condition that the transformation (4.1.6) represent a rigid body motion is

$$\frac{\partial u}{\partial x} = \frac{\partial v}{\partial y} = \frac{\partial w}{\partial z} = 0$$

$$\frac{\partial u}{\partial y} = -\frac{\partial v}{\partial x} \quad \frac{\partial u}{\partial z} = -\frac{\partial w}{\partial x} \quad \frac{\partial v}{\partial z} = -\frac{\partial w}{\partial y} \tag{4.1.10}$$

or

$$u_{i,j} = -u_{j,i}$$

That is, for rigid body motion the tensor $u_{i,j}$ of equation (4.1.7) is skew-symmetric.

Now every second-order tensor can be decomposed into a symmetric tensor and a skew-symmetric tensor in one and only one way (see Problem 4, Chapter 3). It follows therefore that if we decompose the tensor $u_{i,j}$ into symmetric and skew-symmetric parts, the skew-symmetric part will represent rigid body motions, whereas the symmetric part will represent *pure deformation*. Therefore, let

$$u_{i,j} = \tfrac{1}{2}(u_{i,j} + u_{j,i}) + \tfrac{1}{2}(u_{i,j} - u_{j,i}) \tag{4.1.11}$$

or

$$u_{i,j} = \varepsilon_{ij} + \omega_{ij}$$

where

$$\varepsilon_{ij} = \begin{bmatrix} \dfrac{\partial u}{\partial x} & \tfrac{1}{2}\left(\dfrac{\partial u}{\partial y} + \dfrac{\partial v}{\partial x}\right) & \tfrac{1}{2}\left(\dfrac{\partial u}{\partial z} + \dfrac{\partial w}{\partial x}\right) \\ \tfrac{1}{2}\left(\dfrac{\partial u}{\partial y} + \dfrac{\partial v}{\partial x}\right) & \dfrac{\partial v}{\partial y} & \tfrac{1}{2}\left(\dfrac{\partial v}{\partial z} + \dfrac{\partial w}{\partial y}\right) \\ \tfrac{1}{2}\left(\dfrac{\partial u}{\partial z} + \dfrac{\partial w}{\partial x}\right) & \tfrac{1}{2}\left(\dfrac{\partial v}{\partial z} + \dfrac{\partial w}{\partial y}\right) & \dfrac{\partial w}{\partial z} \end{bmatrix} \tag{4.1.12}$$

$$\omega_{ij} = \begin{bmatrix} 0 & \tfrac{1}{2}\left(\dfrac{\partial u}{\partial y} - \dfrac{\partial v}{\partial x}\right) & \tfrac{1}{2}\left(\dfrac{\partial u}{\partial z} - \dfrac{\partial w}{\partial x}\right) \\ \tfrac{1}{2}\left(\dfrac{\partial v}{\partial x} - \dfrac{\partial u}{\partial y}\right) & 0 & \tfrac{1}{2}\left(\dfrac{\partial v}{\partial z} - \dfrac{\partial w}{\partial y}\right) \\ \tfrac{1}{2}\left(\dfrac{\partial w}{\partial x} - \dfrac{\partial u}{\partial z}\right) & \tfrac{1}{2}\left(\dfrac{\partial w}{\partial y} - \dfrac{\partial v}{\partial z}\right) & 0 \end{bmatrix} \tag{4.1.13}$$

ε_{ij} is called the *strain tensor* and ω_{ij} is called the *rotation tensor*. For pure deformation therefore, equations (4.1.6) become

$$\delta A_i = \varepsilon_{ij} A_j \tag{4.1.14}$$

4–2 PHYSICAL INTERPRETATION OF STRAIN COMPONENTS

The physical meaning of the components of the strain tensor (4.1.12) can readily be determined as follows. Assume the vector $\bar{A}$ to be parallel to the x axis so that $A_y = A_z = 0$ and $A_x = A$. Then equation (4.1.14) gives immediately

$$\varepsilon_{11} = \varepsilon_x = \frac{\delta A_x}{A_x} = \frac{\delta A}{A} \tag{4.2.1}$$

Sec. 4–2] Physical Interpretation of Strain Components

Thus ε_x represents the extension or change in length per unit length of a vector originally parallel to the x axis. ε_y and ε_z obviously have similar interpretations.

To interpret the meaning of the strain component $\varepsilon_{ij} = \varepsilon_{zy}$, use is made of Figure 4.2.1. Consider two vectors $\bar{A}$ and $\bar{B}$ initially parallel to the y and z

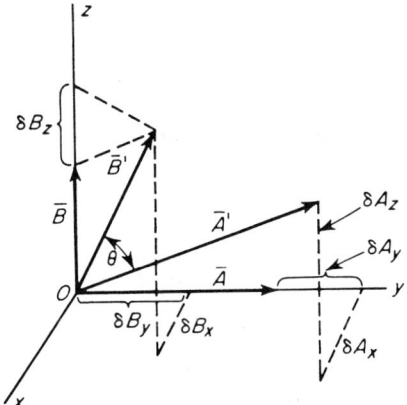

FIGURE 4.2.1 Shear deformation.

axes, respectively. Hence if i, j, and k represent the unit vectors in the x, y, and z directions, then $\bar{A} = jA_y$ and $\bar{B} = kB_z$. After deformation, these vectors are deformed into $\bar{A}'$ and $\bar{B}'$ with components

$$\bar{A}' = i\,\delta A_x + j(A_y + \delta A_y) + k\,\delta A_z$$
$$\bar{B}' = i\,\delta B_x + j\,\delta B_y + k(B_z + \delta B_z) \tag{4.2.2}$$

Denote the angle between $\bar{A}'$ and $\bar{B}'$ by θ. Then from the definition of the dot product

$$A'B' \cos\theta = \bar{A}' \cdot \bar{B}' = \delta A_x\,\delta B_x + (A_y + \delta A_y)\,\delta B_y + (B_z + \delta B_z)\,\delta A_z$$
$$\cong A_y\,\delta B_y + B_z\,\delta A_z \tag{4.2.3}$$

where we have neglected the products of the vector increments, assuming them to be small. Therefore,

$$\cos\theta = \frac{\bar{A}' \cdot \bar{B}'}{A'B'}$$

$$= \frac{A_y\,\delta B_y + B_z\,\delta A_z}{[(\delta A_x)^2 + (A_y + \delta A_y)^2 + (\delta A_z)^2]^{1/2}[(\delta B_x)^2 + (\delta B_y)^2 + (B_z + \delta B_z)^2]^{1/2}}$$

$$\cong \frac{A_y\,\delta B_y + B_z\,\delta A_z}{A_y B_z} = \frac{\delta B_y}{B_z} + \frac{\delta A_z}{A_y} \tag{4.2.4}$$

We have thus neglected all increments except δA_z and δB_y and we can now draw the figure as shown in Figure 4.2.2.

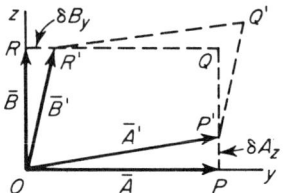

FIGURE 4.2.2   Shear deformation.

Now, from equation (4.1.14),

$$\delta A_z = \varepsilon_{zy} A_y$$
$$\delta B_y = \varepsilon_{yz} B_z \qquad (4.2.5)$$

Also $\cos \theta = \cos(\pi/2 - \alpha)$, where α is the decrease in the right angle between $\bar{A}$ and $\bar{B}$. Or

$$\cos \theta = \sin \alpha \cong \alpha$$

for small angles. From (4.2.4) and (4.2.5),

$$\alpha = \varepsilon_{yz} + \varepsilon_{zy} = 2\varepsilon_{yz} \qquad (4.2.6)$$

Hence ε_{yz} represents half the decrease in the right angle between two vectors that were initially directed along the positive y and z axes. Also, from Figure 4.2.2 and equation (4.2.5),

$$\angle POP' \cong \tan POP' = \frac{\delta A_z}{A_y} = \varepsilon_{zy}$$
$$\angle ROR' \cong \tan ROR' = \frac{\delta B_y}{B_z} = \varepsilon_{yz} \qquad (4.2.7)$$

Thus angles POP' and ROR' are equal. If we now rotate the parallelogram $R'OP'Q'$ through an angle ε_{yz} about the origin, we obtain a figure such as that of Figure 4.2.3. This represents a sliding or shear of the elements parallel to the xy plane, the amount of the sliding being proportional to the distance z of the element from the xy plane. Hence ε_{yz} is called a *shear strain*. Similar derivations can be made for ε_{xy} and ε_{xz}.

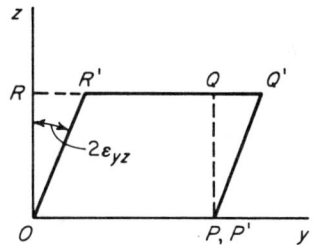

FIGURE 4.2.3   Shear deformation.

4-3 FINITE DEFORMATIONS

In the previous discussion we assumed that the displacements and their derivatives were small; i.e., we considered only infinitesimal strains. If the strains become too large, it is obvious that the previous formulation can no longer be correct. In this case the strains are no longer linearly related to the derivatives of the displacement. Furthermore, the equilibrium equations must be satisfied in the deformed body and must therefore be written in terms of the deformed coordinates if these are considerably different from the undeformed ones.

There are two methods of describing the deformation of a continuous body, when the deformations are large, the *Lagrangian* and the *Eulerian*. The Lagrangian method uses the initial coordinates of each particle to describe the deformation. The Eulerian method uses the coordinates of the particles in the deformed state to describe the deformation. We shall derive the elements of the Eulerian strain tensor using tensor notation for brevity. The power of this notation for derivation purposes, as was mentioned earlier, will be evident.

Let the coordinates of two neighboring particles before deformation be a_i and $a_i + da_i$. After deformation has taken place, let the coordinates of these particles be x_i and $x_i + dx_i$. The initial distance between the particles is

$$ds_0^2 = da_i\, da_i \tag{4.3.1}$$

and the final distance between them is given by

$$ds^2 = dx_i\, dx_i \tag{4.3.2}$$

In the Eulerian description we wish to describe everything in terms of the final coordinates x_i. We therefore write

$$\begin{aligned} a_i &= a_i(x_1, x_2, x_3) \\ da_i &= a_{i,j}\, dx_j = a_{i,k}\, dx_k \end{aligned} \tag{4.3.3}$$

Therefore,
$$ds_0^2 = a_{i,j} a_{i,k}\, dx_j\, dx_k$$

We can also write
$$ds^2 = dx_i\, dx_i = dx_j\, dx_k\, \delta_{jk}$$

Therefore,
$$ds^2 - ds_0^2 = (\delta_{kj} - a_{i,j} a_{i,k})\, dx_j\, dx_k \tag{4.3.4}$$

Now the displacements are given by
$$u_i = x_i - a_i$$
$$a_i = x_i - u_i$$
or
$$a_{i,j} = \delta_{ij} - u_{i,j}$$

Hence
$$\begin{aligned}
ds^2 - ds_0^2 &= [\delta_{jk} - (\delta_{ij} - u_{i,j})(\delta_{ik} - u_{i,k})]\, dx_j\, dx_k \\
&= [\delta_{jk} - (\delta_{jk} - u_{k,j} - u_{j,k} + u_{i,j} u_{i,k})]\, dx_j\, dx_k \\
&= (u_{k,j} + u_{j,k} - u_{i,j} u_{i,k})\, dx_j\, dx_k \\
&= 2\epsilon_{jk}\, dx_j\, dx_k
\end{aligned} \tag{4.3.5}$$

The tensor
$$\epsilon_{jk} = \tfrac{1}{2}(u_{k,j} + u_{j,k} - u_{i,j} u_{i,k}) \tag{4.3.6}$$

is called the *Eulerian strain tensor*. Some typical terms in engineering notation are

$$\begin{aligned}
\epsilon_x &= \frac{\partial u}{\partial x} - \tfrac{1}{2}\left[\left(\frac{\partial u}{\partial x}\right)^2 + \left(\frac{\partial v}{\partial x}\right)^2 + \left(\frac{\partial w}{\partial x}\right)^2\right] \\
\epsilon_{xy} &= \tfrac{1}{2}\left(\frac{\partial u}{\partial y} + \frac{\partial v}{\partial x}\right) - \tfrac{1}{2}\left(\frac{\partial u}{\partial x}\frac{\partial u}{\partial y} + \frac{\partial v}{\partial x}\frac{\partial v}{\partial y} + \frac{\partial w}{\partial x}\frac{\partial w}{\partial y}\right)
\end{aligned} \tag{4.3.7}$$

We see that if the derivatives are small, so that their products can be neglected, these reduce to the same expressions as previously obtained for the infinitesimal strains. The simple physical interpretations given in Section 4.2 for small strains are, however, no longer applicable in this case. Thus if E_1 is the change in length per unit length of an element finally parallel to the x axis, it can be shown [2] that

$$E_1 = \sqrt{1 + 2\epsilon_x} - 1$$

Sec. 4–4] Principal Strains. Strain Invariants

which reduces to $E_1 \cong \epsilon_x$ only if $\epsilon_x \ll 1$. Similarly, the decrease in the right angle between two elements finally directed along the y and z axes is no longer $2\epsilon_{yz}$ but is given by

$$\sin \alpha = \frac{2\epsilon_{yz}}{\sqrt{1 + 2\epsilon_y}\,\sqrt{1 + 2\epsilon_z}}$$

which again reduces to $\alpha = 2\epsilon_{yz}$ only if ϵ_y, ϵ_z, and α are small.

In spite of the foregoing, however, it is possible sometimes to treat problems involving large strains using the equations for infinitesimal strains. This is possible if we do the problem incrementally a small step at a time and after each step change the coordinates to correspond to the deformed body [4]. We are then solving, in essence, a successive series of small strain problems. We will elaborate on this concept more fully in the treatment of creep problems.

4–4 PRINCIPAL STRAINS. STRAIN INVARIANTS

In the study of the stress at a point we found that there exist at least three mutually orthogonal planes which have no shear stress acting on them, i.e., the principal planes. We now ask ourselves, do there exist in a similar fashion planes with no shear strain? By this we mean planes whose normals will not change orientation when the body is strained. Thus a vector $\bar{A}$, which was originally normal to such a plane, will either shorten or lengthen but will not change direction. The answer to this question is yes. Such planes, as in the case of stress, are called *principal planes*, the normal directions to these planes are the *principal directions*, and the corresponding strains are called *principal strains*. To find these directions and the corresponding strains we proceed as follows.

Consider a vector $\bar{A}$ normal to the plane ABC as shown in Figure 4.4.1. Upon straining it is assumed that $\bar{A}$ changes length by an amount $\delta\bar{A}$ but its direction remains the same; i.e., $\delta\bar{A}$ is in the same direction as $\bar{A}$. Since $\bar{A}$ and $\delta\bar{A}$ are parallel, the components of $\bar{A}$ and of $\delta\bar{A}$ are proportional; i.e.,

$$\frac{\delta A}{A} = \frac{\delta A_x}{A_x} = \frac{\delta A_y}{A_y} = \frac{\delta A_z}{A_z} \tag{4.4.1}$$

Now by definition the strain ε in the direction of $\bar{A}$ is

$$\varepsilon = \frac{\delta A}{A} \tag{4.4.2}$$

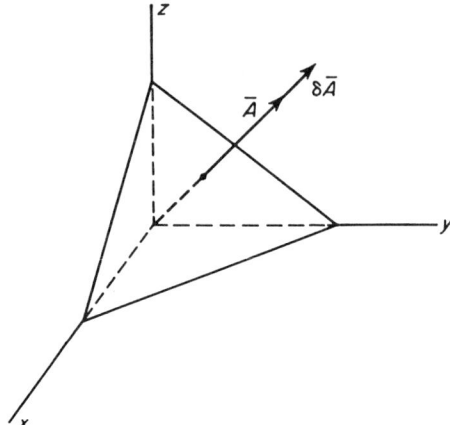

FIGURE 4.4.1 Principal strain vector.

Hence from (4.4.1),

$$\delta A_x = \varepsilon A_x \quad \delta A_y = \varepsilon A_y \quad \delta A_z = \varepsilon A_z \tag{4.4.3}$$

Writing out equation (4.1.14) we have

$$\begin{aligned} \delta A_x &= \varepsilon_x A_x + \varepsilon_{xy} A_y + \varepsilon_{xz} A_z \\ \delta A_y &= \varepsilon_{xy} A_x + \varepsilon_y A_y + \varepsilon_{yz} A_z \\ \delta A_z &= \varepsilon_{xz} A_x + \varepsilon_{yz} A_y + \varepsilon_z A_z \end{aligned} \tag{4.4.4}$$

Substituting (4.4.3) into (4.4.4) gives

$$\begin{aligned} (\varepsilon_x - \varepsilon) A_x + \varepsilon_{xy} A_y + \varepsilon_{xz} A_z &= 0 \\ \varepsilon_{xy} A_x + (\varepsilon_y - \varepsilon) A_y + \varepsilon_{yz} A_z &= 0 \\ \varepsilon_{xz} A_x + \varepsilon_{yz} A_y + (\varepsilon_z - \varepsilon) A_z &= 0 \end{aligned} \tag{4.4.5}$$

To illustrate again the power of the tensor notation, equations (4.4.3) can be written

$$\delta A_i = \varepsilon A_i$$

and substituting into (4.1.14) gives immediately

$$\varepsilon A_i = \varepsilon_{ij} A_j$$

or

$$(\varepsilon_{ij} - \delta_{ij} \varepsilon) A_j = 0$$

which is the same as (4.4.5).

Sec. 4–5] Maximum and Octahedral Shear Strains

The three homogeneous equations (4.4.5) will have a nonvanishing solution if, and only if, the determinant of the coefficients vanish. Thus

$$\begin{vmatrix} \varepsilon_x - \varepsilon & \varepsilon_{xy} & \varepsilon_{xz} \\ \varepsilon_{xy} & \varepsilon_y - \varepsilon & \varepsilon_{yz} \\ \varepsilon_{xz} & \varepsilon_{yz} & \varepsilon_z - \varepsilon \end{vmatrix} = 0 \qquad (4.4.6)$$

Equation (4.4.6) is of exactly the same form as (3.3.8) with stresses replaced by strains. All the remarks and derivations made there, therefore, apply here just as well. Thus (4.4.6) will have three real roots ε_1, ε_2, and ε_3, corresponding to the three principal strains. The invariants appearing in the cubic equation

$$\varepsilon^3 - I_1' \varepsilon^2 - I_2' \varepsilon - I_3' = 0 \qquad (4.4.7)$$

are

$$\begin{aligned} I_1' &= \varepsilon_x + \varepsilon_y + \varepsilon_z \\ I_2' &= \varepsilon_{xy}^2 + \varepsilon_{yz}^2 + \varepsilon_{zx}^2 - (\varepsilon_x \varepsilon_y + \varepsilon_y \varepsilon_z + \varepsilon_z \varepsilon_x) \\ I_3' &= \varepsilon_x \varepsilon_y \varepsilon_z + 2\varepsilon_{xy}\varepsilon_{yz}\varepsilon_{zx} - (\varepsilon_x \varepsilon_{yz}^2 + \varepsilon_y \varepsilon_{zx}^2 + \varepsilon_z \varepsilon_{xy}^2) \end{aligned} \qquad (4.4.8)$$

or in terms of the principal strains

$$\begin{aligned} I_1' &= \varepsilon_1 + \varepsilon_2 + \varepsilon_3 \\ I_2' &= -(\varepsilon_1 \varepsilon_2 + \varepsilon_2 \varepsilon_3 + \varepsilon_3 \varepsilon_1) \\ I_3' &= \varepsilon_1 \varepsilon_2 \varepsilon_3 \end{aligned} \qquad (4.4.9)$$

We note that if equations (4.4.5) are divided by A, then, since $A_x/A = l$, $A_y/A = m$, and $A_z/A = n$, A_x, A_y, and A_z can be replaced in these equations by l, m, and n, respectively. By substituting into these equations the principal strains as obtained by solving the cubic (4.4.7), the principal directions can be obtained, just as was done for the stress tensor.

4–5 MAXIMUM AND OCTAHEDRAL SHEAR STRAINS

Just as in the case of stress, there exists a direction at every point in a strained body for which the shear strain is a maximum. To find this direction we can proceed as follows. Let the coordinate axes be taken in the directions of the principal shear strains and consider a vector $\bar{A}$, as shown in Figure 4.5.1, having direction cosines l, m, and n with respect to these axes, which

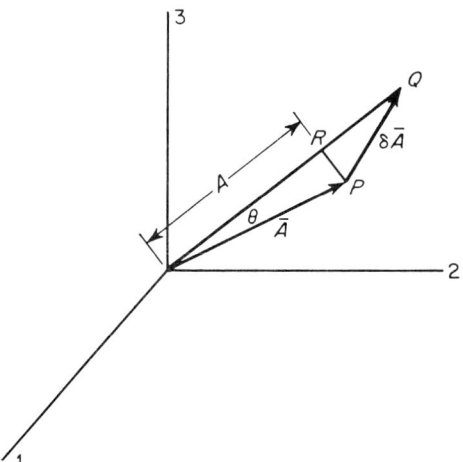

FIGURE 4.5.1 Maximum shear vectors.

are designated by 1, 2, and 3. Let the vector $\bar{A}$ suffer a small strain so that the point P moves to Q. The strain is thus composed of two parts: a linear strain of amount $\varepsilon = RQ/A$, and a rotation or shear of amount $\theta = RP/A$, for small θ. To calculate the linear strain, ε, we have

$$(A + RQ)^2 = (A_1 + \delta A_1)^2 + (A_2 + \delta A_2)^2 + (A_3 + \delta A_3)^2$$
$$A^2 + 2A(RQ) \simeq A_1^2 + A_2^2 + A_3^2 + 2(A_1\,\delta A_1 + A_2\,\delta A_2 + A_3\,\delta A_3) \quad (4.5.1)$$

where squares of small quantities have been neglected. Therefore,

$$\frac{RQ}{A} = \varepsilon = \frac{A_1\,\delta A_1}{A^2} + \frac{A_2\,\delta A_2}{A^2} + \frac{A_3\,\delta A_3}{A^2} \quad (4.5.2)$$

From the basic equation (4.1.14), which is given in expanded form in (4.4.4), we have

$$\delta A_1 = \varepsilon_1 A_1$$
$$\delta A_2 = \varepsilon_2 A_2 \quad (4.5.3)$$
$$\delta A_3 = \varepsilon_3 A_3$$

Making use of the relations $A_1/A = l$, $A_2/A = m$, and $A_3/A = n$ and substituting (4.5.3) into (4.5.2) results in

$$\varepsilon = l^2\varepsilon_1 + m^2\varepsilon_2 + n^2\varepsilon_3 \quad (4.5.4)$$

Sec. 4–5]   Maximum and Octahedral Shear Strains

Equation (4.5.4) gives the linear strain for a vector having direction cosines l, m, and n with respect to the principal strain directions, in terms of the principal strains. To determine the shear $\theta = RP/A$, we have

$$\varepsilon^2 + \theta^2 = \frac{(PQ)^2}{A^2} = \left(\frac{\delta A_1}{A}\right)^2 + \left(\frac{\delta A_2}{A}\right)^2 + \left(\frac{\delta A_3}{A}\right)^2$$

$$= \frac{\varepsilon_1^2 A_1^2}{A^2} + \frac{\varepsilon_2^2 A_2^2}{A^2} + \frac{\varepsilon_3^2 A_3^2}{A^2}$$

or $\quad \varepsilon^2 + \theta^2 = l_1^2 \varepsilon_1^2 + l_2^2 \varepsilon_2^2 + l_3^2 \varepsilon_3^2 \quad$ (4.5.5)

or $\quad \theta^2 = l_1^2 \varepsilon_1^2 + l_2^2 \varepsilon_2^2 + l_3^2 \varepsilon_3^2 - (l_1^2 \varepsilon_1 + l_2^2 \varepsilon_2 + l_3^2 \varepsilon_3)^2 \quad$ (4.5.6)

Comparing equations (4.5.6) with (3.4.2), it is seen that they are of identical forms, with S_s replaced by θ and the principal stresses by the principal strains. The maximum shear strains and the corresponding directions can therefore be obtained in the exact same way as for the stresses. Thus designating *the maximum shear strains* by γ_1, γ_2, and γ_3, we can write directly, by analogy to (3.4.4),

$$\gamma_1 = \pm \tfrac{1}{2}(\varepsilon_2 - \varepsilon_3)$$
$$\gamma_2 = \pm \tfrac{1}{2}(\varepsilon_1 - \varepsilon_3) \quad (4.5.7)$$
$$\gamma_3 = \pm \tfrac{1}{2}(\varepsilon_1 - \varepsilon_2)$$

and the same table of direction cosines is applicable as given on page 35. Thus the maximum shearing strain acts on a plane bisecting the angle between the maximum and minimum principal strain directions and is equal to half the difference of these strains.

If we consider the octahedral planes for which $l = m = n = \pm 1/\sqrt{3}$, we see from (4.5.6) that the shear strain on these planes, which we designate γ_{oct}, is given by

$$\gamma_{\text{oct}}^2 = \tfrac{1}{3}(\varepsilon_1^2 + \varepsilon_2^2 + \varepsilon_3^2) - \tfrac{1}{9}(\varepsilon_1 + \varepsilon_2 + \varepsilon_3)^2$$
$$= \tfrac{1}{9}[(\varepsilon_1 - \varepsilon_2)^2 + (\varepsilon_2 - \varepsilon_3)^2 + (\varepsilon_3 - \varepsilon_1)^2] \quad (4.5.8)$$

in exact analogy to the octahedral shear stress. In terms of the invariants of the strain tensor in analogy to (3.4.10) we have

$$\gamma_{\text{oct}}^2 = \tfrac{2}{9}[I_1'^2 + 3I_2'] \quad (4.5.9)$$

and in terms of nonprincipal strains this becomes

$$\gamma_{\text{oct}}^2 = \tfrac{1}{9}[(\varepsilon_x - \varepsilon_y)^2 + (\varepsilon_y - \varepsilon_z)^2 + (\varepsilon_z - \varepsilon_x)^2 + 6(\varepsilon_{xy}^2 + \varepsilon_{yz}^2 + \varepsilon_{zx}^2)] \quad (4.5.10)$$

It is also apparent by the same analogy [comparing equations (4.5.4) and (4.5.5) with (3.5.1)] that a Mohr's diagram can be constructed for the strains in an identical fashion to the stresses.

4–6 STRAIN DEVIATOR TENSOR

As in the case of the stress tensor, the strain tensor can be separated into two parts, a spherical part q_{ij} and deviator part e_{ij}. The spherical part is given by

$$q_{ij} = \varepsilon_m \delta_{ij} = \begin{bmatrix} \varepsilon_m & 0 & 0 \\ 0 & \varepsilon_m & 0 \\ 0 & 0 & \varepsilon_m \end{bmatrix} \quad (4.6.1)$$

where $\varepsilon_m = \frac{1}{3}(\varepsilon_1 + \varepsilon_2 + \varepsilon_3)$ is the mean strain. The deviator strain then becomes

$$e_{ij} = \begin{bmatrix} \varepsilon_x - \varepsilon_m & \varepsilon_{xy} & \varepsilon_{xz} \\ \varepsilon_{xy} & \varepsilon_y - \varepsilon_m & \varepsilon_{yz} \\ \varepsilon_{xz} & \varepsilon_{yz} & \varepsilon_z - \varepsilon_m \end{bmatrix}$$

$$= \begin{bmatrix} \dfrac{2\varepsilon_x - \varepsilon_z - \varepsilon_y}{3} & \varepsilon_{xy} & \varepsilon_{xz} \\ \varepsilon_{xy} & \dfrac{2\varepsilon_y - \varepsilon_z - \varepsilon_x}{3} & \varepsilon_{yz} \\ \varepsilon_{xz} & \varepsilon_{yz} & \dfrac{2\varepsilon_z - \varepsilon_x - \varepsilon_y}{3} \end{bmatrix} \quad (4.6.2)$$

or, in terms of the principal strains,

$$e_{ij} = \begin{bmatrix} \dfrac{2\varepsilon_1 - \varepsilon_2 - \varepsilon_3}{3} & 0 & 0 \\ 0 & \dfrac{2\varepsilon_2 - \varepsilon_3 - \varepsilon_1}{3} & 0 \\ 0 & 0 & \dfrac{2\varepsilon_3 - \varepsilon_1 - \varepsilon_2}{3} \end{bmatrix} \quad (4.6.3)$$

If we consider a rectangular parallelepiped with sides equal to a, b, and c, so that its initial volume is abc, then after straining its volume will be

$$a(1 + \varepsilon_1)b(1 + \varepsilon_2)c(1 + \varepsilon_3) \quad \text{or} \quad abc + abc(\varepsilon_1 + \varepsilon_2 + \varepsilon_3)$$

neglecting products of the strains. The change in volume per unit volume is then

$$\frac{\Delta V}{V} = \varepsilon_1 + \varepsilon_2 + \varepsilon_3$$

The spherical strain tensor is thus proportional to the volume change. The deviator strain tensor then represents a pure distortional strain.

By analogy with the stress deviator tensor discussed in Section 3.6, the invariants of the strain deviator tensor are

$$J_1' = 0$$
$$J_2' = \tfrac{1}{3}(I_1'^2 + 3I_2')$$
$$J_3' = \tfrac{1}{27}(2I_1'^3 + 9I_1'I_2' + 27I_3')$$
(4.6.4)

or

$$J_2' = \tfrac{1}{6}[(\varepsilon_1 - \varepsilon_2)^2 + (\varepsilon_2 - \varepsilon_3)^2 + (\varepsilon_3 - \varepsilon_1)^2]$$
$$= \tfrac{1}{6}[(\varepsilon_x - \varepsilon_y)^2 + (\varepsilon_y - \varepsilon_z)^2 + (\varepsilon_z - \varepsilon_x)^2 + 6(\varepsilon_{xy}^2 + \varepsilon_{yz}^2 + \varepsilon_{zx}^2)]$$
$$= -(e_1 e_2 + e_2 e_3 + e_3 e_1)$$
(4.6.5)
$$J_3' = e_1 e_2 e_3$$

where
$$e_i = \varepsilon_i - \varepsilon_m$$

Also it follows that

$$J_2' = \tfrac{3}{2}\gamma_{oct}^2$$
(4.6.6)

4-7 COMPATIBILITY OF STRAIN

The strain tensor ε_{ij} defines six strain components in terms of three displacements; i.e.,

$$\varepsilon_x = \frac{\partial u}{\partial x} \qquad \varepsilon_y = \frac{\partial v}{\partial y} \qquad \varepsilon_z = \frac{\partial w}{\partial z}$$
$$\varepsilon_{xy} = \tfrac{1}{2}\left(\frac{\partial u}{\partial y} + \frac{\partial v}{\partial x}\right) \quad \varepsilon_{yz} = \tfrac{1}{2}\left(\frac{\partial v}{\partial z} + \frac{\partial w}{\partial y}\right) \quad \varepsilon_{xz} = \tfrac{1}{2}\left(\frac{\partial u}{\partial z} + \frac{\partial w}{\partial x}\right)$$
(4.7.1)

It is obvious that if the displacements are specified as continuous functions of x, y, and z, then one can use equations (4.7.1) to compute the strain components uniquely. Let us now consider the inverse problem, i.e., to calculate the displacements u, v, and w in a body given the strain components ε_{ij}. Here several problems are encountered. In the first place, it is apparent that the solution will not be unique, for the strains represent pure deformation, whereas the displacements include rigid body motion which have no effect on the strains. The problem can, however, be made unique by specifying the rigid body motion, i.e., specifying the displacement and rotation at some point in the body.

However, a more difficult problem is encountered in calculating the displacements from the strains. Equations (4.7.1) are six equations for the three unknowns u, v, and w. It is evident, therefore, that these equations will not have a solution for any arbitrarily chosen strains, but that some restrictions must be placed on the strains in order that equations (4.7.1) have a solution. This can also be seen from the following physical considerations. Assume the body is divided into infinitesimal cubes. Let all these cubes be separated from each other and let each of the cubes be subjected to some arbitrary strains. It is obvious that if we now try to put all the cubes together, we will, in general, no longer be able to fit all the cubes together the way they were before, to produce a continuous body. Between some of the cube boundaries there will be gaps; others will overlap. This shows that there must be some relationships between the strains at the different points of a body in order that the body remain continuous after straining, i.e., that the displacements be continuous functions of the coordinates. These relationships are called the *compatibility relations*, or, sometimes, the *continuity relations*.

The compatibility equations are found in all standard texts on elasticity. They are derived, for example, in a particularly elegant fashion in reference [3]. In engineering notation these equations are

$$\frac{\partial^2 \varepsilon_x}{\partial y^2} + \frac{\partial^2 \varepsilon_y}{\partial x^2} = 2 \frac{\partial^2 \varepsilon_{xy}}{\partial x \, \partial y}$$

$$\frac{\partial^2 \varepsilon_y}{\partial z^2} + \frac{\partial^2 \varepsilon_z}{\partial y^2} = 2 \frac{\partial^2 \varepsilon_{yz}}{\partial y \, \partial z}$$

$$\frac{\partial^2 \varepsilon_z}{\partial x^2} + \frac{\partial^2 \varepsilon_x}{\partial z^2} = 2 \frac{\partial^2 \varepsilon_{xz}}{\partial x \, \partial z}$$

$$\frac{\partial}{\partial x}\left(-\frac{\partial \varepsilon_{yz}}{\partial x} + \frac{\partial \varepsilon_{zx}}{\partial y} + \frac{\partial \varepsilon_{xy}}{\partial z}\right) = \frac{\partial^2 \varepsilon_x}{\partial y \, \partial z}$$

$$\frac{\partial}{\partial y}\left(-\frac{\partial \varepsilon_{zx}}{\partial y} + \frac{\partial \varepsilon_{xy}}{\partial z} + \frac{\partial \varepsilon_{yz}}{\partial x}\right) = \frac{\partial^2 \varepsilon_y}{\partial z \, \partial x}$$

$$\frac{\partial}{\partial z}\left(-\frac{\partial \varepsilon_{xy}}{\partial z} + \frac{\partial \varepsilon_{yz}}{\partial x} + \frac{\partial \varepsilon_{zx}}{\partial y}\right) = \frac{\partial^2 \varepsilon_z}{\partial x \, \partial y}$$

(4.7.2)

These equations were first derived by Saint-Venant.

Problems

Equations (4.7.2) are *necessary* and *sufficient* conditions that the strain components give single-valued displacements for a simply connected region. For a multiply connected region, however, these conditions are *necessary* but generally *not* sufficient.

Just as for the stress equations of Chapter 3, it should also be emphasized here that all the relations presented in this chapter, including the compatibility relations (4.7.2), are independent of the material properties and therefore hold for both elastically and plastically behaving materials.

Note on Shear Notation

The shear strains defined herein differ from the shear strains defined by many authors by a factor of 1/2. Thus Timoshenko, for example, defines the shear strains by

$$\gamma_{xy} = \frac{\partial u}{\partial y} + \frac{\partial v}{\partial x}$$

$$\gamma_{xz} = \frac{\partial u}{\partial z} + \frac{\partial w}{\partial x}$$

$$\gamma_{yz} = \frac{\partial v}{\partial z} + \frac{\partial w}{\partial y}$$

The factor 1/2 used herein is necessary for the tensor definition of strain as exemplified by equation (4.1.12). Care must be exercised in reading the literature to determine which definition of shear strain is used.

Problems

1. Obtain the equations representing rigid body displacements; i.e., show that

$$u = a + \omega_y z - \omega_z y$$
$$v = b + \omega_z x - \omega_x z$$
$$w = c + \omega_x y - \omega_y x$$

where a, b, and c are constants and

$$\omega_x = \tfrac{1}{2}\left(\frac{\partial w}{\partial y} - \frac{\partial v}{\partial z}\right)$$

$$\omega_y = \tfrac{1}{2}\left(\frac{\partial u}{\partial z} - \frac{\partial w}{\partial x}\right)$$

$$\omega_z = \tfrac{1}{2}\left(\frac{\partial v}{\partial x} - \frac{\partial u}{\partial y}\right)$$

2. Let the rectangle of Figure 4.2.3 be a square. Show that the shear strain is equal to the extension of the diagonal of this square.
3. The displacement vector at a point in a body is given by

$$\bar{u} = C[i(10x + 3y) + j(3x + 2y) + k(6z)]$$

where C is a constant. Show that there is no rotation and compute the principal strains.

4. Given the strain tensor

$$\varepsilon_{ij} = \begin{bmatrix} -0.005 & -0.004 & 0 \\ -0.004 & 0.001 & 0 \\ 0 & 0 & 0.001 \end{bmatrix}$$

Determine:
(a) The principal strains.
(b) The direction cosines of the principal directions.
(c) The largest shearing strain.
(d) The octahedral shear strain.

5. If a rectangular parallelepiped with initial dimensions of 1 in. in the x direction, 2 in. in the y direction, and 3 in. in the z direction is strained to the condition of Problem 4, determine the final dimensions.
6. Derive the Lagrangian strain tensor.
7. Show that in plane polar coordinates the infinitesimal strains are given by

$$\varepsilon_r = \frac{\partial u}{\partial r} \qquad \varepsilon_\theta = \frac{1}{r}\frac{\partial v}{\partial \theta} + \frac{u}{r}$$

$$\varepsilon_{r\theta} = \frac{\partial v}{\partial r} + \frac{1}{r}\frac{\partial u}{\partial \theta} - \frac{v}{r}$$

where u and v are the displacements in the radial and tangential directions, respectively.

8. Assuming axial symmetry, derive the compatibility equation in plane polar coordinates.
9. Verify equations (4.5.9) and (4.5.10).
10. Verify equations (4.6.4) and (4.6.5).

References

1. I. S. Sokolnikoff, *Mathematical Theory of Elasticity*, McGraw-Hill, New York, 1956, p. 21.
2. Ibid., p. 31.
3. Ibid., p. 25.
4. W. Prager and P. G. Hodge, *Theory of Perfectly Plastic Solids*, Wiley, New York, 1951, p. 118.

References

General References

Novozhilov, V. V., *Theory of Elasticity*, Pergamon Press, London, 1961.
Sokolnikoff, I. S., *Mathematical Theory of Elasticity*, McGraw-Hill, New York, 1956.
Timoshenko, S., and J. N. Goodier, *Theory of Elasticity*, McGraw-Hill, New York, 1951.

CHAPTER 5

ELASTIC STRESS–STRAIN RELATIONS

In the previous chapters the states of stress and of strain at a point in a body have been defined and some of their properties discussed. Equations (3.2.2) give the conditions of equilibrium that must be satisfied by the stresses at every point of a body, whereas equations (4.7.2) represent the compatibility conditions that must be satisfied by the strains. The boundary conditions in terms of surface forces are given by equations (3.3.1). There remains to discuss the relations between the stresses and the strains at every point of the body. We shall discuss briefly these relations for an elastic body, as a prelude to the general elastoplastic stress–strain relations to be subsequently discussed.

5–1 EQUATIONS OF ELASTICITY

Hooke first proposed a linear relation between stress and strain for a load applied in one direction. The generalization of Hooke's law to three dimensions is given in Problem 3, Chapter 3. For an isotropic material this becomes, using the tensor subscript notation and including thermal strains,

$$\varepsilon_{ij} = \frac{1}{2G} \sigma_{ij} - \delta_{ij}\left(\frac{\mu}{E} \Theta - \alpha T\right) \qquad (5.1.1)$$

where E is the elastic modulus, μ Poisson's ratio, α the coefficient of linear

Sec. 5–1] Equations of Elasticity

thermal expansion, T the temperature above some arbitrary reference temperature, G the shear modulus related to E and μ by the well-known relation

$$G = \frac{E}{2(1 + \mu)} \tag{5.1.2}$$

and

$$\Theta = \sigma_x + \sigma_y + \sigma_z = I_1 \tag{5.1.3}$$

Equation (5.1.1) can be solved for the stresses to give

$$\sigma_{ij} = 2G(\varepsilon_{ij} - \delta_{ij}\alpha T) + \delta_{ij}\lambda(\theta - 3\alpha T) \tag{5.1.4}$$

where

$$\theta = \varepsilon_x + \varepsilon_y + \varepsilon_z = I_1'$$
$$\lambda = \frac{\mu E}{(1 + \mu)(1 - 2\mu)} \tag{5.1.5}$$

In engineering notation equations (5.1.1) become

$$\varepsilon_x = \frac{1}{E}[\sigma_x - \mu(\sigma_y + \sigma_z)] + \alpha T$$

$$\varepsilon_y = \frac{1}{E}[\sigma_y - \mu(\sigma_x + \sigma_z)] + \alpha T$$

$$\varepsilon_z = \frac{1}{E}[\sigma_z - \mu(\sigma_x + \sigma_y)] + \alpha T$$

$$\varepsilon_{xy} = \frac{1}{2G}\tau_{xy} = \frac{1 + \mu}{E}\tau_{xy} \tag{5.1.6}$$

$$\varepsilon_{yz} = \frac{1}{2G}\tau_{yz} = \frac{1 + \mu}{E}\tau_{yz}$$

$$\varepsilon_{zx} = \frac{1}{2G}\tau_{zx} = \frac{1 + \mu}{E}\tau_{zx}$$

It readily follows from (5.1.6) that

$$\theta = \frac{1 - 2\mu}{E}\Theta + 3\alpha T \tag{5.1.7}$$

or

$$\varepsilon_m = \frac{1 - 2\mu}{E}\sigma_m + \alpha T \tag{5.1.8}$$

where ε_m and σ_m are the mean strain and mean stress, respectively. Finally, combining (5.1.8) and (5.1.1) results in

$$e_{ij} = \frac{1}{2G} S_{ij} \qquad (5.1.9)$$

where e_{ij} and S_{ij} are the strain deviator and stress deviator tensors, respectively. Thus the deviators of the stress and strain tensors are related to each other, in the elastic case, by the simple equation (5.1.9), whereas the spherical stress components are related to the spherical strain components by equation (5.1.8).

It also follows that

$$J'_1 = J_1 = 0 \qquad (5.1.10)$$
$$J'_2 = \frac{1}{4G^2} J_2$$

$$\gamma^2_{oct} = \frac{1}{4G^2} \tau^2_{oct} \qquad (5.1.11)$$

It should be noted that nothing in the foregoing discussion requires that E, G, μ, or α be constant throughout the body. They may, for example, be functions of temperature, so that if the body is not at a uniform temperature, these constants may have different values at different points in the body.

It is probably worthwhile to briefly discuss the origin of the αT terms appearing in equations (5.1.1). We start by defining the *free thermal expansion* of a material as that part of the expansion which is uninfluenced by stress but is due to temperature rise alone. We also make the following four assumptions:

1. The material is isotropic and, therefore, the free thermal expansion is the same in all directions.
2. The free thermal expansion is directly proportional to the temperature.
3. The principle of superposition of strains holds.
4. The thermal expansion does not influence the shear strains.

Let us now denote the strains due to temperature rise by primes and the strains due to stresses by double primes. Then assumptions 1 and 2 enable us to write

$$\varepsilon'_x = \varepsilon'_y = \varepsilon'_z = \alpha T$$

Sec. 5-2] Elastic Strain Energy Functions

where α is the coefficient of linear thermal expansion. From Hooke's law we of course have

$$\varepsilon_x'' = \frac{1}{E}[\sigma_x - \mu(\sigma_y + \sigma_z)]$$

$$\varepsilon_y'' = \frac{1}{E}[\sigma_y - \mu(\sigma_z + \sigma_x)]$$

$$\varepsilon_z'' = \frac{1}{E}[\sigma_z - \mu(\sigma_x + \sigma_y)]$$

and now making use of assumption 3, the total strains are as given by the first three of equations (5.1.6), whereas Hooke's law and assumption 4 give the last three of equations (5.1.6). Note that assumption 2 may be modified to the extent that α may be a function of temperature.

5-2 ELASTIC STRAIN ENERGY FUNCTIONS

If a body is deformed by the action of external forces, then the latter do work on the body. If the body is in equilibrium so that none of this work goes into kinetic energy, then this work is stored as strain energy of deformation in the body. The elastic strain energy can be written as follows [1]:

$$U = \tfrac{1}{2}\sigma_{ij}\varepsilon_{ij} \qquad (5.2.1)$$

If we substitute into (5.2.1) equation (5.1.4), there results

$$U = G\varepsilon_{ij}\varepsilon_{ij} + \frac{\lambda}{2}\theta^2 - \frac{2G + 3\lambda}{2}\alpha T\theta \qquad (5.2.2)$$

Making use of the definitions of the invariants of the strain and strain deviator tensors, equation (5.2.2) can be written after some algebraic manipulations as

$$U = \frac{2G + 3\lambda}{6}(I_1'^2 - 3\alpha T I_1') + 2GJ_2' \qquad (5.2.3)$$

Now I_1' represents the spherical state of strain which is proportional to the volume change, whereas J_2' is an invariant of the deviator strain tensor which represents a pure distortional strain. Equation (5.2.3) can therefore be considered as the sum of two energy terms. The first term is the energy involved in changing the volume; the second term is the energy of distortion, which is designated by U_d and can be written

$$U_d = 2GJ'_2 = 3G\gamma^2_{\text{oct}}$$

or, from (5.1.10),

$$U_d = \frac{1}{2G}J_2 = \frac{3}{4G}\tau^2_{\text{oct}} \qquad (5.2.4)$$

5-3 SOLUTION OF ELASTIC PROBLEMS

To solve an elastic problem we have to find the stresses and the strains which will satisfy the previously derived equations. Thus the stresses must satisfy the equilibrium equations, (3.2.2) and (3.2.3), as well as the boundary conditions (3.3.5). The strains must satisfy the compatibility equations (4.7.2). Finally, the stresses must be related to the strains through the stress–strain relations (5.1.1) or their equivalent. The problem of finding a set of stresses and strains satisfying the above relations is known as the *first boundary-value problem* of elasticity. Alternatively, it is possible to reduce the above set of equations to three equations, called the *Navier equations*, involving only the displacement u_i [2].

$$G\nabla^2 u_i + (\lambda + G)\theta_{,i} + F_i = 0$$

where ∇^2 is the Laplacian operator, λ and G are Lamé's constants previously defined, F_i are the body forces per unit volume, and θ is the invariant, defined in (5.1.5) and which can also be written

$$\theta = \frac{\partial u}{\partial x} + \frac{\partial v}{\partial y} + \frac{\partial w}{\partial z}$$

This last formulation is most convenient if the displacements are specified on the boundary rather than the forces. The problem is then known as the *second boundary-value problem* of elasticity.

From the previous derivations it is apparent that both the equilibrium equations and compatibility relations are independent of the properties of the material under consideration. The equilibrium equations express the static equilibrium of an element of the body and the compatibility relations express the continuity of the body. These equations will therefore hold whether the body behaves elastically or whether plastic flow occurs. The same holds for the boundary conditions. The difference between the elastic problem

References

and the plastic problem occurs in the stress–strain relations. If plastic flow occurs, the linear generalized Hooke's law no longer holds and the relations between stress and strain become nonlinear in a very complicated manner. What happens to the relations between stress and strain when plastic flow occurs will be the subject of Chapters 6 and 7.

Problems

1. Show how equation (5.1.4) can be obtained from equation (5.1.1).
2. Derive equation (5.1.9).
3. Given the following stress tensor at a point:

$$\begin{matrix} 10{,}000 & 1{,}000 & -8{,}000 \\ 1{,}000 & -6{,}000 & 6{,}000 \\ -8{,}000 & 6{,}000 & 20{,}000 \end{matrix}$$

   Assume the material behaves elastically with an elastic modulus of 30×10^6 and Poisson's ratio of 0.3. Determine the strain deviator tensor. Calculate the mean strain assuming the temperature rise at the point is 300°F and that $\alpha = 7.5 \times 10^{-6}$ in./in./°F.
4. Verify equations (5.1.10) and (5.1.11).
5. Prove that the axes of principal stress coincide with the axes of principal strain for an isotropic material obeying Hooke's law.
6. Prove that for the material of Problem 5, Mohr's stress diagram will be similar to Mohr's strain diagram.
7. Derive equation (5.2.3).

References

1. I. S. Sokolnikoff, *Mathematical Theory of Elasticity*, McGraw-Hill, New York, 1956, p. 84.
2. *Ibid.*, p. 73.

General References

Hoffman, O., and G. Sachs, *Introduction to the Theory of Plasticity for Engineers*, McGraw-Hill, New York, 1953.

Sokolnikoff, I. S., *Mathematical Theory of Elasticity*, McGraw-Hill, New York, 1956.

CHAPTER 6

CRITERIA FOR YIELDING

6-1 EXAMPLES OF MULTIAXIAL STRESS

In Chapter 2 the stress–strain curve for a material in simple tension was discussed and it was shown that there exists a yield point at which the material will begin to deform plastically. In this case the stress is uniaxial and this point can readily be determined. But what if there are several stresses acting at a point in different directions? What combination of these stresses will cause yielding? We know, for example, that for a hydrostatic stress, i.e., equal stresses in all directions, yielding does not occur even for very large values of stress. As another example involving a test which can be performed without too much difficulty, consider a thin-walled cylinder which is being pulled axially by a load P, is being twisted by a twisting moment T, and is pressurized internally (see Figure 6.1.1) by a pressure p.

By varying the pressure p, the axial pull P, and the torque T, it is possible to get various combinations of stresses, which will also of course result in

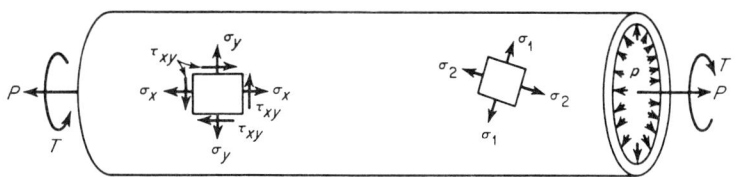

FIGURE 6.1.1 Combined stresses in thin-walled cylinder.

different principal directions. The question here is: For what combinations of loads will the cylinder begin to yield plastically?

Another simple example is the plane stress problem of a thin rotating disk with or without temperature gradients. At every point of the disk (except possibly at the rim) there exists a state of biaxial stress. The question again arises: For which states of biaxial stress will the disk deform plastically? The criteria for deciding which combination of multiaxial stresses will cause yielding are called *yield criteria*. The first step of any plastic flow analysis is to decide on a yield criterion. The next step is to decide how to describe the behavior of the material after yielding has started. In this chapter we shall discuss the choice of a yield criterion.

6-2 EXAMPLES OF YIELD CRITERIA

Numerous criteria have been proposed for the yielding of solids, going as far back as Coulomb in 1773. Many of these were originally suggested as criteria for failure of brittle materials and were later adopted as yield criteria for ductile materials. Some of the more common ones will be briefly discussed. Although some of these theories are no longer in use, they are included here both for their historic interest and to give the reader a feeling for the type of approach used in promulgating yield criteria.

Maximum Stress Theory, or Rankine Theory

This theory assumes that yielding occurs when one of the principal stresses becomes equal to the yield stress in simple tension σ_0, or the yield stress in compression $\sigma_{0,c}$. Thus if σ_1 is the maximum principal stress and σ_2 is the minimum principal stress, yielding will occur in tension when $\sigma_1 = \sigma_0$ and it will occur in compression when $\sigma_2 = \sigma_{0,c}$. For a material with the same yield in tension and compression, this criterion becomes

$$\sigma_1 = \sigma_0$$

or (6.2.1)

$$\sigma_2 = -\sigma_0$$

A simple plot illustrating this criterion for the case of biaxial stress with $\sigma_3 = 0$ is shown in Figure 6.2.1. The coordinates are the remaining principal stresses σ_1 and σ_2. Yielding occurs when the state of stress is on the boundary of the rectangle, for then one of the stresses is at the yield point in tension or compression. For example, consider a thin-walled cylinder subjected to an

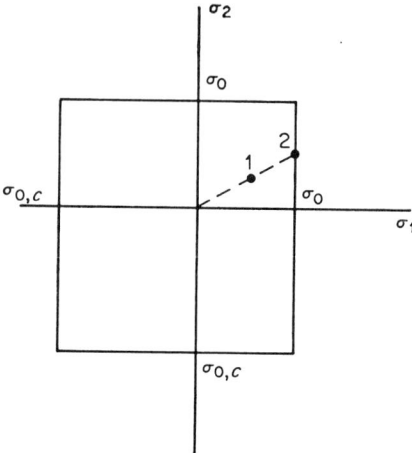

FIGURE 6.2.1   Maximum stress theory.

increasing internal pressure p. Let σ_1 be the circumferential stress and σ_2 the axial stress. Then $\sigma_1 = 2\sigma_2$. As the pressure is increased from zero, the stresses follow the dashed line of Figure 6.2.1, as shown, σ_1 always being equal to twice σ_2. At point 1 the cylinder is still elastic, neither stress having reached the value of σ_0. At point 2, σ_1 is equal to σ_0 and yielding begins even though σ_2 is only $\tfrac{1}{2}\sigma_0$. This maximum stress criterion, however, shows very poor agreement with experiment and is rarely used.

Maximum Strain Theory, or Saint-Venant Theory

This theory assumes yielding will occur when the maximum value of the principal strain equals the value of the yield strain in simple tension (or compression), $\varepsilon_0 = \sigma_0/E$. Thus if ε_1 is assumed to be the largest strain in absolute value, yielding will occur when

$$E\varepsilon_1 = \sigma_1 - \mu(\sigma_2 + \sigma_3) = \pm\sigma_0 \qquad (6.2.2)$$

or, for the biaxial case with $\sigma_3 = 0$,

$$\begin{aligned} E\varepsilon_1 &= \sigma_1 - \mu\sigma_2 = \pm\sigma_0 \quad \text{for } |\sigma_1| \geq |\sigma_2| \\ E\varepsilon_2 &= \sigma_2 - \mu\sigma_1 = \pm\sigma_0 \quad \text{for } |\sigma_2| \geq |\sigma_1| \end{aligned} \qquad (6.2.3)$$

A plot in the $\sigma_1\sigma_2$ plane showing the boundary at which yielding begins is shown in Figure 6.2.2. This theory also does not agree well with most experiments. It has, however, been used in the design of guns, since some experimental results on thick-walled cylinders are in agreement with this theory [1].

Sec. 6-2] Examples of Yield Criteria

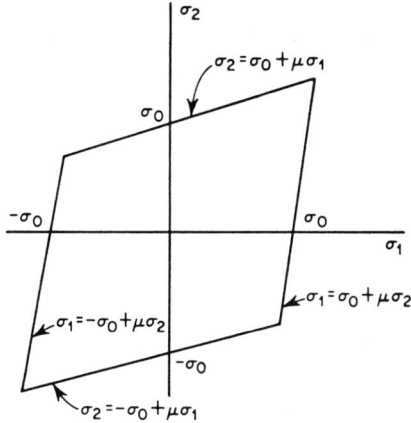

FIGURE 6.2.2 Maximum strain theory.

Maximum Shear Theory, or Tresca Criterion

This theory (sometimes called the Coulomb theory) assumes that yielding will occur when the maximum shear stress reaches the value of the maximum shear stress occurring under simple tension. The maximum shear stress is given by equation (3.4.4) and is equal to half the difference between the maximum and minimum principal stresses. For simple tension, therefore, since $\sigma_2 = \sigma_3 = 0$, the maximum shear stress at yield is $\frac{1}{2}\sigma_0$. The Tresca criterion then asserts that yielding will occur when any one of the following six conditions is reached:

$$\sigma_1 - \sigma_2 = \pm \sigma_0$$
$$\sigma_2 - \sigma_3 = \pm \sigma_0 \qquad (6.2.4)$$
$$\sigma_3 - \sigma_1 = \pm \sigma_0$$

For the biaxial case with $\sigma_3 = 0$, we have

$$\begin{aligned}
\sigma_1 - \sigma_2 &= \sigma_0 & &\text{if } \sigma_1 > 0,\ \sigma_2 < 0 \\
\sigma_1 - \sigma_2 &= -\sigma_0 & &\text{if } \sigma_1 < 0,\ \sigma_2 > 0 \\
\sigma_2 &= \sigma_0 & &\text{if } \sigma_2 > \sigma_1 > 0 \\
\sigma_1 &= \sigma_0 & &\text{if } \sigma_1 > \sigma_2 > 0 \\
\sigma_1 &= -\sigma_0 & &\text{if } \sigma_1 < \sigma_2 < 0 \\
\sigma_2 &= -\sigma_0 & &\text{if } \sigma_2 < \sigma_1 < 0
\end{aligned} \qquad (6.2.5)$$

A plot in the $\sigma_1\sigma_2$ plane for this yield criterion is shown in Figure 6.2.3. It is to be noted that one limitation of this theory is the requirement that the yield

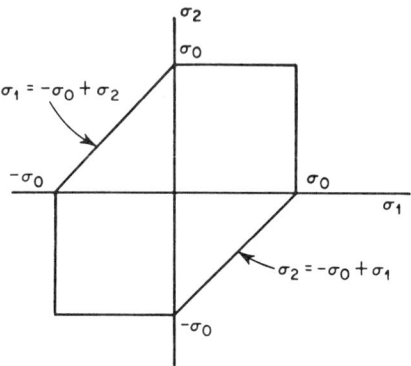

FIGURE 6.2.3 Maximum shear stress theory.

stresses in tension and compression be equal. The Tresca criterion is in fair agreement with experiment and is used to a considerable extent by designers. It suffers, however, from one major difficulty—it is necessary to know in advance which are the maximum and minimum principal stresses. For the case of pure shear (see Section 3.7),

$$\sigma_1 = -\sigma_2 = k \qquad \sigma_3 = 0$$

the Tresca criterion predicts yielding to occur when

$$\sigma_1 - \sigma_2 = 2k = \sigma_0$$

or
$$k = \tfrac{1}{2}\sigma_0$$

That is, *the yield stress in pure shear is $\tfrac{1}{2}$ the yield stress in simple tension.*

Maximum Strain Energy Theory, or Beltrami's Energy Theory

This theory assumes that yielding will occur when the total strain energy per unit volume equals the total strain energy per unit volume at yielding in uniaxial tension or compression. The total strain energy at yield in the tensile test is

$$\tfrac{1}{2}\sigma_0\varepsilon_0 = \frac{1}{2E}\sigma_0^2$$

Sec. 6-2] Examples of Yield Criteria

and the total strain energy U is given by

$$U = \tfrac{1}{2}(\sigma_1 \varepsilon_1 + \sigma_2 \varepsilon_2 + \sigma_3 \varepsilon_3)$$

$$= \frac{1}{2E}[\sigma_1^2 + \sigma_2^2 + \sigma_3^2 - 2\mu(\sigma_1\sigma_2 + \sigma_2\sigma_3 + \sigma_3\sigma_1)]$$

The yield criterion becomes

$$\sigma_1^2 + \sigma_2^2 + \sigma_3^2 - 2\mu(\sigma_1\sigma_2 + \sigma_2\sigma_3 + \sigma_3\sigma_1) = \sigma_0^2 \tag{6.2.6}$$

For the biaxial case this becomes

$$\sigma_1^2 + \sigma_2^2 - 2\mu\sigma_1\sigma_2 = \sigma_0^2 \tag{6.2.7}$$

This is the equation of an ellipse in the $\sigma_1\sigma_2$ plane. It is apparent from (6.2.6) that yielding can occur under sufficiently high hydrostatic pressure $\sigma_1 = \sigma_2 = \sigma_3$, which, as has been shown, is contrary to experiment. This is also apparent from equation (5.2.4), where it is shown that only the distortional strain energy can contribute to yielding. This theory has therefore been superseded by the theory described next.

Distortion Energy Theory, or the von Mises Yield Criterion

The distortion energy theory (also associated with Hencky) assumes that yielding begins when the distortion energy equals the distortion energy at yield in simple tension. Thus from equation (5.2.4),

$$U_d = \frac{1}{2G} J_2 = \frac{3}{4G} \tau_{\text{oct}}^2$$

At the yield point in simple tension, from (3.6.8),

$$J_2 = \tfrac{1}{3}\sigma_0^2$$

Therefore the yield condition becomes

$$\tfrac{1}{2}[(\sigma_1 - \sigma_2)^2 + (\sigma_2 - \sigma_3)^2 + (\sigma_3 - \sigma_1)^2] = \sigma_0^2 \tag{6.2.8}$$

and, for the biaxial case,

$$\sigma_1^2 - \sigma_1\sigma_2 + \sigma_2^2 = \sigma_0^2 \tag{6.2.9}$$

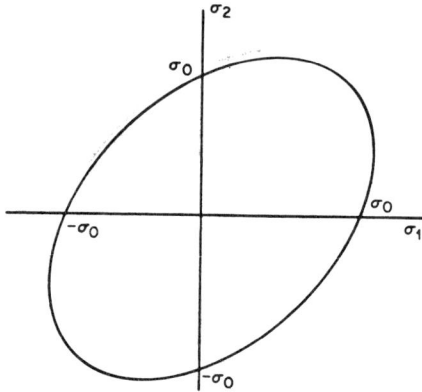

FIGURE 6.2.4 Distortion energy theory.

This plots as an ellipse, called the *von Mises ellipse*, in the $\sigma_1 \sigma_2$ plane, as shown in Figure 6.2.4. For the case of pure shear

$$\sigma_1 = -\sigma_2 = k \qquad \sigma_3 = 0$$
$$J_2 = \tfrac{1}{6}[(\sigma_1 - \sigma_2)^2 + (\sigma_2 - \sigma_3)^2 + (\sigma_3 - \sigma_1)^2]$$
$$= \sigma_1^2 = k^2$$

and the von Mises criterion would predict yielding to occur when

$$k^2 = \tfrac{1}{3}\sigma_0^2$$

or
$$k = \frac{\sigma_0}{\sqrt{3}}$$

That is, *the yield stress in pure shear is $1/\sqrt{3}$ times the yield stress in simple tension.* Thus the von Mises criterion predicts a pure shear yield stress which is about 15 per cent higher than predicted by the Tresca criterion. It will subsequently be shown that this is the maximum difference between the two criteria.

The von Mises yield criterion usually fits (but not always) the experimental data better than the other theories, and it is usually easier to apply than the Tresca criterion because no knowledge is needed regarding the relative magnitudes of the principal stresses. For these reasons, this criterion is widely used at the present time. If, however, the relative magnitudes of the principal stresses are known, as, for example, in the case of the thick-walled tube discussed in Chapter 8, the Tresca criterion is easier to apply.

Sec. 6–2] Examples of Yield Criteria

Von Mises originally proposed his criterion because of mathematical convenience. Hencky later showed that it was equivalent to assuming that yielding will take place when the distortion or shear strain energy reaches a critical value, as shown above. Also, since the octahedral shear stress is equal to

$$\tau_{oct} = \tfrac{1}{3}\sqrt{(\sigma_1 - \sigma_2)^2 + (\sigma_1 - \sigma_3)^2 + (\sigma_3 - \sigma_2)^2}$$

which for simple tension at yield becomes

$$\tau_{oct,0} = \frac{\sqrt{2}}{3}\sigma_0$$

then equation (6.2.8) can be written

$$\tau_{oct} = \tau_{oct,0}$$

That is, yielding will occur when the octahedral shear stress reaches the octahedral shear stress at yield in simple tension.

Alternatively, the criterion (6.2.8) can be looked upon as stating that yielding will occur when the second invariant J_2 of the stress deviator tensor reaches a critical value, i.e., the value of J_2 at yield in simple tension. The assumption that the yield criterion should depend on the invariants of the stress deviator tensor is generally accepted, as will be discussed in the next section.

Mohr's Theory of Yielding

Mohr extended the maximum shear stress theory by assuming that the critical shear stress is not necessarily equal to the maximum shear stress but depends also on the normal stress acting on the shearing plane. In general, the greater the normal stress, the lower is the critical shear stress. Mohr's theory therefore takes into account the effect of mean stress, which has experimentally been shown to be important in some cases, particularly with regard to fracture. Mohr's theory can also take into account differences in the yield points in tension and compression, as exemplified by the *internal friction theory*.

Internal Friction Theory

This is a special case of Mohr's theory, in which the critical shear stress is assumed to be a linear function of the normal stress acting on the plane of maximum shear. This results in the following conditions for the biaxial case [1a]: If $\sigma_1 > \sigma_2 > \sigma_3$, $\sigma_2 = 0$,

$$\sigma_1 - \frac{\sigma_0}{\sigma_{0,c}} \sigma_3 = \sigma_0 \tag{6.2.10}$$

where $\sigma_{0,c}$ is the yield stress in compression. If $\sigma_1 > \sigma_3 > \sigma_2$, $\sigma_2 = 0$, then

$$\sigma_1 = \sigma_0$$

If $\sigma_1 > \sigma_3$, both negative, then (6.2.11)

$$\sigma_3 = -\sigma_{0,c}$$

A plot in the $\sigma_1\sigma_3$ plane is shown in Figure 6.2.5.

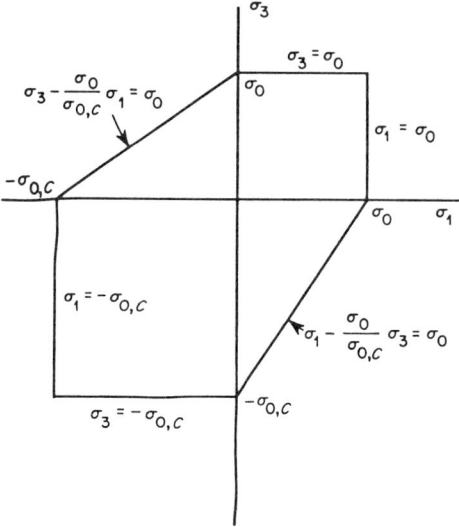

FIGURE 6.2.5 Internal friction theory.

The various yield criteria listed by no means exhaust the available list. Besides, there are other, more complicated theories which attempt to take into account anisotropy. For example, a yield criterion proposed by Marin [2] for anisotropic materials with different yields in tension and compression has the form (for biaxial stress)

$$\sigma_1^2 + \left[2 - \frac{\sigma_{c1}\sigma_{t1} - \tau_0(\sigma_{c1} - \sigma_{t1} - (\sigma_{c1}\sigma_{t1}/\sigma_{t2}) + \sigma_{t2})}{\tau_0^2}\right]\sigma_1\sigma_2$$

$$+ \sigma_2^2 + (\sigma_{c1} - \sigma_{t1})\sigma_1 + \left(\frac{\sigma_{c1}\sigma_{t1}}{\sigma_{t2}} - \sigma_{t2}\right)\sigma_2 = \sigma_{c1}\sigma_{t1}$$

where σ_{t1} and σ_{t2} are the tensile yields in the σ_1 and σ_2 directions, σ_{c1} is the compressive yield in the σ_1 direction, and τ_0 is the shear yield strength.

6-3 YIELD SURFACE. HAIGH-WESTERGAARD STRESS SPACE

In Section 6.2 we discussed several yield criteria and also plotted several two-dimensional plots for biaxial stress cases, showing the curves at which yielding takes place. In the most general case, the yield criterion will depend on the complete state of stress at the point under consideration and will therefore be a function of the nine components of stress at the point. Since the stress tensor is symmetric, we can reduce this function to a function of the six independent components of the stress tensor. This yield criterion for a virgin material is then essentially the extension of the single yield point of the uniaxial tensile test to the six-component stress tensor. For a material loaded to the initial yield, it can be expressed by the relationship

$$F(\sigma_{ij}) = K \tag{6.3.1}$$

where K is a known function, or if desired,

$$F_1(\sigma_{ij}) = 0 \tag{6.3.2}$$

Equation (6.3.1) represents a hypersurface in the six-dimensional stress space and any point on this surface represents a point at which yielding can begin. For example, for the simple tensile test at the yield point σ_0, the point $\sigma_x = \sigma_0$, $\sigma_y = \sigma_z = \tau_{xy} = \tau_{xz} = \tau_{yz} = 0$, must lie on this surface, and the point $\sigma_x = \sigma_y = \sigma_z = \tau_{xz} = \tau_{yz} = 0$, $\tau_{xy} = \tau_0$, which represents a thin-walled tube loaded in torsion to the torsional yield τ_0, must also lie on this surface.

The function appearing in (6.3.1) is called the *yield function* and the surface described by (6.3.1) or (6.3.2) in the stress space is called the *yield surface*. Without specifying yet any particular form for this surface, the equation describing it can be simplified somewhat by making use of some of the previously discussed assumptions regarding the yielding of metals. If, as usual, isotropy is assumed so that rotating the axes does not affect the yielding, we can choose the principal axes for the coordinates, and then (6.3.1) can be written

$$F_2(\sigma_1, \sigma_2, \sigma_3) = K \tag{6.3.3}$$

Furthermore, since it is always assumed that hydrostatic tension or compression does not influence yielding, we can assume that only the stress deviators enter into the yield function and write

$$f_1(S_1, S_2, S_3) = K \qquad (6.3.4)$$

Alternatively, since S_1, S_2, and S_3 can be written in terms of the invariants J_1, J_2, and J_3, where

$$J_1 = S_1 + S_2 + S_3 = 0$$
$$J_2 = \tfrac{1}{2}(S_1^2 + S_2^2 + S_3^2) = -(S_1 S_2 + S_2 S_3 + S_3 S_1) \qquad (6.3.5)$$
$$J_3 = \tfrac{1}{3}(S_1^3 + S_2^3 + S_3^3) = S_1 S_2 S_3$$

we can write

$$f(J_2, J_3) = K \qquad (6.3.6)$$

Subject therefore to the above two assumptions, the yield criterion has been reduced to a function of the two nonzero invariants of the stress deviator tensor. We note in passing that $f(J_2, J_3)$ is symmetric in the principal stresses, which is to be expected since in an isotropic material all of the principal stresses must play the same role in yielding. Thus whatever yield function is chosen, it must be symmetric in the principal stresses.

The two most widely used yield criteria, the Tresca maximum shear criterion and the von Mises yield criterion, discussed in Section 6.2, are specific cases of (6.3.6). The von Mises criterion is by far the simplest one that can be associated with equation (6.3.6),

$$J_2 = \tfrac{1}{3}\sigma_0^2 = k^2 \qquad (6.3.7)$$

where k is the yield in pure shear.

The Tresca criterion can be written in the form

$$4J_2^3 - 27J_3^2 - 36k^2 J_2^2 + 96k^4 J_2 - 64k^6 = 0 \qquad (6.3.8)$$

and we see the great complexity of the Tresca criterion in its general form compared to the von Mises criterion. Only in the case, as previously discussed, where the maximum and minimum principal stresses are known a priori can the Tresca criterion be reduced to the simple form

$$\sigma_1 - \sigma_3 = 2k \qquad (6.3.9)$$

Sec. 6-3] Yield Surface. Haigh-Westergaard Stress Space

We can learn a great deal about the yield surface defined by equation (6.3.3) from simple geometric considerations. We introduce the $(\sigma_1, \sigma_2, \sigma_3)$ coordinate system, which represents a stress space called the *Haigh-Westergaard stress space* [3]. Every point in this space having coordinates σ_1, σ_2, and σ_3 is a possible stress state. Consider a line ON as shown in Figure 6.3.1

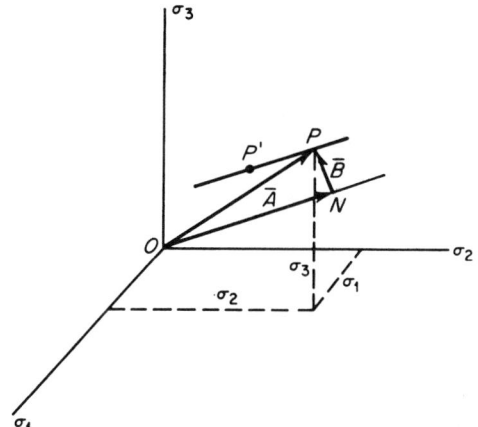

FIGURE 6.3.1 Haigh-Westergaard stress space.

passing through the origin, and having equal angles with the coordinate axes. Then for every point on this line the stress state is one for which

$$\sigma_1 = \sigma_2 = \sigma_3 = \sigma_m \quad (6.3.10)$$

Thus every point on this line corresponds to a hydrostatic or spherical stress state, the deviatoric stresses $S_1 = (2\sigma_1 - \sigma_2 - \sigma_3)/3$, etc., being equal to zero. Furthermore, if we consider any plane perpendicular to ON, the equation of this plane will be

$$\sigma_1 + \sigma_2 + \sigma_3 = \sqrt{3}\,\rho \quad (6.3.11)$$

where ρ is the distance along the normal from the origin to the plane. Hence the spherical component of the stress tensor increases linearly with the distance of the plane from the origin. On the plane passing through the origin, the spherical stress is zero, the equation being $\sigma_1 + \sigma_2 + \sigma_3 = 0$. This plane is called the π *plane*.

Now consider any arbitrary stress state such as at point P with stress components σ_1, σ_2, and σ_3. The stress vector $\overline{OP}$ can be decomposed into two components, the component $\overline{A}$ parallel to ON and the component $\overline{B}$ perpendicular to ON, the latter parallel to the π plane. The component $\overline{A}$ can readily be determined by projecting the components of $\overline{OP}$—σ_1, σ_2, and σ_3—on to ON. Thus

$$A = \frac{1}{\sqrt{3}} \sigma_1 + \frac{1}{\sqrt{3}} \sigma_2 + \frac{1}{\sqrt{3}} \sigma_3$$

or
$$A = \sqrt{3}\, \sigma_m \qquad (6.3.12)$$

To determine $\bar{B}$ we have

$$\begin{aligned}
B^2 &= \overline{OP}^2 - A^2 \\
&= \sigma_1^2 + \sigma_2^2 + \sigma_3^2 - 3\sigma_m^2 \\
&= (\sigma_1 - \sigma_m)^2 + (\sigma_2 - \sigma_m)^2 + (\sigma_3 - \sigma_m)^2 \\
&= S_1^2 + S_2^2 + S_3^2 \\
&= 2J_2 \qquad (6.3.13)
\end{aligned}$$

The components of $\bar{B}$ are therefore the stress deviators S_1, S_2, and S_3. We have thus decomposed the arbitrary stress state at P into a spherical part $\bar{A}$ and a deviatoric part $\bar{B}$, the latter parallel to the π plane. Furthermore, if we take a different stress state P' lying on a line through P parallel to ON, then obviously the projection of the vector $\overline{OP'}$ onto the π plane will be the same as the projection of the vector $\overline{OP}$ onto this plane. The two states of stress, at P and at P', will therefore have the same deviatoric components and will differ only in their spherical parts. In fact, all points on the line through P parallel to ON will correspond to the same deviatoric states of stress.

Since it is assumed that yielding is determined by the deviatoric state of stress only, it follows that if one of the points on the line through P parallel to ON lies on the yield surface, they must all lie on the yield surface, since they all have the same deviatoric stress components. Hence the yield surface must be composed of lines parallel to ON; i.e., it must be a cylinder with generators parallel to ON. The only assumption needed to arrive at this conclusion is the dependence of the yield surface on the deviatoric stress components only.

The intersection of this *yield cylinder* with any plane perpendicular to it will produce a curve called the *yield locus*. Since this curve will be the same for all planes perpendicular to the cylinder, we need consider only the yield locus on one such plane. For this purpose we choose the π plane, on which, as was pointed out previously, the spherical stress state is zero. Figure 6.3.2 shows the π plane as the plane of the paper and the projections upon this plane of the coordinate axes σ_1, σ_2, and σ_3. Since this plane makes equal angles with the coordinate axes, their projections upon this plane must make

Sec. 6–3] Yield Surface. Haigh-Westergaard Stress Space

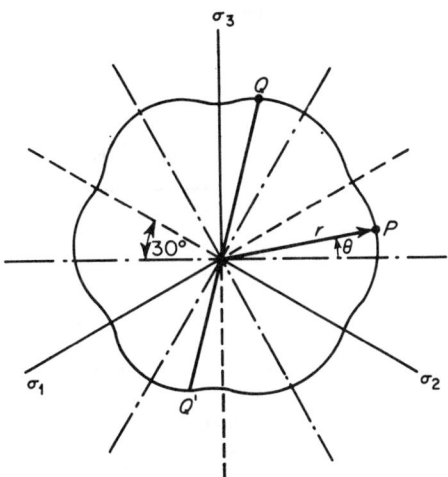

FIGURE 6.3.2 π plane.

equal angles with each other, 120°, as shown. It can be shown from symmetry considerations that the yield locus must have the same shape in each of the twelve 30° sectors dividing the π plane as shown in the figure. This follows from the following considerations. The yield surface must be symmetric in the principal stresses since it certainly does not matter, for example, if we interchange the values of σ_2 and σ_3. It follows, therefore, that the lines bisecting the angle between any two principal axes in the plane must be lines of symmetry. The lines σ_1, $-\sigma_1$; σ_2, $-\sigma_2$; and σ_3, $-\sigma_3$ are therefore lines of symmetry and we now have six symmetric sectors. Furthermore, if we assume equal yielding in tension and compression, then if we go from a point Q on the yield locus to the point Q', where all the stresses have reversed signs, we should again be on the yield locus. Therefore, if σ_1 goes to $-\sigma_1$, we must have symmetry about a line perpendicular to the σ_1 axis; if σ_2 goes to $-\sigma_2$, we must have symmetry about a line perpendicular to the σ_2 axis; and if σ_3 goes to $-\sigma_3$, we must have symmetry about the line perpendicular to the σ_3 axis. We have thus divided the yield locus into 12 symmetric sectors, each of 30°, and we need only consider the stress states lying in one of these sectors.

To determine the location on the π plane of the projection of the point P lying on the yield cylinder, we proceed as follows. Let r and θ be the polar coordinates of the point P in the π plane, θ being measured from the horizontal axis, and let a and b be the horizontal and vertical components of r, as shown in Figure 6.3.3. P has components σ_1, σ_2, and σ_3, and in the π plane they will project as $\sqrt{2/3}\,\sigma_1$, $\sqrt{2/3}\,\sigma_2$, and $\sqrt{2/3}\,\sigma_3$, respectively. This can be seen in Figure 6.3.4, which shows the plane containing the vector σ_1 the

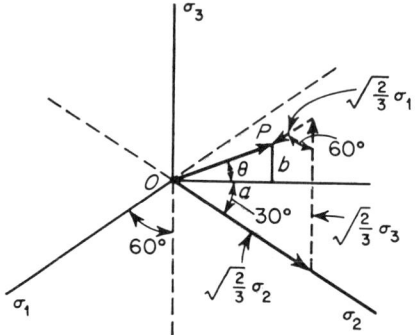

FIGURE 6.3.3 Projection of point P on π plane.

normal to the plane, ON, and the projection of σ_1 on the π plane. It is obvious from the figure that the projection of σ_1 on the π plane is equal to $\sqrt{2/3}\,\sigma_1$. From Figure 6.3.3:

$$a = \sqrt{2/3}\,\sigma_2 \cos 30° - \sqrt{2/3}\,\sigma_1 \cos 30° = (\sigma_2 - \sigma_1)/\sqrt{2}$$
$$b = \sqrt{2/3}\,\sigma_3 - \sqrt{2/3}\,\sigma_2 \sin 30° - \sqrt{2/3}\,\sigma_1 \sin 30°$$
$$= 1/\sqrt{6}\,(2\sigma_3 - \sigma_2 - \sigma_1) \tag{6.3.14}$$

$$r^2 = a^2 + b^2 = \frac{(\sigma_2 - \sigma_1)^2}{2} + \tfrac{1}{6}(2\sigma_3 - \sigma_2 - \sigma_1)^2$$
$$= \tfrac{1}{3}[(\sigma_1 - \sigma_2)^2 + (\sigma_2 - \sigma_3)^2 + (\sigma_3 - \sigma_1)^2]$$
$$= [(\sigma_1 - \sigma_m)^2 + (\sigma_2 - \sigma_m)^2 + (\sigma_3 - \sigma_m)^2]$$
$$= 2J_2 \tag{6.3.15}$$

$$\theta = \tan^{-1}\frac{b}{a} = \tan^{-1}\frac{2\sigma_3 - \sigma_2 - \sigma_1}{\sqrt{3}\,(\sigma_2 - \sigma_1)}$$

$$\sqrt{3}\tan\theta = \frac{2\sigma_3 - \sigma_2 - \sigma_1}{\sigma_2 - \sigma_1}$$

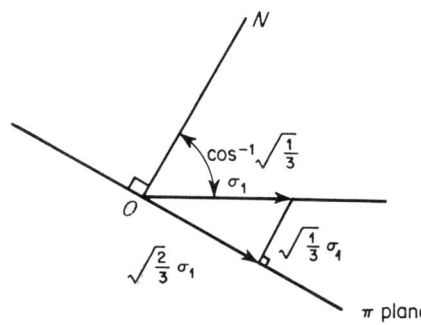

FIGURE 6.3.4 Projection of σ_1 on π plane.

Sec. 6-3] Yield Surface. Haigh-Westergaard Stress Space

Let us now consider the yield locus for the von Mises yield criterion [equation (6.3.7)]:

$$J_2 = \tfrac{1}{3}\sigma_0^2$$

From (6.3.15) we have

$$r^2 = \tfrac{2}{3}\sigma_0^2 \qquad (6.3.16)$$

The yield locus is therefore a circle of radius $r = \sqrt{2/3}\,\sigma_0$, as shown in Figure 6.3.5. The yield surface in the Haigh-Westergaard stress space will

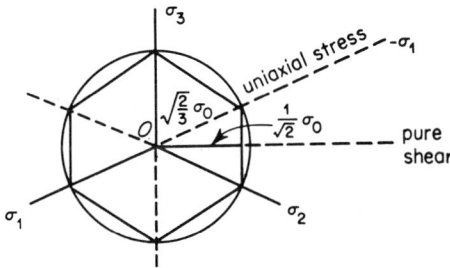

FIGURE 6.3.5   von Mises circle and Tresca hexagon.

then be a circular cylinder whose axis is the line ON equally inclined to the stress axes.

The yield locus for the Tresca maximum shear criterion is a regular hexagon inscribed in the von Mises circle, as shown in Figure 6.3.5. This can be proved as follows.

Consider the sector between the σ_2 and $-\sigma_1$ axes in Figure 6.3.5. For any stress state in this sector,

$$\sigma_2 \geq \sigma_3 \geq \sigma_1$$

The Tresca yield condition for this sector is then $\sigma_2 - \sigma_1 = \sigma_0$. This will be represented by a line parallel to the σ_3 axis, since it is independent of σ_3, the distance of this line from the σ_3 axis being equal to $1/\sqrt{2}\,(\sigma_2 - \sigma_1) = 1/\sqrt{2}\,\sigma_0$ [by equation (6.3.14)]. In a similar manner, a straight line is obtained for each sector, thus forming the hexagon shown in the figure. The corners of the hexagon will touch the von Mises circle, as can be seen again from the first sector, since the radius to the corner is equal to

$$r = \frac{(1/\sqrt{2})\sigma_0}{\cos 30°} = \sqrt{\tfrac{2}{3}}\,\sigma_0$$

which is the radius of the von Mises circle. In the stress space the Tresca yield surface is a regular hexagonal cylinder inscribed in the von Mises cylinder, as shown in Figure 6.3.6.

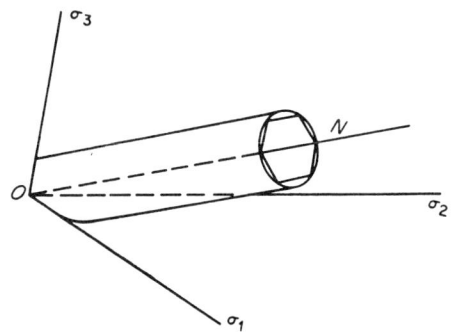

FIGURE 6.3.6 Tresca and von Mises cylinders.

It is now evident from Figure 6.3.5 that the maximum difference between the von Mises and Tresca criteria occurs at $\theta = 0°$ and is equal to

$$\frac{\sqrt{\tfrac{2}{3}}\,\sigma_0 - (1/\sqrt{2})\sigma_0}{(1/\sqrt{2})\sigma_0} = \frac{2}{\sqrt{3}} - 1 = 0.156$$

as previously mentioned.

The line at $\theta = 0°$ corresponds to pure shear stress states. This follows from (6.3.15), for $\theta = 0$:

$$\sigma_3 = \tfrac{1}{2}(\sigma_1 + \sigma_2)$$

and the mean stress is

$$\sigma_m = \tfrac{1}{3}(\sigma_1 + \sigma_2 + \sigma_3) = \tfrac{1}{2}(\sigma_1 + \sigma_2)$$

Therefore, if we subtract the mean stress from the stress components σ_1, σ_2, and σ_3, we get a stress state $\tfrac{1}{2}(\sigma_1 - \sigma_2)$, $\tfrac{1}{2}(\sigma_2 - \sigma_1)$, 0, which corresponds to a state of pure shear (see p. 41). In the π plane itself, since $\sigma_1 + \sigma_2 + \sigma_3 = 0$, it follows that $\sigma_3 = 0$ and $\sigma_1 = -\sigma_2$. For any other plane parallel to the π plane, we have a case of pure shear, as indicated, with the addition of a spherical state of stress of amount $\tfrac{1}{2}(\sigma_1 + \sigma_2)$.

The line at $\theta = 30°$ corresponds to a uniaxial stress, since

$$\frac{2\sigma_3 - \sigma_2 - \sigma_1}{\sigma_2 - \sigma_1} = 1$$

or

$$\sigma_3 = \sigma_2$$

If we subtract a hydrostatic stress of σ_2, we get a stress state $\sigma_1 - \sigma_2$, 0, 0.

Sec. 6-3] Yield Surface. Haigh-Westergaard Stress Space

It can now readily be shown that if the yield locus is assumed to be *convex* and one circumscribes the von Mises circle by a regular hexagon, then all possible yield loci must lie between the two regular hexagons inscribed in, and circumscribing, the von Mises circle. By the locus being convex, we mean simply that any straight line in the π plane may pierce the locus in at most two points. It will be proved later that the yield surface must indeed be convex; for the present we shall merely assume it. Let the point A on the σ_3 axis in Figure 6.3.7 lie on the yield locus. By the symmetry conditions previously discussed, the points B and C must also lie on the yield locus. The

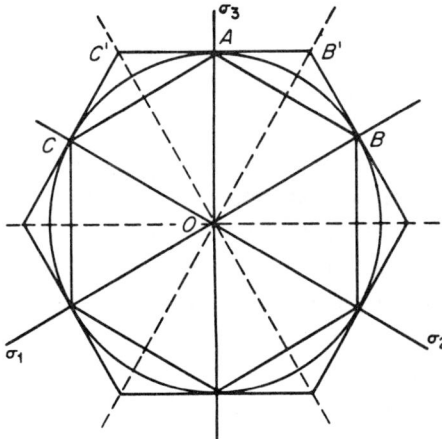

FIGURE 6.3.7 Bounds on yield loci.

curve CAB is a convex piecewise smooth curve passing through CAB and having the proper symmetries. Any other curve through these points passing inside CAB will obviously not be convex. CAB is therefore a lower bound for the yield loci.

Now draw a horizontal line symmetric about the σ_3 axis intersecting the adjacent axes of symmetry at C' and B'. Then it follows that no piecewise smooth curve passing through $C'AB'$ can lie outside $C'AB'$, since then it would not be convex. Thus $C'AB'$ represents an upper bound to the yield locus. The rest of the upper and lower bounds can be constructed from symmetry and are shown as the two regular hexagons in Figure 6.3.7. Thus it has been shown that all conceivable yield loci satisfying the conditions of isotropy, equal yield in tension and compression, independence of hydrostatic stress, and convexity must lie between two regular hexagons as shown; only convex curves are, of course, admissible. The usual Tresca yield locus is represented by the inner hexagon, and the von Mises circle circumscribes the inner hexagon and is circumscribed by the outer one. If the von Mises yield surface is taken as a reference, then the maximum deviation of any admissible yield surface is about 15.5 per cent.

6–4 LODE'S STRESS PARAMETER. EXPERIMENTAL VERIFICATION OF YIELD CRITERIA

The first investigation of a yield criterion was performed by Tresca in 1864, in which he measured the loads required to extrude metals through dies of various shapes. On the basis of these experiments he arrived at the maximum shear stress criterion, previously discussed. It is evident that according to the Tresca criterion the intermediate stress has no effect on yielding. The von Mises criterion, on the other hand, gives equal weight to all three principal stresses. The simplest and most common type test specimen used to check these criteria is the thin-walled tube shown in Figure 6.1.1. The first such experiments were run by Lode [5], who tested tubes of steel, copper, and nickel under various combinations of longitudinal tension and internal hydrostatic pressure. Lode devised a very sensitive method of differentiating between the Tresca and von Mises criteria by determining the effect of the intermediate principal stress on yielding. According to the Tresca criterion, if $\sigma_1 \geq \sigma_2 \geq \sigma_3$ the yield criterion is given by

$$\frac{\sigma_1 - \sigma_3}{\sigma_0} = 1 \qquad (6.4.1)$$

To account for the influence of the intermediate stress in the von Mises criterion, Lode introduced the parameter μ, called *Lode's stress parameter* (not to be confused with Poisson's ratio, μ):

$$\begin{aligned}\mu &= -\frac{2\sigma_2 - \sigma_3 - \sigma_1}{\sigma_1 - \sigma_3} \\ &= -\frac{\sigma_2 - \tfrac{1}{2}(\sigma_1 + \sigma_3)}{\tfrac{1}{2}(\sigma_1 - \sigma_3)}\end{aligned} \qquad (6.4.2)$$

μ is thus the ratio of the difference between the intermediate stress and the average of the largest and smallest stresses to half the difference between the largest and smallest stresses, and is therefore a measure of the effect of the intermediate stress. Comparing (6.4.2) to (6.3.15) for the case $\sigma_1 \geq \sigma_2 \geq \sigma_3$, we see that

$$\mu = -\sqrt{3} \tan \theta \qquad (6.4.3)$$

Sec. 6–4] Lode's Stress Parameter. Verification of Yield Criteria

Thus if θ varies from $0°$ to $30°$, μ will vary from 0 to -1. All combinations of loading from pure shear to simple tension are therefore included in the range 0 to -1.

By means of (6.4.2), the von Mises criterion can be written

$$\frac{\sigma_1 - \sigma_3}{\sigma_0} = \frac{2}{(3 + \mu^2)^{1/2}} \quad (6.4.4)$$

The difference between the Tresca and von Mises criteria is then determined by how much the right side of (6.4.4) differs from 1. For $\mu = -1$, equation (6.4.4) agrees with (6.4.1). This is the case of simple tension, and the two criteria agree as expected. The maximum difference in the range 0 to -1 obviously occurs when $\mu = 0$, which is the case of pure shear. The difference is then $2/\sqrt{3}$, which agrees again with the previous results.

Lode ran a series of tests covering the range of μ from 1 to -1 ($-30° \leq \theta \leq 30°$). A μ of 1 corresponds to uniaxial compression. The results are shown in Figure 6.4.1. It is seen that the data favor the von Mises criterion even

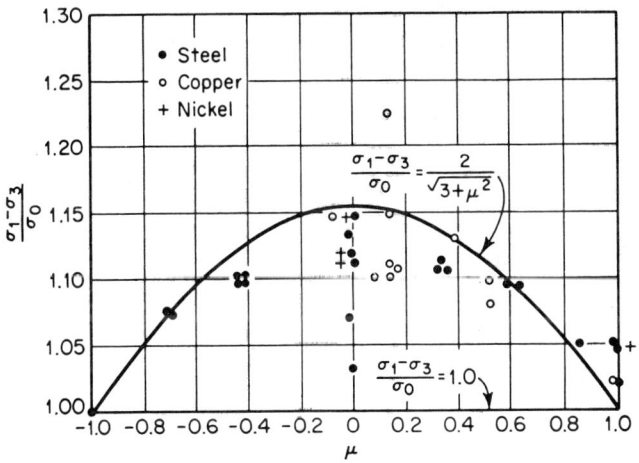

FIGURE 6.4.1 Lode's test results.

though appreciable deviation occurs. These deviations were partially attributed to lack of isotropy of the material. Some of the results are also plotted in Figure 6.4.2 in the $\sigma_1 \sigma_2$ plane, since σ_3 is zero for these tests. The intersection of the von Mises circular cylinder with the $\sigma_1 \sigma_2$ plane is obviously an ellipse, and the intersection of the Tresca cylinder is a hexagon (no longer regular), as shown in Figures 6.2.3 and 6.2.4. Figure 6.4.2 shows that the data plot between the hexagon and the ellipse, although generally closer to the ellipse.

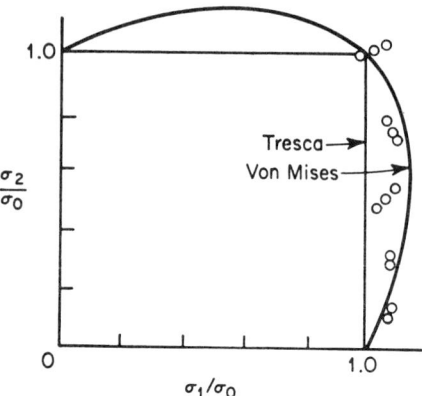

FIGURE 6.4.2   Lode's test results.

In 1931 Taylor and Quinney [6] published their classical experiments, which were intended to settle this question. They used copper and steel tubings which were very nearly isotropic and tested them very carefully in combined tension and torsion. They concluded that the deviations from the von Mises criterion were real and could not be explained on the basis of lack of experimental accuracy or isotropy. Their results are shown in Figure 6.4.3.

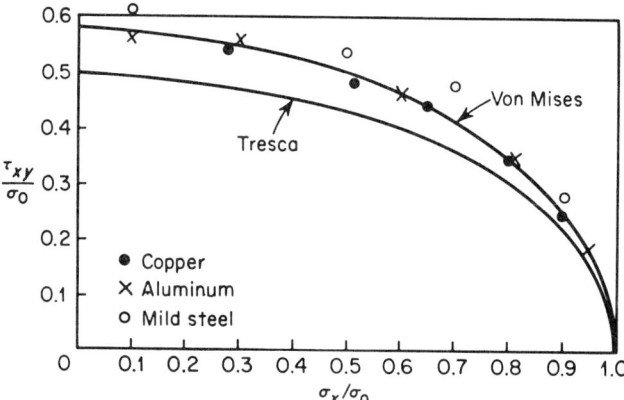

FIGURE 6.4.3   Results of Taylor and Quinney.

The theoretical curves are obtained as follows. Using the coordinate system of Figure 6.1.1, for tension and torsion loads only, the stresses are σ_x and τ_{xy}, all the others being zero. The Tresca criterion then becomes

$$2\sqrt{\frac{\sigma_x^2}{4} + \tau_{xy}^2} = \sigma_0$$

or

$$\left(\frac{\sigma_x}{\sigma_0}\right)^2 + 4\left(\frac{\tau_{xy}}{\sigma_0}\right)^2 = 1 \tag{6.4.5}$$

Sec. 6–4] Lode's Stress Parameter. Verification of Yield Criteria

The von Mises criterion, on the other hand, becomes

$$\left(\frac{\sigma_x}{\sigma_0}\right)^2 + 3\left(\frac{\tau_{xy}}{\sigma_0}\right)^2 = 1 \tag{6.4.6}$$

Both (6.4.5) and (6.4.6) plot as ellipses in the $\sigma_x \tau_{xy}$ plane, as shown in Figure 6.4.3. It is seen in the figure that the data fit the von Mises criterion considerably better than the Tresca criterion, although appreciable deviations from the von Mises criterion do sometimes occur. Figure 6.4.4. shows a replot of

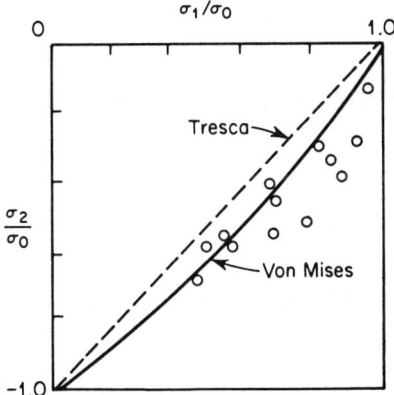

FIGURE 6.4.4 Results of Taylor and Quinney.

some of these results in the $\sigma_1 \sigma_2$ plane, and Figure 6.4.5 shows similar results obtained by Ros and Eichinger [7]. Other tests of similar nature can be found in the literature.

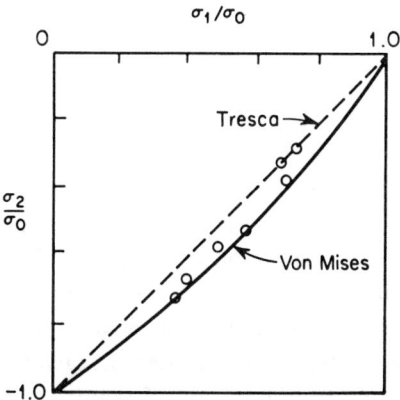

FIGURE 6.4.5 Results of Ros and Eichinger.

Attempts have been made to improve the correlation of the data by including the effect of the third invariant J_3 into the yield criterion [8]. It seems, however, that from an engineering viewpoint the accuracy of the von Mises criterion for yielding is amply sufficient, considering the general scatter and lack of uniformity in the properties of nominally the same material obtained from different batches from the manufacturer. Therefore, the search for more accurate theories, particularly since they are bound to be more complex, seems to be a rather thankless task. In what follows, therefore, we shall generally use the von Mises yield criterion and occasionally the Tresca criterion.

6-5 SUBSEQUENT YIELD SURFACES. LOADING AND UNLOADING

So far we have discussed the *initial* yield surface at which a material will first start yielding. For a perfectly plastic material, this yield surface remains fixed, as is seen in the uniaxial tensile test, where the stress after yielding remains constant at the yield stress [Figure 2.6.1(d)]. However, for a material that *strain hardens* the yield surface must change for continued straining beyond the initial yield. We know that for the uniaxial case, if a material is strained beyond the yield point to some point such as B' in Figure 2.1.2, the load removed so that the stress state moves to C', and then the load is increased again, yielding will not take place until the point B' is reached again. Thus the yield point has been raised in the work-hardened material. In the same way the yield surface in the case of multiaxial stress must "move out" in some way, at least at the point where yielding initially took place.

In equation (6.3.1) we defined a yield function by the relation

$$F(\sigma_{ij}) = K \quad (6.5.1)$$

such that whenever the function F became equal to the constant K, yielding would begin. K then represented an initial yield surface in the stress space. We can now generalize this type of relation to subsequent yield surfaces. After yielding has occurred, K takes on a new value (or values), depending on the strain-hardening properties of the material. If the material is unloaded and then loaded again, additional yielding will not occur until the new value of K is reached. The function F can then be looked upon as a *loading function*, or loading surface, which represents the load being applied, and the function K is a *yield function*, or strain-hardening function, and will depend on the complete previous stress and strain history of the material and its strain-

Sec. 6–5] Subsequent Yield Surfaces. Loading and Unloading

hardening properties. We can now distinguish three cases for a strain-hardening material:

$$(1) \quad F = K \quad dF = \frac{\partial F}{\partial \sigma_{ij}} d\sigma_{ij} > 0$$

This constitutes *loading*.

$$(2) \quad F = K \quad dF = \frac{\partial F}{\partial \sigma_{ij}} d\sigma_{ij} = 0 \qquad (6.5.2)$$

This is called *neutral loading*.

$$(3) \quad F = K \quad dF = \frac{\partial F}{\partial \sigma_{ij}} d\sigma_{ij} < 0$$

This constitutes *unloading*.

Geometrically the conditions (6.5.2) are readily visualized. $F = K$ means the stress state is on the yield surface. $dF > 0$ means the stress state is "moving out" from the yield surface and plastic flow is occurring. $dF < 0$ means the stress state is "moving in" from the yield surface and unloading is therefore taking place. $dF = 0$ corresponds to the case of the stress state moving on the yield surface and is called *neutral loading*. For a strain-hardening material no plastic flow occurs. If $F < K$, the stress state is an elastic one. For perfectly plastic materials plastic flow occurs for

$$F = K \quad dF = 0 \qquad (6.5.3)$$

The case $dF > 0$ does not exist.

The above discussion is illustrated in Figure 6.5.1. Point P represents the

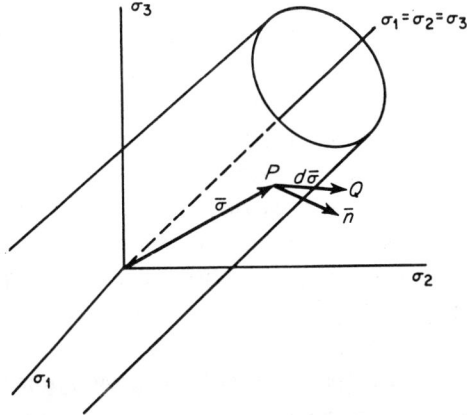

FIGURE 6.5.1 Stress increment vector for loading.

existing state of stress lying on the yield surface. If the stress state changes to Q, so that the vector $d\bar{\sigma}$ points "outward" from the cylinder, then loading is taking place. If Q lies on the surface, we have neutral loading, and if Q is inside the surface, we have unloading. As loading continues, the point representing the new existing state of stress will again lie on the surface, which will move and/or change its shape correspondingly. A similar picture can be drawn in the π plane.

Assume the material obeys the von Mises criterion, so that the initial yield locus is a circle of radius $\sqrt{\tfrac{2}{3}}\,\sigma_0$ in the π plane. Suppose that straining takes place to some point $\sigma_0' > \sigma_0$ and the material is then unloaded. If we now assume that the material remains isotropic, just as it was originally, then the new yield locus is a circle of radius $\sqrt{\tfrac{2}{3}}\,\sigma_0'$, which is larger than, but concentric with, the original yield circle. We have thus assumed that the material strain hardens isotropically, as shown for the uniaxial case by the curve $ABCFG$ of Figure 2.3.1. For *isotropic hardening*, therefore, the yield cylinder will expand with stress and strain history but will retain the same shape as initially. For a Tresca material, the subsequent yield loci will be a series of concentric regular hexagons. This is shown in Figure 6.5.2.

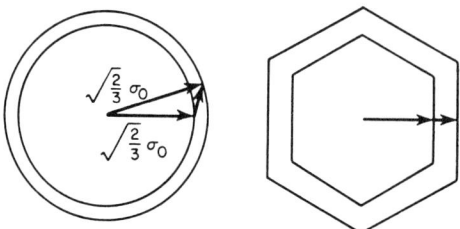

FIGURE 6.5.2 Subsequent yield loci.

The assumption of isotropic hardening is the simplest one, mathematically, to use. However, it does not take into account a Bauschinger effect. The Bauschinger effect would tend to reduce the size of the locus on one side as that on the other side is increased. The yield surface would thus change shape as the yielding progresses. Experiments to verify this effect are described in reference [9]. Several tests were carried out with aluminum alloy tubes with various ratios or torsion and tension to obtain an initial yield surface. By unloading and loading again in a specified manner, subsequent yield surfaces were obtained. The results plotted in the $\tau\sigma$ plane are shown in Figure 6.5.3. Without going into a detailed discussion of this data, it is evident that the initial von Mises yield ellipse does not just grow symmetrically but that a definite Bauschinger effect exists.

To account for the Bauschinger effect, Prager (reference [14] of Chapter 2)

Sec. 6-5] Subsequent Yield Surfaces. Loading and Unloading

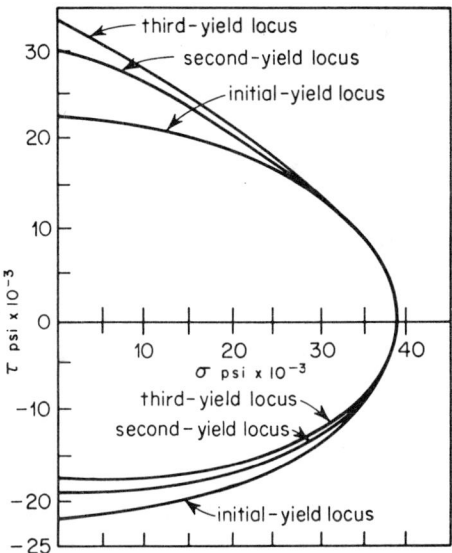

FIGURE 6.5.3   Initial and subsequent yield loci.

introduced the *kinematic model*, discussed for the one-dimensional case in Section 2.6 and illustrated for that case in Figures 2.6.2 and 2.6.3. In this model the total elastic range is maintained constant by translating the initial yield surface without deforming it. The model is represented by a rigid frame having the shape of the yield surface, as shown for the Tresca criterion in Figure 6.5.4. The state of stress before yield occurs is represented by the position of a pin free to move within the frame. As the pin contacts the side of the frame, yielding occurs. The frame is assumed to be constrained against rotation and to be perfectly smooth, so that only forces normal to the frame can be transmitted to it. As the pin pushes against the frame it causes it to translate in a direction normal to the surface at the point of contact. At corners, if the motion of the pin engages both sides, the frame translates in the direction of the motion of the pin. Depending on the material being described by the model, i.e., rigid perfectly plastic, rigid strain hardening, or elastic perfectly plastic, the state of stress and the state of strain are represented in the model in different ways. For example, for a rigid strain-hardening material, the displacement of the center of the frame relative to the origin is proportional to the total strain, and the state of stress is represented by the position of the pin relative to the origin (compare to Figure 2.6.2).

Figure 6.5.4 (from reference [4]) shows the translating of the Tresca hexagon by means of this model for a stress path OP. This type of hardening is called *kinematic hardening* because of the type of model used to represent it. It takes

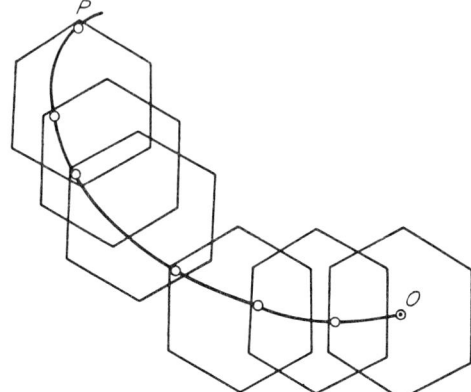

FIGURE 6.5.4 Kinematic model for Tresca criterion.

into account a Bauschinger effect but, because it maintains the total elastic range constant, it probably overcorrects somewhat for the Bauschinger effect, as discussed in Section 2.3. Attempts have therefore been made to improve on the kinematic hardening model. This type of model, however, is much more difficult to handle mathematically, and therefore the isotropic hardening assumption is still generally used. For small plastic strains it probably gives answers that are sufficiently accurate.

Problems

1. Compare the values of the pure shear yield strengths based on the six theories of yielding listed (exclude Mohr's theory). Assume $\mu = 0.3$ and $\sigma_{0,c} = 1.25\sigma_0$. For this case, $\tau = \sigma_1 = -\sigma_2$.
2. A circular shaft 10 in. in diameter has a tensile yield strength of 90,000 psi and a compressive yield strength of 120,000 psi. Determine the twisting moment M_t required to produce yielding based on:
   (a) The maximum stress theory.
   (b) The maximum shear theory.
   (c) The distortion energy theory.
   (d) The internal friction theory.
3. Derive equation (6.2.10).
4. It was shown in Section 6.3 that if the von Mises and Tresca criteria are assumed to agree for the case of uniaxial tension, then they will disagree for the case of pure shear and the von Mises yield circle will circumscribe the Tresca hexagon. Show that if it is assumed that the von Mises and Tresca criteria agree for the case of pure shear, then they will disagree for the case of uniaxial tension and the Tresca hexagon will circumscribe the von Mises ellipse.
5. Derive the general form of the Tresca yield criteria as given by equation (6.3.8).

References

6. Derive the equation for the Tresca yield criterion in the sector between the $-\sigma_1$ and σ_3 axes as was done in the text for the sector between the σ_2 and $-\sigma_1$ axes.
7. Derive equation (6.4.4).
8. Show that Lode's stress parameter $\mu = 1$ corresponds to uniaxial compression, $\mu = -1$ to uniaxial tension, and $\mu = 0$ to pure shear.
9. Derive equations (6.4.5) and (6.4.6).

References

1. J. Marin, *Mechanical Behavior of Engineering Materials*, Prentice-Hall, Englewood Cliffs, N.J., 1962, p. 117.
1a. *Ibid.*, p. 122.
2. J. Marin, Theories of Strength for Combined Stresses and Nonisotropic Materials, *J. Aeronautical Sci.*, 24, No. 4, 1956, pp. 265–269.
3. H. M. Westergaard, On the Resistance of Ductile Materials to Combined Stresses, *J. Franklin Inst.*, 189, 1920, pp. 627–640.
4. P. M. Naghdi, Stress–Strain Relations in Plasticity and Thermoplasticity, *Inst. Eng. Res., Univ. Calif. (Berkeley), Ser. 131, Issue 9*, March 1960, p. 18.
5. W. Lode, Versuche ueber den Einfluss der mittleren Hauptspannung auf das Fliessen der Metalle Eisen Kupfer und Nickel, *Z. Physik.*, 36, 1926, pp. 913–939.
6. G. I. Taylor and H. Quinney, The Plastic Distortion of Metals, *Phil. Trans. Roy. Soc., London*, A230, 1931, pp. 323–362.
7. M. Ros and A. Eichinger, Versuche Zur Klaerung der Frage der Bruchgefahr III, Metalle, *Eidgenoss. Material pruf. und Versuchsanstalt Industriell Bauwerk und Gewerbe, Diskussionsbericht No. 34*, Zurich, 1929, pp. 3–59.
8. F. D. Stockton and D. C. Drucker, Fitting Mathematical Theories of Plasticity to Experimental Results, *J. Coll. Sci. (Rheology Issue)*, 5, 1950, pp. 239 250.
9. P. M. Naghdi, F. Essenberg and W. Koff, An Experimental Study of Initial and Subsequent Yield Surfaces in Plasticity, *J. Appl. Mech.*, 25, 1958, pp. 201–209.

General References

Hill, R., *The Mathematical Theory of Plasticity*, Oxford Univ. Press, London, 1950.
Hoffman, O., and G. Sachs, *Introduction to the Theory of Plasticity for Engineers*, McGraw-Hill, New York, 1953.
Marin, J., *The Mechanical Behavior of Engineering Materials*, Prentice-Hall, Englewood Cliffs, N.J., 1962.

CHAPTER 7

PLASTIC STRESS–STRAIN RELATIONS

7-1 DISTINCTION BETWEEN ELASTIC AND PLASTIC STRESS–STRAIN RELATIONS

In the previous sections the relations between stress and strain in the elastic range were discussed and also the stress states at which plastic flow or yielding will begin. There is one more ingredient necessary in constructing a plasticity theory—the relations between stress and strain when plastic flow is occurring. These relations are the subject of this chapter.

Whereas the strains are linearly related to the stresses by Hooke's law in the elastic range, the relation will generally be nonlinear in the plastic range, as is evident from the uniaxial stress–strain curve. A more complicated distinction between elastic and plastic stress–strain relations arises from the fact that whereas in the elastic range the strains are uniquely determined by the stresses, i.e., for a given set of stresses we can compute the strains directly using Hooke's law without any regard as to how this stress state was attained, in the plastic range the strains are in general not uniquely determined by the stresses but depend on the whole history of loading or how the stress state was reached. This can readily be illustrated by considering a thin-walled tube in tension and torsion as in the experiments of Taylor and Quinney.

Consider the initial yield curve to be as shown in Figure 7.1.1. Let the specimen be strained in uniaxial tension beyond the initial yield to some

Sec. 7-1] Distinction Between Elastic and Plastic Stress-Strain Relations

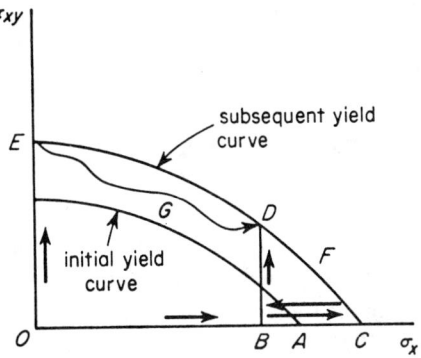

FIGURE 7.1.1 Effect of loading path on plastic strains.

point C, where CDE defines the subsequent yield curve. The plastic strains will then be

$$\varepsilon_x^P = \varepsilon_p$$
$$\varepsilon_y^P = \varepsilon_z^P = -\tfrac{1}{2}\varepsilon_p$$
$$\varepsilon_{xy}^P = \varepsilon_{yz}^P = \varepsilon_{zx}^P = 0$$

Let the specimen now be unloaded to the point B and let us apply a shear stress increasing from B to D on the new yield locus. The plastic strains will still be as given above. Any other path could have been used in arriving at D from C such as $OCFD$, as long as we do not move outside the yield locus. Now suppose that the specimen were first stressed in shear to the point E on the new yield locus and then, by any other path inside EDC, such as EGD, were stressed to the point D. The plastic strains would be

$$\varepsilon_{xy}^P = \gamma_p$$
$$\varepsilon_x^P = \varepsilon_y^P = \varepsilon_z^P = \varepsilon_{xz}^P = \varepsilon_{yz}^P = 0$$

which is obviously completely unrelated to the previous strain state. Thus even though the same stress state at D exists for both loading paths, and therefore the elastic strain states are the same, the plastic strain states are different.

Because of the above illustrated dependence of the plastic strains on the loading path, it becomes necessary, in general, to compute the differentials or increments of plastic strain throughout the loading history and then obtain the total strains by integration or summation. However, there is at least one important class of loading paths for which the plastic strains are independent of the loading path and depend only on the final state of stress. These are the so-called *radial* or *proportional loading paths*, in which all the stresses increase in the same ratio. These will be more fully discussed subsequently.

7-2 PRANDTL-REUSS EQUATIONS

The first approach to plastic stress–strain relations was suggested by Saint-Venant in 1870 [1], who proposed that the principal axes of *strain increment* coincided with the principal stress axes. The general three-dimensional equations relating the increments of total strain to the stress deviations were given by Lévy in 1871 [2] and independently by von Mises in 1913 [3]. These are known as the *Lévy–Mises equations*. These equations are

$$\frac{d\varepsilon_x}{S_x} = \frac{d\varepsilon_y}{S_y} = \frac{d\varepsilon_z}{S_z} = \frac{d\varepsilon_{yz}}{\tau_{yz}} = \frac{d\varepsilon_{zx}}{\tau_{zx}} = \frac{d\varepsilon_{xy}}{\tau_{xy}} = d\lambda \quad (7.2.1)$$

or $\quad d\varepsilon_{ij} = S_{ij}\, d\lambda$

where S_{ij} is the stress deviator tensor and $d\lambda$ is a nonnegative constant which may vary throughout the loading history. In these equations the total strain increments are assumed to be equal to the plastic strain increments, the elastic strains being ignored. Thus these equations can only be applied to problems of large plastic flow and cannot be used in the elastoplastic range. The generalization of equations (7.2.1) to include both elastic and plastic components of strain is due to Prandtl [4] and Reuss [5] and are known as the *Prandtl–Reuss equations*.

Reuss assumed that the *plastic* strain increment is at any instant of loading proportional to the instantaneous stress deviation; i.e.,

$$\frac{d\varepsilon_x^P}{S_x} = \frac{d\varepsilon_y^P}{S_y} = \frac{d\varepsilon_z^P}{S_z} = \frac{d\varepsilon_{xy}^P}{\tau_{xy}} = \frac{d\varepsilon_{yz}^P}{\tau_{yz}} = \frac{d\varepsilon_{zx}^P}{\tau_{zx}} = d\lambda \quad (7.2.2)$$

or $\quad d\varepsilon_{ij}^P = S_{ij}\, d\lambda$

Equations (7.2.1) can then be considered as a special case of (7.2.2) where the elastic strain components are neglected.

Equations (7.2.2) state that the increments of plastic strain depend on the current values of the deviatoric stress state, not on the stress increment required to reach this state. They also imply that the principal axes of stress and of plastic strain increment tensors coincide. The equations themselves merely give a relationship between the ratios of plastic strain increments in the different directions. To determine the actual magnitudes of the increments a yield criterion is required, as will shortly be shown.

Sec. 7-2] Prandtl-Reuss Equations

If the principal directions are considered, equations (7.2.2) can be written

$$\frac{d\varepsilon_1^p}{S_1} = \frac{d\varepsilon_2^p}{S_2} = \frac{d\varepsilon_3^p}{S_3} = d\lambda \tag{7.2.3}$$

or $\quad d\varepsilon_1^p = S_1\, d\lambda \quad d\varepsilon_2^p = S_2\, d\lambda \quad d\varepsilon_3^p = S_3\, d\lambda$

or

$$\frac{d\varepsilon_1^p - d\varepsilon_2^p}{S_1 - S_2} = \frac{d\varepsilon_2^p - d\varepsilon_3^p}{S_2 - S_3} = \frac{d\varepsilon_3^p - d\varepsilon_1^p}{S_3 - S_1} = d\lambda \tag{7.2.4}$$

The numerators of the first three terms of (7.2.4) are the diameters of the three Mohr's circles for the plastic strain increments and the denominators are the diameters of Mohr's stress circles, as shown in Figure 7.2.1. Equations (7.2.4) therefore imply that the Mohr's circles of stress and plastic strain increment are similar. Also from the relations for the principal shears, equations (7.2.4) can be considered as stating that the ratios of the three principal plastic shear strain increments to the principal shear stresses are constant at any instant.

Equations (7.2.2) can be written in terms of the actual stresses as

$$\begin{aligned}
d\varepsilon_x^p &= \tfrac{2}{3}d\lambda\, [\sigma_x - \tfrac{1}{2}(\sigma_y + \sigma_z)] \\
d\varepsilon_y^p &= \tfrac{2}{3}d\lambda\, [\sigma_y - \tfrac{1}{2}(\sigma_z + \sigma_x)] \\
d\varepsilon_z^p &= \tfrac{2}{3}d\lambda\, [\sigma_z - \tfrac{1}{2}(\sigma_x + \sigma_y)] \\
d\varepsilon_{xy}^p &= d\lambda\, \tau_{xy} \\
d\varepsilon_{yz}^p &= d\lambda\, \tau_{yz} \\
d\varepsilon_{zx}^p &= d\lambda\, \tau_{zx}
\end{aligned} \tag{7.2.5}$$

Therefore, if $d\lambda$ were known, we would have the desired stress-strain relations. To determine $d\lambda$ use is made of the yield criterion as follows. By means of equations (7.2.2),

$$(d\varepsilon_x^p - d\varepsilon_y^p)^2 + (d\varepsilon_y^p - d\varepsilon_z^p)^2 + (d\varepsilon_z^p - d\varepsilon_x^p)^2 + 6(d\varepsilon_{xy}^p)^2 + 6(d\varepsilon_{yz}^p)^2 + 6(d\varepsilon_{zx}^p)^2$$
$$= (d\lambda)^2\, [(\sigma_x - \sigma_y)^2 + (\sigma_y - \sigma_z)^2 + (\sigma_z - \sigma_x)^2 + 6\tau_{xy}^2 + 6\tau_{yz}^2 + 6\tau_{zx}^2] \tag{7.2.6}$$

The bracketed quantity on the right side of equation (7.2.6) is seen to be proportional to the square of the octahedral shear stress [equation (3.4.11)], and the left side of equation (7.2.6) is proportional to the square of the increment of octahedral plastic shear strain defined by [see equation (4.5.10)]

$$(d\gamma_0^p)^2 \equiv \tfrac{1}{9}[(d\varepsilon_x^p - d\varepsilon_y^p)^2 + (d\varepsilon_y^p - d\varepsilon_z^p)^2 + (d\varepsilon_z^p - d\varepsilon_x^p)^2$$
$$+ 6(d\varepsilon_{xy}^p)^2 + 6(d\varepsilon_{yz}^p)^2 + 6(d\varepsilon_{zx}^p)^2] \tag{7.2.7}$$

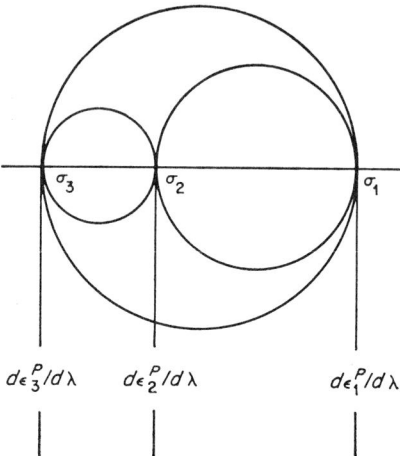

FIGURE 7.2.1 Mohr's circles for stress and plastic strain increments.

The constant $d\lambda$ now becomes

$$d\lambda = \frac{d\gamma_0^p}{\tau_{\text{oct}}}$$

$$= \sqrt{\tfrac{3}{2}}\, \frac{d\gamma_0^p}{\sqrt{J_2}} \qquad (7.2.8)$$

where J_2 is the second invariant of the stress deviator tensor.

It is convenient to define an *equivalent* or *effective* stress and an *equivalent* or *effective* plastic strain increment as

$$\sigma_e \equiv \frac{1}{\sqrt{2}}[(\sigma_x - \sigma_y)^2 + (\sigma_y - \sigma_z)^2 + (\sigma_z - \sigma_x)^2 + 6(\tau_{xy}^2 + \tau_{yz}^2 + \tau_{zx}^2)]^{1/2}$$

$$= \frac{3}{\sqrt{2}} \tau_{\text{oct}}$$

$$= \sqrt{3J_2} \qquad (7.2.9)$$

and

$$d\varepsilon_p \equiv \frac{\sqrt{2}}{3}[(d\varepsilon_x^p - d\varepsilon_y^p)^2 + (d\varepsilon_y^p - d\varepsilon_z^p)^2 + (d\varepsilon_z^p - d\varepsilon_x^p)^2$$

$$+ 6(d\varepsilon_{xy}^p)^2 + 6(d\varepsilon_{yz}^p)^2 + (d\varepsilon_{zx}^p)^2]^{1/2}$$

$$= \sqrt{2}\, d\gamma_0^p \qquad (7.2.10)$$

Sec. 7-2] Prandtl-Reuss Equations

For a uniaxial tensile test in the x direction the equivalent stress and equivalent plastic strain increment reduce to

$$\sigma_e = \sigma_x$$
$$d\varepsilon_p = d\varepsilon_x^P \qquad (7.2.11)$$

The convenience of the above definitions now becomes apparent. The equivalent or effective stress, σ_e, and the equivalent or effective plastic strain increment, $d\varepsilon_p$, will henceforth be used in this text rather than the octahedral shear stress and octahedral plastic shear strain increment.

The constant $d\lambda$ can therefore be written

$$d\lambda = \tfrac{3}{2} \frac{d\varepsilon_p}{\sigma_e} \qquad (7.2.12)$$

and the stress–strain relations (7.2.5) become

$$d\varepsilon_x^P = \frac{d\varepsilon_p}{\sigma_e} [\sigma_x - \tfrac{1}{2}(\sigma_y + \sigma_z)]$$

$$d\varepsilon_y^P = \frac{d\varepsilon_p}{\sigma_e} [\sigma_y - \tfrac{1}{2}(\sigma_z + \sigma_x)]$$

$$d\varepsilon_z^P = \frac{d\varepsilon_p}{\sigma_e} [\sigma_z - \tfrac{1}{2}(\sigma_x + \sigma_y)] = -(d\varepsilon_x^P + d\varepsilon_y^P)$$

$$d\varepsilon_{xy}^P = \frac{3}{2} \frac{d\varepsilon_p}{\sigma_e} \tau_{xy} \qquad (7.2.13)$$

$$d\varepsilon_{yz}^P = \frac{3}{2} \frac{d\varepsilon_p}{\sigma_e} \tau_{yz}$$

$$d\varepsilon_{zx}^P = \frac{3}{2} \frac{d\varepsilon_p}{\sigma_e} \tau_{zx}$$

or
$$d\varepsilon_{ij}^P = \frac{3}{2} \frac{d\varepsilon_p}{\sigma_e} S_{ij}$$

If one compares equation (7.2.9) for the equivalent stress σ_e with equation (6.2.8), which gives the von Mises yield criterion, it is seen that just as yielding begins

$$\sigma_e = \sigma_0 \qquad (7.2.14)$$

where σ_0 is the yield stress in simple tension. The equivalent stress is thus the same as the von Mises yield function, and since equations (7.2.13) make use of this function, the original Prandtl–Reuss assumptions imply the von Mises yield criterion. This will also be shown subsequently from other considerations.

It also follows from (7.2.14) that for a perfectly plastic material the Prandtl–Reuss equations may be written

$$d\varepsilon_{ij}^P = \frac{3}{2}\frac{d\varepsilon_p}{\sigma_0} S_{ij} \qquad (7.2.15)$$

For a material that work hardens, however, σ_e may be greater than σ_0, and it is now necessary to find the relation between the equivalent stress σ_e and the equivalent plastic strain increment, $d\varepsilon_p$. Before this is done, we introduce the concepts of *plastic work* and the *measures* of work hardening.

7-3 PLASTIC WORK. TWO MEASURES OF WORK HARDENING

An important concept that is frequently used in plasticity theory is the concept of *plastic work*. The work done per unit volume on an element during straining is

$$\begin{aligned}dW &= \sigma_{ij}\, d\varepsilon_{ij} \\ &= \sigma_{ij}\,(d\varepsilon_{ij}^e + d\varepsilon_{ij}^P) \\ &= dW^e + dW^P\end{aligned} \qquad (7.3.1)$$

But $dW^e \equiv \sigma_{ij}\, d\varepsilon_{ij}^e$ is recoverable elastic energy, whereas the plastic deformation is an irreversible process from which the energy cannot be recovered. The remainder of the work done is called the *plastic work per unit volume*,

$$dW^P = \sigma_{ij}\, d\varepsilon_{ij}^P \qquad (7.3.2)$$

which in turn can be written

$$dW^P = S_{ij}\, d\varepsilon_{ij}^P \qquad (7.3.3)$$

or, in terms of the principal stresses,

$$dW^P = S_1\, d\varepsilon_1^P + S_2\, d\varepsilon_2^P + S_3\, d\varepsilon_3^P$$

Now let us consider again a plot in the π plane where the axes are taken to be the principal stress deviators S_1, S_2, and S_3 (for a point lying in the π plane, since $\sigma_1 + \sigma_2 + \sigma_3 = 0$, $S_1 = \sigma_1, S_2 = \sigma_2, S_3 = \sigma_3$). Since for plastic strain increments $d\varepsilon_1^P + d\varepsilon_2^P + d\varepsilon_3^P = 0$, the same plot in the π plane can

Sec. 7-3] Plastic Work. Two Measures of Work Hardening

be used for both the stress deviators and the plastic strain increments, provided the latter are multiplied by a constant to give them the dimensions of stress. For this constant we choose $2G$. The stress deviator vector and the plastic strain increment vector can then be plotted on the same plot, as

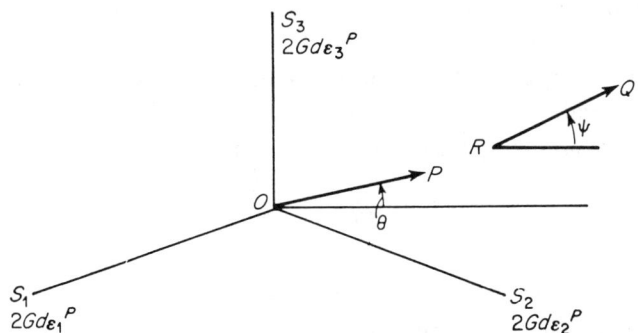

FIGURE 7.3.1 Stress and plastic strain increment vectors in π plane.

shown in Figure 7.3.1. From (7.3.3) the plastic work increment is the scaler product of the two vectors, or

$$dW^P = \frac{\overline{OP} \cdot \overline{RQ}}{2G} = \frac{(OP)(RQ)}{2G} \cos(\theta - \psi)$$

But

$$OP = \sqrt{S_1^2 + S_2^2 + S_3^2} = \sqrt{\tfrac{2}{3}}\, \sigma_e$$

and

$$RQ = 2G\sqrt{(d\varepsilon_1^P)^2 + (d\varepsilon_2^P)^2 + (d\varepsilon_3^P)^2} = 2\sqrt{\tfrac{3}{2}}\, G\, d\varepsilon_p$$

Hence the plastic work increment can be written

$$dW^P = \sigma_e\, d\varepsilon_p \cos(\theta - \psi) \qquad (7.3.4)$$

If $\overline{RQ}$ is parallel to $\overline{OP}$, this reduces to

$$dW^P = \sigma_e\, d\varepsilon_p \qquad (7.3.5)$$

In particular, equation (7.3.5) will be valid for the Prandtl–Reuss relations of Section 7.2. Making use of equation (7.3.5), the Prandtl–Reuss relations (7.2.13) can now also be written

$$d\varepsilon_{ij}^P = \frac{3}{2}\frac{dW^P}{\sigma_e^2} S_{ij} \qquad (7.3.6)$$

We next consider the question of how the amount of work hardening or strain hardening that has taken place in a given material due to plastic flow is measured. For this purpose two work-hardening hypotheses, known as the two measures of work hardening, have been proposed.

The first hypothesis assumes that the amount of hardening depends only on the total plastic work, and is independent of the strain path [6]. This is called the *equivalence of plastic work*. The implication is that the resistance to further yielding depends only on the amount of work which has been done on the material. This amount of work is measured by the yield criterion. Thus, as in (6.5.1), the yield criterion is written

$$F(\sigma_{ij}) = K \tag{7.3.7}$$

where K keeps changing as the material work hardens, and, assuming isotropic hardening, $F(\sigma_{ij})$ remains the same. By the above hypothesis, K is a function of the plastic work done per unit volume and we can write

$$F(\sigma_{ij}) = f(W^P) \tag{7.3.8}$$

where

$$W^P = \int \sigma_{ij} \, d\varepsilon_{ij}^P$$

If the equivalent stress σ_e is taken as the yield function, we can write

$$\sigma_e = f(W^P) \tag{7.3.9}$$

The functional relationship between the yield function and the plastic work can be obtained experimentally and then the plastic strain increments can be calculated using equation (7.3.6).

The second hypothesis uses the equivalent plastic strain previously defined as a measure of work hardening; i.e.,

$$\varepsilon_p = \int d\varepsilon_p \tag{7.3.10}$$

where $d\varepsilon_p$ is given by equations (7.2.10). The yield function is then assumed to be a function of the equivalent plastic strain. Thus

$$F(\sigma_{ij}) = H(\varepsilon_p) \tag{7.3.11}$$

Sec. 7-3] Plastic Work. Two Measures of Work Hardening

Again the functional relationship can be determined by experiment. If, as before, the equivalent stress is used for the yield function, then

$$\sigma_e = H(\varepsilon_p) \tag{7.3.12}$$

and the plastic strain increments can be computed from (7.2.13).

For the case of the von Mises criterion assuming isotropic hardening, these two hypotheses are obviously equivalent. Since

$$W^P = \int \sigma_{ij} \, d\varepsilon_{ij}^P = \int \sigma_e \, d\varepsilon_p$$

equation (7.3.9) can be written

$$\sigma_e = f\left(\int \sigma_e \, d\varepsilon_p\right)$$
$$= H(\varepsilon_p) \tag{7.3.13}$$

which is the same as (7.3.12). In general, however, equations (7.3.11) and (7.3.8) need not be equivalent, because of anisotropy and the Bauschinger effect. A detailed discussion of the conditions under which the two are equivalent is given in reference [7]. Equations (7.3.8) and (7.3.11) are sometimes referred to as the *work-hardening* and *strain-hardening hypotheses*, respectively [8].

The formulation (7.3.11) is simpler to use. For the case of the Prandtl-Reuss equations and the von Mises yield criterion, equation (7.3.12) is used almost exclusively in conjunction with equations (7.2.13). In actual application, the experimental relationship given by equation (7.3.12) is taken from the uniaxial tensile stress–strain curve, as shown in Figure 7.3.2. The abscissa

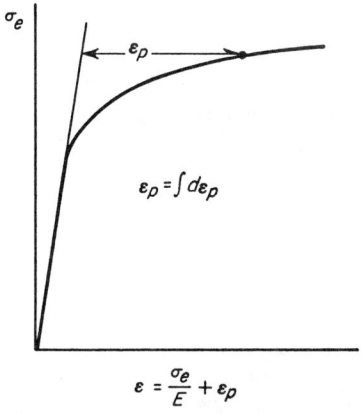

FIGURE 7.3.2 Relation between equivalent stress and equivalent plastic strain.

and ordinate of the uniaxial stress–strain curve are replaced by $\varepsilon_p = \int d\varepsilon_p$ and σ_e, respectively. In terms of the slope of this curve, equations (7.2.13) can be written

$$d\varepsilon_{ij}^P = \frac{3}{2}\frac{d\sigma_e}{H'\sigma_e}S_{ij} \qquad (7.3.14)$$

where

$$H' = \frac{d\sigma_e}{d\varepsilon_p}$$

7-4 STRESS–STRAIN RELATIONS BASED ON TRESCA CRITERION

In Section 7.2 it was shown that the Prandtl–Reuss relations are associated with the von Mises yield criterion. The stress–strain relations generally used with the Tresca criterion are simpler in nature. However, as discussed in Section 6.2, to use the Tresca yield criterion it is necessary to know which is the maximum and which is the minimum principal stress. Assuming that σ_1 is the maximum principal stress and σ_3 is the minimum principal stress, an equivalent stress can be defined for the Tresca criterion by

$$\sigma_T = \sigma_1 - \sigma_3 \qquad (7.4.1)$$

For the uniaxial tensile test in the 1 direction, σ_T becomes equal to σ_1. The Prandtl–Reuss equations can now be written

$$d\varepsilon_{ij}^P = \frac{3}{2}\frac{d\varepsilon_p}{\sigma_T}S_{ij} \qquad (7.4.2)$$

with $d\varepsilon_p$ defined as previously. However, it can easily be shown (see Problem 5) that equations (7.4.2) are inconsistent with the definition (7.4.1). In spite of this inconsistency, equations (7.4.2) are sometimes used to good advantage [9]. Since in using the Tresca criterion it is assumed that the middle stress has no effect on yielding, it is reasonable to assume also that there is no plastic flow in that direction. A consistent set of relations can therefore be obtained by assuming

$$d\varepsilon_1^P = \frac{\sqrt{3}}{2}d\varepsilon_p$$

$$d\varepsilon_2^P = 0$$

$$d\varepsilon_3^P = -\frac{\sqrt{3}}{2}d\varepsilon_p = -d\varepsilon_1^P \qquad (7.4.3)$$

with

$$d\varepsilon_p = \sqrt{\tfrac{2}{3}}\sqrt{(d\varepsilon_1^P)^2 + (d\varepsilon_2^P)^2 + (d\varepsilon_3^P)^2}$$

Equations (7.2.13) and (7.4.3) are known as the *flow rules* associated with the von Mises and Tresca criteria, respectively. In Section 7.6 we will show that these are special cases of a more general flow rule. In that section the results of the previous sections will be essentially rederived based on a more unified approach.

7-5 EXPERIMENTAL VERIFICATION OF PRANDTL-REUSS EQUATIONS

The first experimental investigation to determine the validity of the plastic stress–strain relations was made by Lode (reference [5] of Chapter 6). As explained in Chapter 6, Lode tested tubes of steel, copper, and nickel under combined tension and internal pressure. In addition to the stress parameter μ defined by (6.4.2), Lode introduced the plastic strain parameter ν defined by

$$\nu = \frac{2d\varepsilon_2^P - d\varepsilon_3^P - d\varepsilon_1^P}{d\varepsilon_3^P - d\varepsilon_1^P} = \frac{d\varepsilon_2^P - \tfrac{1}{2}(d\varepsilon_3^P + d\varepsilon_1^P)}{\tfrac{1}{2}(d\varepsilon_3^P - d\varepsilon_1^P)} \tag{7.5.1}$$

From equations (7.2.3),

$$\frac{2d\varepsilon_2^P - d\varepsilon_3^P - d\varepsilon_1^P}{d\varepsilon_3^P - d\varepsilon_1^P} = \frac{2S_2 - S_3 - S_1}{S_3 - S_1} = \frac{2\sigma_2 - \sigma_3 - \sigma_1}{\sigma_3 - \sigma_1}$$

or
$$\nu = \mu \tag{7.5.2}$$

If the Prandtl–Reuss relations are valid, then equation (7.5.2) should be satisfied over the whole range of experiments. The results of Lode's tests are shown schematically in Figure 7.5.1. Although the relation appears to be approximately satisfied, there is a definite deviation which cannot be accounted for by experimental scatter. This deviation was confirmed by Taylor and Quinney (reference [6] of Chapter 6), whose results on tension–torsion tests are shown roughly in Figure 7.5.2. Although some of these deviations can be attributed to lack of isotropy, it appears that the Prandtl–Reuss relations are not quite correct. Prager [10] has shown that the data can be made to fit quite well by including J_3 in the yield function. However, the relations are too cumbersome for general use.

From a practical viewpoint the Prandtl–Reuss relations appear to be sufficiently accurate, just as the von Mises yield criterion is sufficiently accurate, and the small deviations of the experimental data are not large enough to make additional complexity worthwhile. These relations will therefore henceforth be used most frequently, the Tresca criterion and its

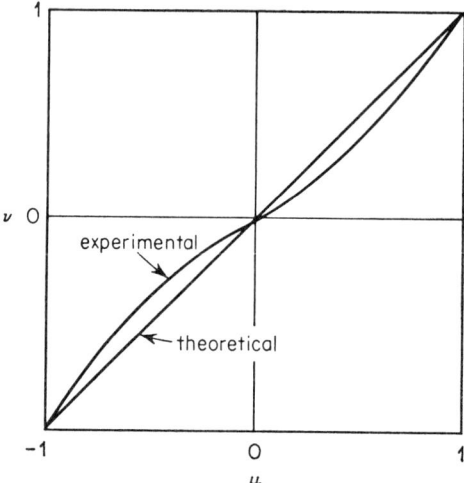

FIGURE 7.5.1 Results of Lode's tests.

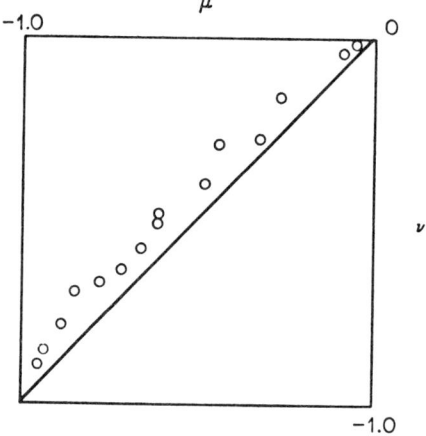

FIGURE 7.5.2 Results of Taylor and Quinney's tests.

associated flow rule being used occasionally. It is to be noted that the assumed coincidence of the principal stress axes and principal plastic strain increment axes are borne out very well by the experiments of Taylor and Quinney.

7-6 GENERAL DERIVATION OF PLASTIC STRESS-STRAIN RELATIONS

In Section 7.5 the Saint-Venant–Lévy–Mises and the Prandtl–Reuss relations were described as originating basically from an assumption that "the

Sec. 7-6] General Derivation of Plastic Stress-Strain Relations

maximum shear and maximum slide velocity are co-directional," as Saint-Venant expressed it (see reference [11]). It was also shown that these relations imply the von Mises yield function. In this section the general equations for determining the plastic stress-strain relations for any yield criterion will be derived based on a unified approach due to Drucker [12, 13].

We start with a more precise definition of work hardening which is due to Drucker, and it will subsequently be shown that together with two additional assumptions it is sufficient to obtain the most general form of the stress-strain relations.

Suppose we have a given state of stress and then some external agency applies an additional set of stresses and then slowly removes them. Work hardening implies that for all such added sets of stresses the material will remain in equilibrium, and

1. Positive work is done by the external agency during the *application* of the set of stresses.
2. The net work performed by it over the cycle of application and removal is zero or positive.

It should be emphasized that the work referred to is not the total work done by all the forces acting; it is only the work done by the added set of forces on the displacements which result. In other words, work hardening means that useful net energy over and above the elastic energy cannot be extracted from the material and the system of forces acting on it.

This definition can be put into mathematical language as follows: Suppose that to a state of stress σ_{ij} and strain ε_{ij} some external agency applies small surface forces so that the stress at each point is changed by an amount $d\sigma_{ij}$ and the strain by an amount $d\varepsilon_{ij}$. Part of $d\varepsilon_{ij}$ is elastic and part may be plastic; i.e., $d\varepsilon_{ij} = d\varepsilon_{ij}^e + d\varepsilon_{ij}^p$. Now suppose these added forces are removed, thus releasing the elastic strain increments, $d\varepsilon_{ij}^e$. It then follows from implication 1 that for work hardening

$$d\sigma_{ij}\,d\varepsilon_{ij} > 0$$

and, from implication 2,

$$d\sigma_{ij}(d\varepsilon_{ij} - d\varepsilon_{ij}^e) \geq 0$$

or
$$\left.\begin{array}{c}d\sigma_{ij}(d\varepsilon_{ij}^e + d\varepsilon_{ij}^p) > 0 \\ d\sigma_{ij}\,d\varepsilon_{ij}^p \geq 0\end{array}\right\} \qquad (7.6.1)$$

Equations (7.6.1) represent the mathematical definition of work hardening. The second of equations (7.6.1) is sometimes referred to as *the uniqueness condition*.

To obtain the general stress–strain relations, we use the above definition plus two basic assumptions. These are:

1. A loading function exists. At each stage of the plastic deformation there exists a function $f(\sigma_{ij})$ so that further plastic deformation takes place only for $f(\sigma_{ij}) > K$. Both f and K may depend on the existing state of stress and on the strain history.

2. The relation between *infinitesimals* of stress and plastic strain is linear; i.e.,

$$d\varepsilon_{ij}^P = C_{ijkl}\, d\sigma_{kl} \qquad (7.6.2)$$

Although equation (7.6.2) seems very reasonable, it should be noted that there appears to be no theoretical justification for it. It is purely an assumption. Although the C_{ijkl} may be functions of stress, strain, and history of loading, (7.6.2) implies that they are independent of the $d\sigma_{kl}$.

From assumption 1 it follows that for plastic deformation to take place

$$df(\sigma_{ij}) > 0$$

or
$$\frac{\partial f}{\partial \sigma_{ij}}\, d\sigma_{ij} > 0 \qquad (7.6.3)$$

and from the linearity assumption 2 it follows that the superposition principle may be applied to the stress and strain *increments*. Thus if $d\sigma'_{ij}$ and $d\sigma''_{ij}$ are two increments producing plastic strain increments, $d\varepsilon_{ij}^{P'}$ and $d\varepsilon_{ij}^{P''}$, then an increment $d\sigma_{ij} = d\sigma'_{ij} + d\sigma''_{ij}$ will produce an increment, $d\varepsilon_{ij}^{P'} + d\varepsilon_{ij}^{P''}$.

Now assume that for a given state of stress σ_{kl}, an increment of stress $d\sigma_{kl}$ producing plastic flow is imposed. This increment $d\sigma_{kl}$ can be decomposed into two parts $d\sigma'_{kl}$ and $d\sigma''_{kl}$ such that $d\sigma'_{kl}$ produces no plastic flow and $d\sigma''_{kl}$ is proportional to the gradient of $f(\sigma_{ij})$. Geometrically this means that the vector $d\sigma_{kl}$ is decomposed into a component tangent to f and a component perpendicular to f, as shown in Figure 7.6.1.

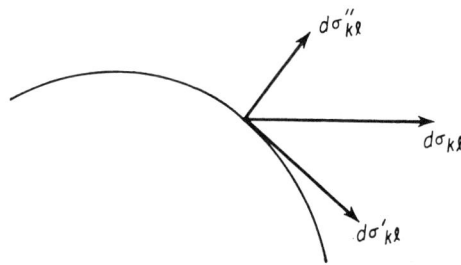

FIGURE 7.6.1   Decomposition of stress increment vector.

Sec. 7-6] General Derivation of Plastic Stress-Strain Relations   113

Since the increment $d\sigma_{kl}$ produces plastic flow, from (7.6.3) we have

$$\frac{\partial f}{\partial \sigma_{kl}} d\sigma_{kl} = \frac{\partial f}{\partial \sigma_{kl}} (d\sigma'_{kl} + d\sigma''_{kl}) > 0 \qquad (7.6.4)$$

But the increment $d\sigma'_{kl}$ produces no plastic flow. Therefore,

$$\frac{\partial f}{\partial \sigma_{kl}} d\sigma'_{kl} = 0 \qquad (7.6.5)$$

Also, $d\sigma''_{kl}$ has been taken proportional to the gradient of f; therefore,

$$d\sigma''_{kl} = a \frac{\partial f}{\partial \sigma_{kl}} \qquad (7.6.6)$$

where a is a scaler > 0 and from (7.6.4), (7.6.5), and (7.6.6) it follows that

$$\frac{\partial f}{\partial \sigma_{kl}} d\sigma_{kl} = \frac{\partial f}{\partial \sigma_{kl}} d\sigma''_{kl} = \frac{\partial f}{\partial \sigma_{kl}} a \frac{\partial f}{\partial \sigma_{kl}} > 0$$

Hence

$$a = \frac{\partial f/\partial \sigma_{kl}}{(\partial f/\partial \sigma_{mn})(\partial f/\partial \sigma_{mn})} d\sigma_{kl} \qquad (7.6.7)$$

Equation (7.6.7) proves that the proposed decomposition is possible.

Comparing (7.6.2) and (7.6.7) [realizing that $d\sigma_{kl}$ in (7.6.2) can be replaced by $d\sigma''_{kl}$ since $d\sigma'_{kl}$ produces no plastic flow] we see that every component of $d\varepsilon^P_{ij}$ must be proportional to a, or

$$d\varepsilon^P_{ij} = h_{ij} a \qquad (7.6.8)$$

or, combining with (7.6.7) gives

$$d\varepsilon^P_{ij} = g_{ij} \frac{\partial f}{\partial \sigma_{kl}} d\sigma_{kl} \qquad (7.6.9)$$

where g_{ij} depends in general on stress, strain, and history of loading.

Now the second of conditions (7.6.1) can be written

$$d\sigma_{ij} d\varepsilon^P_{ij} = (d\sigma'_{ij} + d\sigma''_{ij}) d\varepsilon^P_{ij} \geq 0 \qquad (7.6.10)$$

But $d\sigma'_{ij}$ produces no plastic flow, so that the increment $d\sigma_{ij} = C d\sigma'_{ij} + d\sigma''_{ij}$ for any value of C, positive or negative, will produce the same plastic increment $d\varepsilon^P_{ij}$. We can therefore write the strain-hardening condition as

$$(C d\sigma'_{ij} + d\sigma''_{ij}) d\varepsilon^P_{ij} \geq 0 \qquad (7.6.11)$$

But $d\sigma'_{ij} d\varepsilon^P_{ij}$ must vanish; otherwise C could be chosen (a large negative number) so as to violate (7.6.11). Therefore,

$$d\sigma'_{ij} d\varepsilon^P_{ij} = d\sigma'_{ij} g_{ij} \frac{\partial f}{\partial \sigma_{kl}} d\sigma_{kl} = 0$$

But
$$\frac{\partial f}{\partial \sigma_{kl}} d\sigma_{kl} > 0$$

Hence
$$d\sigma'_{ij} g_{ij} = 0$$

Comparing with (7.6.5) it is seen that

$$g_{ij} = G \frac{\partial f}{\partial \sigma_{ij}} \qquad (7.6.12)$$

where G is a scalar which may depend on stress, strain, and history. Substituting (7.6.12) into (7.6.9) gives

$$d\varepsilon^P_{ij} = G \frac{\partial f}{\partial \sigma_{ij}} \frac{\partial f}{\partial \sigma_{kl}} d\sigma_{kl} \qquad (7.6.13)$$

or
$$d\varepsilon^P_{ij} = G \frac{\partial f}{\partial \sigma_{ij}} df \qquad (7.6.14)$$

which is the general stress–strain relation consistent with the original assumptions.

Let us take some specific examples. For the von Mises yield function, let

$$f = J_2 = \tfrac{1}{6}[(\sigma_1 - \sigma_2)^2 + (\sigma_2 - \sigma_3)^2 + (\sigma_3 - \sigma_1)^2]$$

$$\frac{\partial f}{\partial \sigma_1} = \tfrac{2}{3}[\sigma_1 - \tfrac{1}{2}(\sigma_2 + \sigma_3)]$$

Therefore,

$$d\varepsilon^P_1 = \tfrac{2}{3} d\lambda \, [\sigma_1 - \tfrac{1}{2}(\sigma_2 + \sigma_3)]$$

Sec. 7-6] General Derivation of Plastic Stress–Strain Relations

where $d\lambda = G\,df$, and we recognize the Prandtl–Reuss equations. For the Tresca yield condition, assuming it is known which is the maximum principal stress σ_1 and minimum principal stress σ_3, we have

$$f = \tfrac{1}{2}(\sigma_1 - \sigma_3)$$

$$\frac{\partial f}{\partial \sigma_1} = \frac{1}{2} \qquad \frac{\partial f}{\partial \sigma_2} = 0 \qquad \frac{\partial f}{\partial \sigma_3} = -\frac{1}{2}$$

Then

$$d\varepsilon_1^P = \tfrac{1}{2}\,d\lambda$$
$$d\varepsilon_2^P = 0$$
$$d\varepsilon_3^P = -\tfrac{1}{2}\,d\lambda$$

which are the same as equations (7.4.3). We note that the form of the flow rule or plastic stress–strain relations *associated* with the Tresca criterion is entirely different than that for the von Mises; thus each yield condition has an *associated flow rule*, as was pointed out in Section 7.4. This is sometimes ignored and, for example, the Tresca criterion has been used with the von Mises flow rule. There is, however, no theoretical justification for that type of assumption.

It is worth noting one other important fact from the previous derivation. Since $d\sigma'_{ij}\,d\varepsilon_{ij}^P = 0$ and $d\sigma'_{ij}$ is tangent to the yield surface, it follows that $d\varepsilon_{ij}^P$ *is normal to the yield surface*, for the above equation merely represents a dot product of two vectors. This can also be seen from (7.6.14), since $d\varepsilon_{ij}^P$ is equal to the gradient of f times a scalar. It also follows from the above that the Prandtl–Reuss equations imply the use of the von Mises criterion.

To summarize: Starting with the definition of work hardening and postulating the existence of a loading function and linearity between increments of stress and increments of strain, we can arrive at the general flow rule (7.6.14) for a strain-hardening material. It can also be shown that the plastic strain increment vector must always be normal to the yield surface. The scalar G, which depends in general on the stress, strain, and history, must be determined from experiment, and its derivation will be discussed shortly.

Perfectly Plastic Material

For this case the work done by an external agency which slowly applies and removes a set of stresses is zero over the cycle, or

$$d\sigma_{ij}\,d\varepsilon_{ij}^P = 0 \qquad (7.6.15)$$

It should be remarked that this equation is not the same as the second of (7.6.1) with the equality sign. In (7.6.1) the equality sign is used only when $d\varepsilon_{ij}^P = 0$.

For ideal plasticity it is also assumed that $f(\sigma_{ij})$ exists and is a function of stress only, and that plastic flow takes place without limit when $f(\sigma_{ij}) = K$ and the material behaves elastically when $f(\sigma_{ij}) < K$. For plastic flow, therefore,

$$df = \frac{\partial f}{\partial \sigma_{ij}} d\sigma_{ij} = 0 \tag{7.6.16}$$

Comparing (7.6.15) and (7.6.16) it is seen that

$$d\varepsilon_{ij}^P = d\lambda \frac{\partial f}{\partial \sigma_{ij}} \tag{7.6.17}$$

where $d\lambda$ is a scalar.

Determination of the Function G. Effective Stress and Effective Strain

For (7.6.14) to be of any practical use, it must be related somehow to the experimental uniaxial stress–strain curve. What we are looking for is some function of the stresses, which might be called the *effective stress*, and some function of the strains or strain history, which might be called *effective strain*, so that results obtained by different loading programs can all be correlated by means of a single curve of effective stress versus effective strain. This curve should preferably be the uniaxial tensile curve.

The definition of effective stress can be arrived at rather simply; since it should reduce to the stress in the uniaxial tension test, it is a quantity which will determine whether plastic flow takes place or not, and it must be a positively increasing function of the stresses during plastic flow. Now the loading function $f(\sigma_{ij})$ also, by definition, determines whether additional plastic flow takes place. It is also a positively increasing function as long as plastic flow takes place and, if unloading takes place, plastic flow is not resumed until the highest previous value of f is exceeded. The loading function $f(\sigma_{ij})$ must therefore be some constant times the effective stress to some power; i.e.,

$$f(\sigma_{ij}) = C\sigma_e^n \tag{7.6.18}$$

For example, if we assume again

$$f = J_2$$

then
$$J_2 = C\sigma_e^n$$

Sec. 7-6] General Derivation of Plastic Stress–Strain Relations 117

or $\quad \sigma_e = \left(\dfrac{J_2}{C}\right)^{1/n} = \left\{\dfrac{1}{6C}[(\sigma_1 - \sigma_2)^2 + (\sigma_2 - \sigma_3)^2 + (\sigma_3 - \sigma_1)^2]\right\}^{1/n}$

and for the uniaxial tensile test $\sigma_e = \sigma_1$. Therefore,

$$n = 2 \quad C = 1/3 \quad \sigma_e = \sqrt{3J_2}$$

which agrees with the previous definition in equation (7.2.9).

The definition of effective plastic strain, ε_p, is not quite as simple. There are two methods generally used. One defines the effective strain increment in terms of the plastic work per unit volume; i.e.,

$$dW^P = \sigma_e \, d\varepsilon_p \quad (7.6.19)$$

and since $\quad dW^P = S_{ij} \, d\varepsilon_{ij}^P$

$$d\varepsilon_p = \dfrac{1}{\sigma_e} S_{ij} \, d\varepsilon_{ij}^P \quad (7.6.20)$$

For example, if $f = J_2$, it can readily be shown that

$$d\varepsilon_p = \sqrt{\tfrac{2}{3} d\varepsilon_{ij}^P \, d\varepsilon_{ij}^P} \quad (7.6.21)$$

and, if $f = \sigma_1 - \sigma_3$ with $\sigma_1 > \sigma_2 > \sigma_3$ as for the Tresca criterion, then

$$d\varepsilon_p = d\varepsilon_1^P \quad (7.6.22)$$

Equation (7.6.21) expanded becomes

$$d\varepsilon_p = \sqrt{\tfrac{2}{3}}\,[(d\varepsilon_x^P)^2 + (d\varepsilon_y^P)^2 + (d\varepsilon_z^P)^2 + 2(d\varepsilon_{xy}^P)^2 + 2(d\varepsilon_{yz}^P)^2 + 2(d\varepsilon_{zx}^P)^2]^{1/2} \quad (7.6.23)$$

and, in terms of principal strain increments,

$$d\varepsilon_p = \sqrt{\tfrac{2}{3}}\,[(d\varepsilon_1^P)^2 + (d\varepsilon_2^P)^2 + (d\varepsilon_3^P)^2]^{1/2}$$
$$= \dfrac{2}{\sqrt{3}}\,[(d\varepsilon_1^P)^2 + (d\varepsilon_2^P)^2 + d\varepsilon_1^P \, d\varepsilon_2^P]^{1/2}$$

where the incompressibility condition $d\varepsilon_1^P + d\varepsilon_2^P + d\varepsilon_3^P = 0$ has been used.

A second method for arriving at (7.6.21) is sort of intuitive. One seeks to find a definition of effective plastic strain increment which when integrated is a function of σ_e only. The simplest combination of plastic strain increments which is positive increasing and has the correct "dimension" is

$$d\varepsilon_p = C \sqrt{d\varepsilon_{ij}^P \, d\varepsilon_{ij}^P}$$

To make this definition agree for simple tension we must have

$$d\varepsilon_x^P = d\varepsilon_p = C \sqrt{(d\varepsilon_x^P)^2 + \tfrac{1}{4}(d\varepsilon_x^P)^2 + \tfrac{1}{4}(d\varepsilon_x^P)^2} = C \sqrt{\tfrac{3}{2}} \, d\varepsilon_x^P$$

Therefore,

$$C = \sqrt{\tfrac{2}{3}}$$

$$d\varepsilon_p = \sqrt{\tfrac{2}{3} d\varepsilon_{ij}^P \, d\varepsilon_{ij}^P}$$

and, for $f = J_2$,

$$d\varepsilon_p = H(\sigma_e) d\sigma_e \qquad (7.6.24)$$

so that the integrated effective strain is a function of effective stress only; i.e.,

$$\varepsilon_p = \int d\varepsilon_p = \int H(\sigma_e) \, d\sigma_e \qquad (7.6.25)$$

It should be noted here that the definition (7.6.21) for $d\varepsilon_p$ has been derived for $f = J_2$ only. Drucker has shown that it is reasonably correct for almost any $f(J_2, J_3)$. The second intuitive approach for defining $d\varepsilon_p$ is, of course, not based on any specific loading function.

We are now in a position to determine the function G. It should first be realized that for the previous formulation to agree with the uniaxial tensile curve, $d\sigma_e/d\varepsilon_p$ must be the slope of that curve (in the plastic range). Substituting the basic equation

$$d\varepsilon_{ij}^P = G \frac{\partial f}{\partial \sigma_{ij}} df$$

into (7.6.21) gives

$$d\varepsilon_p = \sqrt{\tfrac{2}{3}} \, G \sqrt{\frac{\partial f}{\partial \sigma_{ij}} \frac{\partial f}{\partial \sigma_{ij}}} \, df \qquad (7.6.26)$$

or

$$G \, df = d\lambda = \sqrt{\tfrac{3}{2}} \frac{d\varepsilon_p}{\sqrt{(\partial f/\partial \sigma_{ij})(\partial f/\partial \sigma_{ij})}} \qquad (7.6.27)$$

and the general plastic stress–strain relation becomes

$$d\varepsilon_{ij}^P = \frac{\sqrt{\tfrac{3}{2}} (\partial f/\partial \sigma_{ij}) \, d\varepsilon_p}{\sqrt{(\partial f/\partial \sigma_{mn})(\partial f/\partial \sigma_{mn})}} \qquad (7.6.28)$$

Sec. 7-7] Incremental and Deformation Theories

or
$$d\varepsilon_{ij}^P = \sqrt{\tfrac{3}{2}} \frac{(\partial f/\partial \sigma_{ij})}{\sqrt{(\partial f/\partial \sigma_{mn})(\partial f/\partial \sigma_{mn})}} \frac{d\sigma_e}{\sigma_e'} \qquad (7.6.29)$$

where $\sigma_e' = d\sigma_e/d\varepsilon_p$ is the slope of the uniaxial stress–plastic strain curve at the current value of σ_e. As an example, for $f = J_2$, equation (7.6.29) gives

$$\begin{aligned} d\varepsilon_{ij}^P &= \frac{3}{2}\frac{S_{ij}}{\sigma_e}\frac{d\sigma_e}{\sigma_e'} \\ &= \frac{3}{2}\frac{S_{ij}}{\sigma_e} d\varepsilon_p \end{aligned} \qquad (7.6.30)$$

Equations (7.6.30) constitute the flow rule (or plastic stress–strain relations) *associated* with the von Mises yield criterion. They are the well-known Prandtl–Reuss relations we obtained previously. If we replace the plastic strain increments in the above equations by total strain increments, the Lévy–Mises relations are obtained which are valid only if the plastic strains are so large that the elastic strains can be neglected.

As a final note, a general flow law such as (7.6.14) can also be obtained on the basis of a hypothesis that there exists a *plastic potential* (similar to the strain energy density function) which is a scalar function of stress, $g(\sigma_{ij})$, from which the plastic strain increments can be obtained by partial differentiation with respect to the stresses. Thus

$$d\varepsilon_{ij}^P = \frac{\partial g}{\partial \sigma_{ij}} d\beta \qquad (7.6.31)$$

where $d\beta$ is a nonnegative constant. The plastic potential $g(\sigma_{ij})$ was first introduced by Melan [14]. By comparison with (7.6.14), it would appear that the plastic potential should play the same role as the yield function, and indeed Bland [15] has proved that they must be the same function, so that g in (7.6.31) can be replaced by f; (7.6.31) and (7.6.14) are then the same.

7-7 INCREMENTAL AND DEFORMATION THEORIES

Equations such as (7.6.30) are called *incremental stress–strain relations* because they relate the increments of plastic strain to the stress. To obtain the total plastic strain components, one must integrate these equations over the whole history of loading. Hencky [16] proposed *total stress–strain relations* whereby the *total* strain components are related to the current stress. Thus, instead of (7.6.30), one would have

$$\varepsilon_{ij}^P = \frac{3}{2} \frac{S_{ij}}{\sigma_e} \varepsilon_p \qquad (7.7.1)$$

The plastic strains then are functions of the current state of stress and are independent of the history of loading. Such theories are called *total* or *deformation theories* in contrast to the *incremental* or *flow theories* previously described. This type of assumption greatly simplifies the problem; however, as was previously shown, the plastic strains cannot in general be independent of the loading path and deformation theories cannot generally be correct. There has often been a tendency therefore to ignore all deformation theory as of little value.

It can easily be shown, however, that for the case of *proportional* or *radial loading*, i.e., if all the stresses are increasing in ratio, the incremental theory reduces to the deformation theory. For if $\sigma_{ij} = K\sigma_{ij}^0$, where σ_{ij}^0 is an arbitrary reference state of stress (nonzero) and K is a monotonically increasing function of time, then $S_{ij} = KS_{ij}^0$ and $\sigma_e = K\sigma_e^0$ and (7.6.30) becomes

$$d\varepsilon_{ij}^P = \frac{3d\varepsilon_p}{2\sigma_e^0} S_{ij}^0$$

which can be immediately integrated to give

$$\varepsilon_{ij}^P = 3\varepsilon_p \frac{S_{ij}^0}{2\sigma_e^0} = \frac{3\varepsilon_p S_{ij}}{2\sigma_e} \qquad (7.7.2)$$

so the plastic strain is a function only of the current state of stress and is independent of the loading path.

Furthermore, it has been proposed by Budiansky [17] that there are ranges of loading paths other than proportional loading for which the basic postulates of plasticity theory are satisfied by deformation theories. Budiansky's theory postulates the occurrence of corners or singular points on the successive yield surfaces and, although the existence of such singular points has as yet not been established experimentally, one cannot rule out the possibility of loading paths other than proportional loading for which total plasticity theories may give satisfactory answers.

From a practical viewpoint, there are a great many engineering problems where the loading path is not far from proportional loading, *provided one is careful when unloading occurs to separate the problem into separate parts, the loading parts, and the unloading parts.*

Problems of plastic flow in thermally stressed disks and cylinders have been handled in this way and good results obtained using deformation theory.

On the other hand, it will subsequently be shown that with the present widespread availability of high-speed computors, many simplifying assumptions heretofore made, including the use of deformation theories under doubtful conditions, are often unnecessary.

7-8 CONVEXITY OF YIELD SURFACE. SINGULAR POINTS

In Section 6.3 the statement was made that the yield surface was convex. A proof, as given in reference [13], will now be presented. Consider some state of stress σ_{ij}^* inside the loading surface, as shown in Figure 7.8.1. Let some

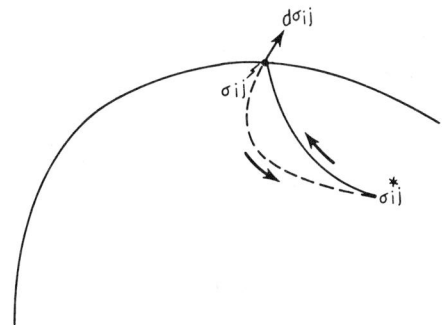

FIGURE 7.8.1 Stress path produced by external agency.

external agency add stresses along some arbitrary path inside the surface until a state of stress σ_{ij} is reached which is on the yield surface. Only elastic changes have taken place so far. Now suppose the external agency to add a very small outward pointing stress increment $d\sigma_{ij}$ which produces small plastic strain increments $d\varepsilon_{ij}^P$, as well as elastic increments. The external agency then releases the $d\sigma_{ij}$ and the state of stress is returned to σ_{ij}^* along an elastic path. The work done by the external agency over the cycle is

$$\delta W = (\sigma_{ij} - \sigma_{ij}^*)\, d\varepsilon_{ij}^P + d\sigma_{ij}\, d\varepsilon_{ij}^P \qquad (7.8.1)$$

If the plastic strain coordinates are superimposed on the stress coordinates, as in Figure 7.8.2, δW may be interpreted as the scalar product of the vector

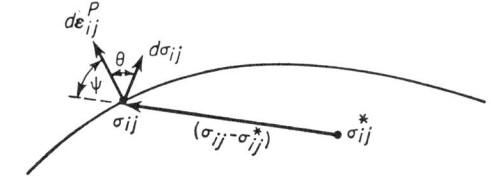

FIGURE 7.8.2 Stress and plastic strain increment vectors.

$\sigma_{ij} - \sigma_{ij}^*$ and the vector $d\varepsilon_{ij}^P$ plus the scalar product of $d\sigma_{ij}$ and $d\varepsilon_{ij}^P$. Now, from the strain-hardening definition equation (7.6.1),

$$d\sigma_{ij}\, d\varepsilon_{ij}^P \geq 0$$

or

$$|d\sigma_{ij}|\, |d\varepsilon_{ij}^P|\cos\theta \geq 0 \qquad (7.8.2)$$

or

$$-\frac{\pi}{2} \leq \theta \leq \frac{\pi}{2}$$

That is, the vectors $d\sigma_{ij}$ and $d\varepsilon_{ij}^P$ make an acute angle with each other. In a similar fashion, since the magnitude of $\sigma_{ij} - \sigma_{ij}^*$ can always be made larger than the magnitude of $d\sigma_{ij}$, it follows that

$$(\sigma_{ij} - \sigma_{ij}^*)d\varepsilon_{ij}^P \geq 0$$

or

$$|\sigma_{ij} - \sigma_{ij}^*|\, |d\varepsilon_{ij}^P|\cos\psi \geq 0$$

Hence

$$-\frac{\pi}{2} \leq \psi \leq \frac{\pi}{2} \qquad (7.8.3)$$

Thus the vector $\sigma_{ij} - \sigma_{ij}^*$ makes an acute angle with the vector $d\varepsilon_{ij}^P$ *for all choices of* σ_{ij}^*. Therefore, all points σ_{ij}^* must lie on one side of a plane perpendicular to $d\varepsilon_{ij}^P$, and, since $d\varepsilon_{ij}^P$ is normal to the yield surface, this plane will be tangent to the yield surface. This must be true for all points σ_{ij} on the yield surface, so that no vector $\sigma_{ij} - \sigma_{ij}^*$ can pass outside the surface intersecting the surface twice, as shown in Figure 7.8.3. The surface must therefore be

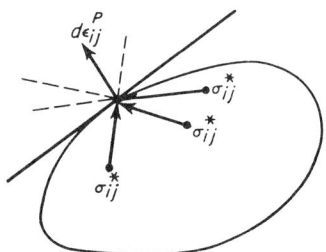

FIGURE 7.8.3 Convex surface, only acute angles possible.

Sec. 7-9] Plastic Strain–Total Strain Plasticity Relations

convex. On the other hand, if the surface is not convex, there exist some points σ_{ij} and σ_{ij}^* such that the vector $\sigma_{ij} - \sigma_{ij}^*$ forms an obtuse angle with the vector $d\sigma_{ij}$, as shown in Figure 7.8.4. This completes the convexity proof.

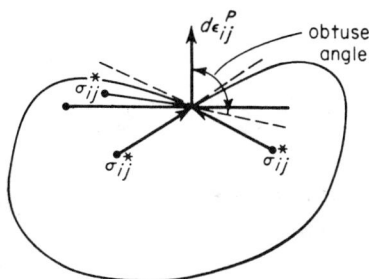

FIGURE 7.8.4 Surface not convex, obtuse angles possible.

Equation (7.6.14) implies that the yield surface has a unique gradient. It may happen, however, that the yield surface has vertices or corners where the gradient is not defined. For example, the Tresca hexagon has no unique normal at the corners, where two of the stresses are equal. Such points are called *singular points* or *singular yield conditions*. Such points can be treated by introducing an auxiliary parameter as described in reference [18].

7-9 PLASTIC STRAIN–TOTAL STRAIN PLASTICITY RELATIONS

The Prandtl–Reuss equations relate the plastic strain increments to the stresses. We shall now derive a similar set of equations involving only strains. These equations enable one to compute the plastic strain increments from the total strains without recourse to the stresses. In effect, they provide a simple method for separating the total strains into their elastic and plastic components. The advantage of this formulation will become evident later when certain iterative methods for solving plasticity problems are discussed.

Assume some loading path to a given state of stress and total plastic strains ε_{ij}^P. Let the load be increased by a small amount, producing additional plastic strains $\Delta\varepsilon_{ij}^P$. The total strains can now be written

$$\varepsilon_{ij} = \varepsilon_{ij}^e + \varepsilon_{ij}^P + \Delta\varepsilon_{ij}^P \qquad (7.9.1)$$

where ε_{ij}^e is the elastic component of the total strain, ε_{ij}^P is the accumulated plastic strain up to (but not including) the current increment of load, and

$\Delta\varepsilon_{ij}^P$ is the increment of plastic strain due to the increment of load. ε_{ij}^P is presumed to be known, $\Delta\varepsilon_{ij}^P$ is to be computed. Define *modified total strains* as follows:

$$\varepsilon'_{ij} \equiv \varepsilon_{ij} - \varepsilon_{ij}^P \qquad (7.9.2)$$

Then

$$\varepsilon'_{ij} = \varepsilon_{ij}^e + \Delta\varepsilon_{ij}^P \qquad (7.9.3)$$

Subtracting the mean strain from the diagonal components of both sides of equation (7.9.3) results in

$$e'_{ij} = e_{ij}^e + \Delta\varepsilon_{ij}^P \qquad (7.9.4)$$

where e_{ij}^e is the elastic strain deviator tensor and e'_{ij} is the modified strain deviator tensor. From Hooke's law and the Prandtl–Reuss relations,

$$e_{ij}^e = \frac{1}{2G} S_{ij} = \frac{1}{2G\,\Delta\lambda} \Delta\varepsilon_{ij}^P$$

Hence

$$e'_{ij} = \left(1 + \frac{1}{2G\,\Delta\lambda}\right) \Delta\varepsilon_{ij}^P \qquad (7.9.5)$$

$$\frac{2}{3} e'_{ij} e'_{ij} = \frac{2}{3}\left(1 + \frac{1}{2G\,\Delta\lambda}\right)^2 \Delta\varepsilon_{ij}^P \Delta\varepsilon_{ij}^P \qquad (7.9.6)$$

We now *define* an *equivalent modified total strain* by

$$\varepsilon_{et} = \sqrt{\tfrac{2}{3} e'_{ij} e'_{ij}} \qquad (7.9.7)$$

so that, from (7.9.6),

$$1 + \frac{1}{2G\,\Delta\lambda} = \frac{\varepsilon_{et}}{\Delta\varepsilon_p} \qquad (7.9.8)$$

and, from (7.9.5),

$$\Delta\varepsilon_{ij}^P = \frac{\Delta\varepsilon_p}{\varepsilon_{et}} e'_{ij} \qquad (7.9.9)$$

Sec. 7-9] Plastic Strain–Total Strain Plasticity Relations

or, in expanded form,

$$\Delta \varepsilon_x^P = \frac{\Delta \varepsilon_p}{3\varepsilon_{et}} (2\varepsilon_x' - \varepsilon_y' - \varepsilon_z')$$

$$\Delta \varepsilon_y^P = \frac{\Delta \varepsilon_p}{3\varepsilon_{et}} (2\varepsilon_y' - \varepsilon_z' - \varepsilon_x')$$

$$\Delta \varepsilon_z^P = \frac{\Delta \varepsilon_p}{3\varepsilon_{et}} (2\varepsilon_z' - \varepsilon_x' - \varepsilon_y')$$

$$= -(\Delta \varepsilon_x^P + \Delta \varepsilon_y^P) \qquad (7.9.10)$$

$$\Delta \varepsilon_{xy}^P = \frac{\Delta \varepsilon_p}{\varepsilon_{et}} \varepsilon_{xy}'$$

$$\Delta \varepsilon_{yz}^P = \frac{\Delta \varepsilon_p}{\varepsilon_{et}} \varepsilon_{yz}'$$

$$\Delta \varepsilon_{zx}^P = \frac{\Delta \varepsilon_p}{\varepsilon_{et}} \varepsilon_{zx}'$$

with ε_{et} given by (7.9.7) or alternatively by

$$\varepsilon_{et} = \frac{\sqrt{2}}{3} [(\varepsilon_x' - \varepsilon_y')^2 + (\varepsilon_y' - \varepsilon_z')^2 + (\varepsilon_z' - \varepsilon_x')^2 + 6(\varepsilon_{xy}')^2 + 6(\varepsilon_{yz}')^2 + 6(\varepsilon_{zx}')^2]^{1/2}$$
(7.9.11)

and the primed quantities are the modified total strains as given by equation (7.9.2).

Equations (7.9.10) are equivalent to the Prandtl–Reuss equations. The stresses do not appear in these equations and the increments of plastic strain can be computed from the total strains. Note that since they have been derived by use of the Prandtl–Reuss equations, they implicitly make use of the von Mises yield function. It should also be emphasized that the equivalent total strain defined by (7.9.7) is a purely mathematically defined quantity without any *direct* physical meaning, even in the uniaxial case. However, it can be *related* to the uniaxial stress–strain curve as follows: From equation (7.2.12),

$$\Delta \lambda = \frac{3}{2} \frac{\Delta \varepsilon_p}{\sigma_e} \qquad (7.9.12)$$

Substituting this value for $\Delta\lambda$ into equation (7.9.8) gives

$$\left(1 + \frac{\sigma_e}{3G\,\Delta\varepsilon_p}\right) = \frac{\varepsilon_{et}}{\Delta\varepsilon_p}$$

or
$$\varepsilon_{et} = \Delta\varepsilon_p + \frac{1}{3G}\,\sigma_e$$

$$= \Delta\varepsilon_p + \frac{2(1+\mu)}{3E}\,\sigma_e \qquad (7.9.13)$$

Referring to the uniaxial stress-strain curve as shown in Figure 7.9.1, let

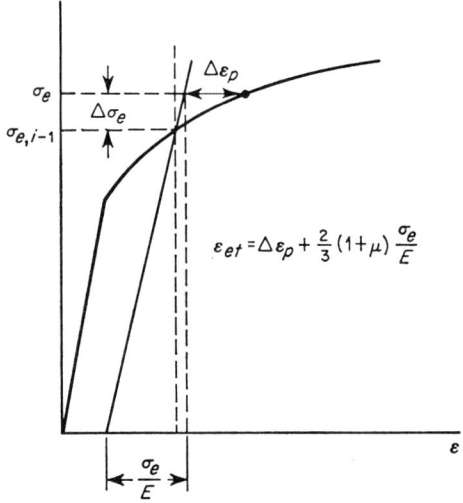

FIGURE 7.9.1 Relation between ε_{et}, σ_e, and $\Delta\varepsilon_p$.

$\Delta\sigma_e$ be the increment in stress to which corresponds a plastic strain increment $\Delta\varepsilon_p$ and let σ_e be the stress at the end of the increment. Then ε_{et} is the sum of the plastic strain increment plus the total elastic strain multiplied by $2/3(1 + \mu)$. Solving equation (7.9.13) for $\Delta\varepsilon_p$ results in

$$\Delta\varepsilon_p = \varepsilon_{et} - \frac{2}{3}\frac{(1+\mu)}{E}\,\sigma_e \qquad (7.9.14)$$

σ_e can now readily be eliminated from equations (7.9.13) or (7.9.14) as follows. Let the stress preceding the increment of load be $\sigma_{e,i-1}$, i.e., $\sigma_e = \sigma_{e,i-1} + \Delta\sigma_e$. Then expanding σ_e in a Taylor series about $\sigma_{e,i-1}$ gives approximately

$$\sigma_e = \sigma_{e,i-1} + \left(\frac{d\sigma_e}{d\varepsilon_p}\right)_{i-1}\Delta\varepsilon_p + \cdots \qquad (7.9.15)$$

where higher-order terms in $\Delta\varepsilon_p$ have been neglected. Substituting into (7.9.14) and solving for $\Delta\varepsilon_p$ gives

$$\Delta\varepsilon_p = \frac{\varepsilon_{et} - \frac{2}{3}[(1+\mu)/E]\sigma_{e,i-1}}{1 + \frac{2}{3}[(1+\mu)/E](d\sigma_e/d\varepsilon_p)_{i-1}} \qquad (7.9.16)$$

For linear strain hardening, equations (7.9.15) and (7.9.16) are obviously exact. Equation (7.9.16) shows how ε_{et} is related to $\Delta\varepsilon_p$ through the geometry of the uniaxial stress–strain curve. We shall use this relationship subsequently together with equations (7.9.10) and (7.9.11) to solve specific problems. For want of a better name we will refer to equations (7.9.10) as the plastic strain–total strain equations.

If one desires to use the total or deformation theory of plasticity, it can be shown [19] that it is only necessary to remove the primes and increment symbols from equations (7.9.9) through (7.9.11); i.e.,

$$\varepsilon_{ij}^p = \frac{\varepsilon_p}{\varepsilon_{et}} e_{ij}$$

where

$$\varepsilon_{et} = \sqrt{\tfrac{2}{3} e_{ij} e_{ij}}$$

$$= \frac{\sqrt{2}}{3}[(\varepsilon_x - \varepsilon_y)^2 + (\varepsilon_y - \varepsilon_z)^2 + (\varepsilon_z - \varepsilon_x)^2 + 6(\varepsilon_{xy}^2 + \varepsilon_{yz}^2 + \varepsilon_{zx}^2)]^{1/2}$$

$$\varepsilon_p = \varepsilon_{et} - \frac{2}{3}\frac{1+\mu}{E}\sigma_e \qquad (7.9.17)$$

7-10 COMPLETE STRESS–STRAIN RELATIONS. SUMMARY

In the previous sections the relations between the increments of plastic strain and the stresses at any instant were discussed in some detail. The fundamental problem in applying plasticity theory is to determine the total plastic strain as a function of the history of loading or history of stress. Suppose a body is loaded along some specified load path to some final load condition. To calculate the plastic strains at this final load condition it is theoretically necessary, in general, to integrate the infinitesimal plastic strain increments over the actual loading path. Although this can be done in relatively simple cases, it is usually more expeditious to assume the load applied in small finite increments and calculate the finite increments of plastic strain for each of the load increments. All these increments of plastic strain are then added to give the total plastic strain. The integration is thus replaced by a summation.

Let the total loading path be divided into N increments of load. Assume that the plastic strains have been computed for the first $i - 1$ increments of load and we now wish to compute them for the ith increment of load. The total strains at the end of the ith increment can be written with thermal strains included, as

$$\varepsilon_{ij} = \frac{1}{2G}\sigma_{ij} - \delta_{ij}\left(\frac{\mu}{E}\Theta - \alpha T\right) + \sum_{k=1}^{i-1}\Delta\varepsilon_{ij,k}^P + \Delta\varepsilon_{ij,i}^P \quad (7.10.1)$$

where $\Theta = \sigma_{ii} = \sigma_x + \sigma_y + \sigma_z$. The first two terms on the right side of equation (7.10.1) represent the elastic part of the total strain, the third term is the thermal strain, the fourth term is the plastic strain accumulated in the first $i - 1$ increments of load, and the fifth term is the plastic strain due to the ith increment of load. In expanded form these equations are

$$\varepsilon_x = \frac{1}{E}[\sigma_x - \mu(\sigma_y + \sigma_z)] + \alpha T + \sum_{k=1}^{i-1}\Delta\varepsilon_{x,k}^P + \Delta\varepsilon_{x,i}^P$$

$$\varepsilon_y = \frac{1}{E}[\sigma_y - \mu(\sigma_x + \sigma_z)] + \alpha T + \sum_{k=1}^{i-1}\Delta\varepsilon_{y,k}^P + \Delta\varepsilon_{y,i}^P$$

$$\varepsilon_z = \frac{1}{E}[\sigma_z - \mu(\sigma_x + \sigma_y)] + \alpha T + \sum_{k=1}^{i-1}\Delta\varepsilon_{z,k}^P + \Delta\varepsilon_{z,i}^P$$

$$\varepsilon_{xy} = \frac{1+\mu}{E}\tau_{xy} + \sum_{k=1}^{i-1}\Delta\varepsilon_{xy,k}^P + \Delta\varepsilon_{xy,i}^P \quad (7.10.2)$$

$$\varepsilon_{yz} = \frac{1+\mu}{E}\tau_{yz} + \sum_{k=1}^{i-1}\Delta\varepsilon_{yz,k}^P + \Delta\varepsilon_{yz,i}^P$$

$$\varepsilon_{zx} = \frac{1+\mu}{E}\tau_{zx} + \sum_{k=1}^{i-1}\Delta\varepsilon_{xz,k}^P + \Delta\varepsilon_{zx,i}^P$$

In the above equations the sums are known and the problem is to calculate the plastic strain increments for the current or ith increment of load, and the corresponding stresses. To do this it is necessary to use one or another of the plastic stress–strain relations discussed in previous sections. A yield criterion must be chosen and the associated flow rule as given by equation (7.6.14). In particular, since we shall concern ourselves only with the von Mises and

Sec. 7-10] Complete Stress–Strain Relations. Summary

Tresca yield criteria, the following relations previously derived will be used. For the von Mises criterion:

$$\sigma_e = \sqrt{3J_2} = \frac{3}{\sqrt{2}} \tau_{\text{oct}}$$

$$= \sqrt{\tfrac{3}{2} S_{ij} S_{ij}}$$

$$= \frac{1}{\sqrt{2}} [(\sigma_x - \sigma_y)^2 + (\sigma_y - \sigma_z)^2 + (\sigma_z - \sigma_x)^2 + 6(\tau_{xy}^2 + \tau_{yz}^2 + \tau_{zx}^2)]^{1/2}$$

(7.10.3)

$$\Delta \varepsilon_p = \sqrt{\tfrac{2}{3} \Delta \varepsilon_{ij}^P \Delta \varepsilon_{ij}^P}$$

$$= \sqrt{\tfrac{2}{3}} [(\Delta \varepsilon_x^P)^2 + (\Delta \varepsilon_y^P)^2 + (\Delta \varepsilon_z^P)^2 + 2(\Delta \varepsilon_{xy}^P)^2 + 2(\Delta \varepsilon_{yz}^P)^2 + 2(\Delta \varepsilon_{zx}^P)^2]^{1/2}$$

$$= \frac{2}{\sqrt{3}} [(\Delta \varepsilon_x^P)^2 + (\Delta \varepsilon_y^P)^2 + (\Delta \varepsilon_{xy}^P)^2 + (\Delta \varepsilon_{yz}^P)^2 + (\Delta \varepsilon_{zx}^P)^2 + \Delta \varepsilon_x^P \Delta \varepsilon_y^P]^{1/2}$$

$$= \frac{\sqrt{2}}{3} [(\Delta \varepsilon_x^P - \Delta \varepsilon_y^P)^2 + (\Delta \varepsilon_y^P - \Delta \varepsilon_z^P)^2 + (\Delta \varepsilon_z^P - \Delta \varepsilon_x^P)^2$$

$$+ 6(\Delta \varepsilon_{xy}^P)^2 + 6(\Delta \varepsilon_{yz}^P)^2 + 6(\Delta \varepsilon_{zx}^P)^2]^{1/2} \quad (7.10.4)$$

$$\Delta \varepsilon_{ij}^P = \frac{3}{2} \frac{\Delta \varepsilon_p}{\sigma_e} S_{ij}$$

or

$$\Delta \varepsilon_x^P = \frac{\Delta \varepsilon_p}{2\sigma_e} (2\sigma_x - \sigma_y - \sigma_z)$$

$$\Delta \varepsilon_y^P = \frac{\Delta \varepsilon_p}{2\sigma_e} (2\sigma_y - \sigma_z - \sigma_x)$$

$$\Delta \varepsilon_z^P = \frac{\Delta \varepsilon_p}{2\sigma_e} (2\sigma_z - \sigma_x - \sigma_y) = -(\Delta \varepsilon_x^P + \Delta \varepsilon_y^P) \quad (7.10.5)$$

$$\Delta \varepsilon_{xy}^P = \frac{3}{2} \frac{\Delta \varepsilon_p}{\sigma_e} \tau_{xy}$$

$$\Delta \varepsilon_{yz}^P = \frac{3}{2} \frac{\Delta \varepsilon_p}{\sigma_e} \tau_{yz}$$

$$\Delta \varepsilon_{zx}^P = \frac{3}{2} \frac{\Delta \varepsilon_p}{\sigma_e} \tau_{zx}$$

and $\Delta \varepsilon_p$ is related to σ_e through the uniaxial tensile stress–strain curve as shown in Figures 7.3.2 or 7.9.1.

Alternatively we define

$$e'_{ij} = e_{ij} - \sum_{k=1}^{i-1} \Delta\varepsilon^P_{ij,k} = \varepsilon'_{ij} - \delta_{ij}\varepsilon_m$$

$$\varepsilon'_{ij} = \varepsilon_{ij} - \sum_{k=1}^{i-1} \Delta\varepsilon^P_{ij,k}$$

$$\varepsilon_m = \tfrac{1}{3}\varepsilon_{ij} = \tfrac{1}{3}(\varepsilon_x + \varepsilon_y + \varepsilon_z)$$

$$\varepsilon_{et} = \sqrt{\tfrac{2}{3}e'_{ij}e'_{ij}}$$

$$= \frac{\sqrt{2}}{3}[(\varepsilon'_x - \varepsilon'_y)^2 + (\varepsilon'_y - \varepsilon'_z)^2 + (\varepsilon'_z - \varepsilon'_x)^2$$

$$+ 6(\varepsilon'_{xy})^2 + 6(\varepsilon'_{yz})^2 + 6(\varepsilon'_{zx})^2]^{1/2} \quad (7.10.6)$$

Then

$$\Delta\varepsilon^P_{ij} = \frac{\Delta\varepsilon_p}{\varepsilon_{et}} e'_{ij}$$

or

$$\Delta\varepsilon^P_x = \frac{\Delta\varepsilon_p}{3\varepsilon_{et}}(2\varepsilon'_x - \varepsilon'_y - \varepsilon'_z)$$

$$\Delta\varepsilon^P_y = \frac{\Delta\varepsilon_p}{3\varepsilon_{et}}(2\varepsilon'_y - \varepsilon'_x - \varepsilon'_z)$$

$$\Delta\varepsilon^P_z = \frac{\Delta\varepsilon_p}{3\varepsilon_{et}}(2\varepsilon'_z - \varepsilon'_x - \varepsilon'_y)$$

$$\Delta\varepsilon^P_{xy} = \frac{\Delta\varepsilon_p}{\varepsilon_{et}}\varepsilon'_{xy} \quad (7.10.7)$$

$$\Delta\varepsilon^P_{yz} = \frac{\Delta\varepsilon_p}{\varepsilon_{et}}\varepsilon'_{yz}$$

$$\Delta\varepsilon^P_{zx} = \frac{\Delta\varepsilon_p}{\varepsilon_{et}}\varepsilon'_{zx}$$

and $\Delta\varepsilon_p$, ε_{et}, and σ_e are related to each other by

$$\Delta\varepsilon_p = \varepsilon_{et} - \frac{2}{3}\frac{1+\mu}{E}\sigma_e \quad (7.10.8)$$

Furthermore, for small increments

$$\Delta\varepsilon_p \simeq \frac{\varepsilon_{et} - \tfrac{2}{3}[(1+\mu)/E]\,\sigma_{e,i-1}}{1 + \tfrac{2}{3}[(1+\mu)/E](d\sigma_e/d\varepsilon_p)_{i-1}} \quad (7.10.9)$$

Sec. 7-10] Complete Stress-Strain Relations. Summary

where $\sigma_{e,i-1}$ is the value of the equivalent stress at the end of the $(i-1)$st increment of load and $(d\sigma_e/d\varepsilon_p)_{i-1}$ is the slope of the uniaxial tensile curve replotted as true stress versus true plastic strain. Equation (7.10.9) is exact for linear strain hardening. The above relations are shown graphically in Figure 7.9.1.

If the deformation or total theory of plasticity is used, all the above relations are valid if the Δ's are removed from all the previous equations and the primes are removed from equations (7.10.6) through (7.10.7). Equation (7.10.8) becomes

$$\varepsilon_p = \varepsilon_{et} - \frac{2}{3}\frac{1+\mu}{E}\sigma_e \qquad (7.10.10)$$

and by the use of (7.10.10), the uniaxial stress–strain curve can be replotted as a curve of ε_p versus ε_{et} as shown in Figure 7.10.1. This curve can then be used instead of the original stress–strain curve.

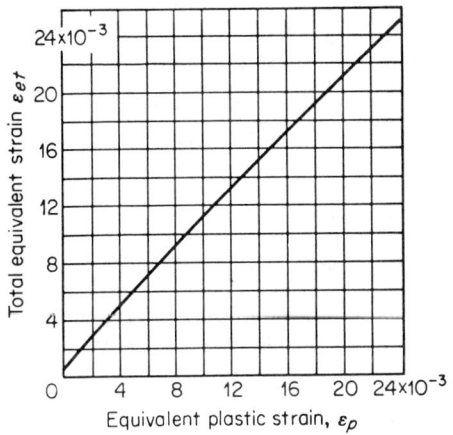

FIGURE 7.10.1 Equivalent total strain–equivalent plastic strain curve.

For the Tresca criterion, assume $\sigma_1 > \sigma_2 > \sigma_3$. Then

$$\sigma_T = \sigma_1 - \sigma_3$$

$$\Delta\varepsilon_p = \frac{2}{\sqrt{3}}\Delta\varepsilon_1^p = -\frac{2}{\sqrt{3}}\Delta\varepsilon_3^p \qquad (7.10.11)$$

or, alternatively,

$$\Delta\varepsilon_p = \Delta\varepsilon_1^p$$

and the relation between σ_T and $\Delta\varepsilon_p$ is taken from the uniaxial tensile curve.

132 Plastic Stress–Strain Relations [Ch. 7

The stress–strain relations discussed in this chapter are just one of four sets of relations that must generally be satisfied in solving an elastoplastic problem. The other three sets of relations are the same as for any elasticity problem. These are

1. The equations of equilibrium of stresses.
2. The strain-displacement or compatibility relations.
3. The boundary conditions.

To obtain a complete solution we must find a set of stresses and strains which satisfy these four sets of relations. In the next several chapters it will be shown how these relations are adapted to specific problems and how solutions to these problems can be obtained. In all that follows, as in the preceding, it is assumed that the material is homogeneous, isotropic, and strain hardens isotropically.

Problems

1. Show that the equivalent stress σ_e defined by equation (7.2.9) can also be written

$$\sigma_e = \sqrt{\tfrac{3}{2} S_{ij} S_{ij}} = \sqrt{\tfrac{3}{2}(S_1^2 + S_2^2 + S_3^2)}$$

2. Show that the equivalent plastic strain increment $d\varepsilon_p$ defined by equation (7.2.10) can also be written

$$d\varepsilon_p = \sqrt{\tfrac{2}{3} d\varepsilon_{ij}^P d\varepsilon_{ij}^P} = \sqrt{\tfrac{2}{3}[(d\varepsilon_1^P)^2 + (d\varepsilon_2^P)^2 + (d\varepsilon_3^P)^2]}$$

3. Show that equation (7.3.5) is valid for the Prandtl–Reuss relations.
4. Show that equation (7.3.3) follows directly from equation (7.3.2).
5. Show that the stress–strain relations (7.4.2) are inconsistent with the definition (7.4.1).
6. Derive equations (7.6.21) and (7.6.22) making use of equations (7.6.20).
7. Show that the following expressions for the effective plastic strain increment are equivalent:

$$d\varepsilon_p = \sqrt{\tfrac{2}{3}}\,[(d\varepsilon_x^P)^2 + (d\varepsilon_y^P)^2 + (d\varepsilon_z^P)^2 + 2(d\varepsilon_{xy}^P)^2 + 2(d\varepsilon_{yz}^P)^2 + 2(d\varepsilon_{zx}^P)^2]^{1/2}$$

$$= \frac{2}{\sqrt{3}}\,[(d\varepsilon_x^P)^2 + (d\varepsilon_y^P)^2 + d\varepsilon_x^P\, d\varepsilon_y^P + (d\varepsilon_{xy}^P)^2 + (d\varepsilon_{yz}^P)^2 + (d\varepsilon_{zx}^P)^2]^{1/2}$$

$$= \frac{\sqrt{2}}{3}\,[(d\varepsilon_x^P - d\varepsilon_y^P)^2 + (d\varepsilon_y^P - d\varepsilon_z^P)^2 + (d\varepsilon_z^P - d\varepsilon_x^P)^2$$
$$+ 6(d\varepsilon_{xy}^P)^2 + 6(d\varepsilon_{yz}^P)^2 + 6(d\varepsilon_{zx}^P)^2]^{1/2}$$

8. From the fact that the plastic strain increment vector is normal to the yield surface, prove that the Prandtl–Reuss equations imply the use of the von Mises yield criterion.
9. Derive equation (7.6.30) from equation (7.6.28).

10. Show that the Prandtl–Reuss relations imply that the principal axes of stress and of plastic strain increment coincide.
11. Derive equations (7.2.12) and (7.2.13) using tensor notation only.
12. Determine the equivalent stress σ_T for the Tresca criterion by means of equation (7.6.18). Assume $\sigma_1 > \sigma_2 > \sigma_3$. Also determine the effective plastic strain increment by the two methods described by equation (7.6.20) and what follows.
13. Prove that

$$\frac{\partial J_2}{\partial \sigma_{ij}} = S_{ij}$$

where S_{ij} is the stress deviator tensor.
14. Show that equations (7.9.7) and (7.9.11) are equivalent.

References

1. B. Saint-Venant, Mémoire sur l'établissement des équations differentielles des mouvements intérieurs opérés dans les corps solides ductiles au delá des limites où l'élasticité pourrait les ramener à leur premier état, *Compt. Rend.*, **70**, 1870, pp. 473–480.
2. M. Lévy, Mémoire sur les équations géneralés des mouvements intérieurs des corps solides ductile au delá limites où l'élasticité pourrait les ramener à leur premier état, *Compt. Rend.*, **70**, 1870, pp. 1323–1325.
3. R. von Mises, Mechanik der festen Koerper in Plastisch deformablem Zustand, *Goettinger Nachr. Math. Phys., Kl.*, **1913**, pp. 582–592.
4. L. Prandtl, Spannungsverteilung in plastischen Koerpern, *Proceedings of the 1st International Congress on Applied Mechanics, Delft*, Technische Boekhandel en Druckerij, J. Waltman, Jr., 1925, pp. 43–54.
5. E. Reuss, Beruecksichtigung der elastischen Formaenderungen in der Plastizitaetstheorie, *Z. Angew. Math. Mech.*, **10**, 1930, pp. 266–274.
6. R. Hill, *The Mathematical Theory of Plasticity*, Oxford Univ. Press, London, 1950, p. 25.
7. D. R. Bland, The Two Measures of Work-Hardening, *9th International Congress of Applied Mechanics*, Univ. de Bruxelles, 1957, pp. 45–50.
8. H. Ford, *Advanced Mechanics of Materials*, Wiley, New York, 1963, p. 416.
9. A. M. Wahl, Effect of Transient Period in Evaluating Rotating Disk Tests Under Creep Conditions, *J. Basic Eng.*, **85**, 1963, pp. 66–70.
10. W. Prager, Strain Hardening Under Combined Stress, *J. Appl. Phys.*, **16**, 1945, pp. 837–840.
11. I. Todhunter and K. Pearson, *A History of the Theory of Elasticity and Strength of Materials*, Vol. II, Part 1, Cambridge Univ. Press, 1893, p. 166.
12. D. C. Drucker, Some Implications of Work Hardening and Ideal Plasticity, *Quart. Appl. Math.*, **7**, 1950, pp. 411–418.
13. D. C. Drucker, A More Fundamental Approach to Plastic Stress–Strain Relations, *1st U.S. Congress of Applied Mechanics*, ASME, New York, 1952, pp. 487–491.
14. E. Melan, Zur Plastizitaet des raeumlichen Kontinuums, *Ingr.-Arch.*, **9**, 1938, pp. 116–126.

15. D. R. Bland, The Associated Flow Rule of Plasticity, *J. Mech. Phys. Solids*, **6**, 1957, pp. 71–78.
16. H. Z. Hencky, Zur Theorie Plastischer Deformationen und der hierdurch im Material hervorgerufenen Nachspannungen, *Z. Angew. Math. Mech.*, **4**, 1924, pp. 323–334.
17. B. Budiansky, A Reassessment of Deformation Theories of Plasticity, *J. Appl. Mech.*, **26**, 1959, pp. 259–264.
18. W. T. Koiter, Stress–Strain Relations, Uniqueness, and Variational Theorems for Elastic–Plastic Materials with a Singular Yield Surface, *Quart. Appl. Math.*, **11**, 1953, pp. 350–354.
19. A. Mendelson and S. S. Manson, Practical Solution of Plastic Deformation Problems in Elastic–Plastic Range, *NASA Tech. Rept. R-28*, 1959.

General References

Drucker, D. C., Stress–Strain Relations in the Plastic Range, a Survey of Theory and Experiment, *Office of Naval Research, Contract N7 onr-358, NR–041–032*, Dec. 1950.

Hill, R., *The Mathematical Theory of Plasticity*, Oxford Univ. Press, London, 1950.

Johnson, W., and P. B. Mellor, *Plasticity for Mechanical Engineers*, Van Nostrand, Princeton, N.J., 1962.

Naghdi, P. M., Stress–Strain Relations in Plasticity and Thermoplasticity, Office of Naval Research, Contract Nonr-222 (69), *Tech. Rept. No. 9*, 1960.

CHAPTER 8

ELASTOPLASTIC PROBLEMS OF SPHERES AND CYLINDERS

8-1 GENERAL RELATIONS

Spheres and cylinders are widely used as pressure vessels, in the chemical industry, for example, as well as many other places. The loads involve high pressures and sometimes high temperatures and high temperature gradients. The elastic stress and strain distributions are relatively simple to obtain, particularly since the loading is usually reasonably symmetric. The solutions in the elastoplastic range, however, become complicated, and so simplifying assumptions of various types are made. These usually involve assuming the material to be incompressible in both the elastic and plastic ranges, and assuming it to be perfectly plastic in the plastic range. With these assumptions closed-form solutions can be obtained. We shall first present some of these classical solutions. Subsequently it will be shown how these problems can be solved without the usual simplifying assumptions.

For later use we record here the equilibrium, compatibility, strain-displacement, and stress-strain relations in spherical coordinates and polar coordinates assuming spherical and axial symmetry, respectively.

Spherical Coordinates

The stresses are designated by σ_r and $\sigma_\theta = \sigma_\phi$ and the strains by ε_r and $\varepsilon_\theta = \varepsilon_\phi$. The equilibrium equations reduce to

$$\frac{d\sigma_r}{dr} + 2\frac{\sigma_r - \sigma_\theta}{r} = -F_r \qquad (8.1.1)$$

where F_r is the body force per unit volume. The strains are related to the displacements by

$$\varepsilon_r = \frac{du}{dr} \qquad \varepsilon_\theta = \frac{u}{r} = \varepsilon_\phi \qquad (8.1.2)$$

where u is the radial displacement. Combining both of equations (8.1.2) gives the compatibility equation

$$\frac{d\varepsilon_\theta}{dr} + \frac{\varepsilon_\theta - \varepsilon_r}{r} = 0 \qquad (8.1.3)$$

Because of symmetry the shear stresses and shear strains are zero as well as the tangential displacements.

The stress–strain relations are

$$\begin{aligned}\varepsilon_r &= \frac{1}{E}(\sigma_r - 2\mu\sigma_\theta) + \alpha T + \varepsilon_r^P \\ \varepsilon_\theta &= \frac{1}{E}[(1-\mu)\sigma_\theta - \mu\sigma_r] + \alpha T + \varepsilon_\theta^P\end{aligned} \qquad (8.1.4)$$

where ε_r^P and ε_θ^P are the total plastic strains. From the incompressibility condition it follows that

$$\varepsilon_r^P = -2\varepsilon_\theta^P \qquad (8.1.5)$$

For the von Mises yield criterion, the equivalent stress becomes

$$\sigma_e = |\sigma_r - \sigma_\theta| \qquad (8.1.6)$$

so that the yield criterion is

$$|\sigma_r - \sigma_\theta| = \sigma_0 \qquad (8.1.7)$$

and the equivalent plastic strain increment is

$$d\varepsilon_p = |d\varepsilon_r^P| \qquad (8.1.8)$$

The Prandtl–Reuss relations thereupon reduce to

$$\begin{aligned}d\varepsilon_r^P = -2d\varepsilon_\theta^P &= d\varepsilon_p \frac{\sigma_r - \sigma_\theta}{|\sigma_r - \sigma_\theta|} \\ &= d\varepsilon_p \operatorname{sgn}(\sigma_r - \sigma_\theta)\end{aligned} \qquad (8.1.9)$$

Sec. 8-1] General Relations

where sgn stands for "the sign of." We note that if the plastic strains vary monotonically with the applied load, equation (8.1.9) can be integrated to give

$$\varepsilon_r^P = \varepsilon_p \, \text{sgn}\,(\sigma_r - \sigma_\theta) \tag{8.1.10}$$

Note also that the Tresca yield criterion in this case coincides with the von Mises criterion.

Polar Coordinates

We assume axial symmetry and either plane strain or plane stress. The equilibrium equations then become

$$\frac{d\sigma_r}{dr} + \frac{\sigma_r - \sigma_\theta}{r} = -F_r \tag{8.1.11}$$

where F_r is the body force per unit volume. The strain-displacement relations and corresponding compatibility equation are

$$\varepsilon_r = \frac{du}{dr} \qquad \varepsilon_\theta = \frac{u}{r}$$

$$\frac{d\varepsilon_\theta}{dr} + \frac{\varepsilon_\theta - \varepsilon_r}{r} = 0 \tag{8.1.12}$$

which are the same as equations (8.1.2) and (8.1.3). The stress–strain relations are given by

$$\varepsilon_r = \frac{1}{E}[\sigma_r - \mu(\sigma_\theta + \sigma_z)] + \alpha T + \varepsilon_r^P$$

$$\varepsilon_\theta = \frac{1}{E}[\sigma_\theta - \mu(\sigma_z + \sigma_r)] + \alpha T + \varepsilon_\theta^P \tag{8.1.13}$$

$$\varepsilon_z = \frac{1}{E}[\sigma_z - \mu(\sigma_r + \sigma_\theta)] + \alpha T - (\varepsilon_\theta^P + \varepsilon_r^P)$$

For the case of plane stress, $\sigma_z = 0$, and for the case of plane strain $\varepsilon_z = 0$ or $\varepsilon_z =$ constant for generalized plane strain. In both cases the shear stresses and strains are zero.

The von Mises and Tresca criteria do not coincide in this case as they do for the case of spherical symmetry. The yield criteria and corresponding plasticity relations will be described subsequently as they are used. Several examples will now be discussed beginning with the case of a thick hollow sphere.

8–2 THICK HOLLOW SPHERE WITH INTERNAL PRESSURE AND THERMAL LOADING

Consider a sphere with inner radius a and outer radius b, subjected to an internal pressure p and a radial temperature distribution $T(r)$. It is obvious that complete symmetry about the center will exist so that the radial and any two tangential directions will be principal directions. Equations (8.1.1) through (8.1.10) apply. We start by finding the elastic solution. Substituting the stress–strain relations (8.1.4) (with the plastic strains set to zero) into the compatibility equation (8.1.3) and making use of the equilibrium equation (8.1.1), the following solution for the stresses can readily be obtained:

$$\sigma_r = \frac{-2E}{1-\mu} \frac{1}{r^3} \int_a^r \alpha T r^2 \, dr + \frac{2}{3}\left(1 - \frac{a^3}{r^3}\right) C_1 + \frac{C_2}{r^3}$$

$$\sigma_\theta = -\tfrac{1}{2}\sigma_r - \frac{E\alpha T}{1-\mu} + C_1$$

(8.2.1)

where C_1 and C_2 are integration constants. Note that E and α have been assumed constant in obtaining the above solution. The constants C_1 and C_2 can be obtained using the boundary conditions

$$\sigma_r(a) = -p$$
$$\sigma_r(b) = 0$$

(8.2.2)

resulting in

$$C_2 = -pa^3$$
$$C_1 = \frac{3E}{1-\mu} \frac{1}{b^3 - a^3} \int_a^b \alpha T r^2 \, dr + \frac{3}{2} \frac{pa^3}{b^3 - a^3}$$

(8.2.3)

For convenience the following dimensionless quantities are now introduced:

$$\beta \equiv \frac{b}{a} \qquad \rho \equiv \frac{r}{a} \qquad P \equiv \frac{p}{\sigma_0}$$

$$\tau \equiv \frac{E\alpha T}{(1-\mu)\sigma_0} \qquad S_r \equiv \frac{\sigma_r}{\sigma_0} \qquad S_\theta \equiv \frac{\sigma_\theta}{\sigma_0}$$

(8.2.4)

where σ_0 is the yield stress in uniaxial tension. Equation (8.2.1) can now be written in the dimensionless form as

$$S_r = 2\frac{\rho^3 - 1}{\rho^3(\beta^3 - 1)} \int_1^\beta \tau\rho^2 \, d\rho - \frac{2}{\rho^3} \int_1^\rho \tau\rho^2 \, d\rho + \frac{\rho^3 - \beta^3}{\rho^3(\beta^3 - 1)} P$$

$$S_\theta = \frac{2\rho^3 + 1}{\rho^3(\beta^3 - 1)} \int_1^\beta \tau\rho^2 \, d\rho + \frac{1}{\rho^3} \int_1^\rho \tau\rho^2 \, d\rho - \tau + \frac{2\rho^3 + \beta^3}{2\rho^3(\beta^3 - 1)} P$$

(8.2.5)

Sec. 8–2] Hollow Sphere with Internal Pressure and Thermal Loading

The strains can be computed from (8.1.4) and the displacement from (8.1.2).

We note that in the case of pressure loads only, the stress distribution is independent of Poisson's ratio [see equations (8.2.5) and (8.2.4)]. The assumption that is often made that the material is incompressible in the elastic range ($\mu = \frac{1}{2}$), as well as the plastic range, therefore leads to no error in the elastic stress distribution. In the case of temperature loads, however, assuming $\mu = \frac{1}{2}$ instead of 0.3, for example, results in approximately a 30 per cent error in the elastic thermal stresses. The strains are not independent of Poisson's ratio even for the case of pressure loading. In what follows, the effect of Poisson's ratio is always taken into account.

The conditions for the onset of yielding in the sphere can now be investigated. In terms of the dimensionless stresses defined in (8.2.4), the yield criterion (8.1.7) is written

$$|S_r - S_\theta| = 1$$

From equations (8.2.5) the yield condition becomes

$$\left| \frac{3}{\rho^3(\beta^3 - 1)} \int_1^\beta \tau\rho^2 \, d\rho + \frac{3}{\rho^3} \int_1^\rho \tau\rho^2 \, d\rho - \tau + \frac{3\beta^3}{2\rho^3(\beta^3 - 1)} P \right| = 1 \quad (8.2.6)$$

As a specific example, assume a temperature distribution resulting from an outward flow of heat due to an inner surface temperature of T_0 and outer surface temperature of zero. This steady-state temperature distribution will be given by

$$T = \frac{T_0 a}{b - a}\left(\frac{b}{r} - 1\right)$$

or

$$\tau = \frac{\tau_0}{\beta - 1}\left(\frac{\beta}{\rho} - 1\right) \quad (8.2.7)$$

where

$$\tau_0 = \frac{E\alpha T_0}{(1 - \mu)\sigma_0}$$

Evaluating the integrals and substituting into the yield condition (8.2.6) results in

$$\left| \frac{3\beta^3}{2\rho^3(\beta^3 - 1)} P - \frac{\beta\tau_0}{2\rho^3(\beta^3 - 1)} [3\beta^2 - \rho^2(\beta^2 + \beta + 1)] \right| = 1 \quad (8.2.8)$$

Consider first the case of pressure only. Then the yield condition becomes

$$\frac{3\beta^3 P}{2\rho^3(\beta^3 - 1)} = 1 \quad (8.2.9)$$

Yielding will first occur at the smallest value of ρ, i.e., $\rho = 1$, and the dimensionless pressure necessary to first cause yielding, the *critical pressure*, will be

$$P_{\text{crit}} = \frac{2(\beta^3 - 1)}{3\beta^3} \quad (8.2.10)$$

A plot of the ratio of this critical pressure as a function of the ratio of the outer to the inner radii β is shown in Figure 8.2.1. For a given value of β, yielding will start at the inner surface at a pressure as given by equation (8.2.10) or Figure 8.2.1. As P is increased, the plastic zone will spread from

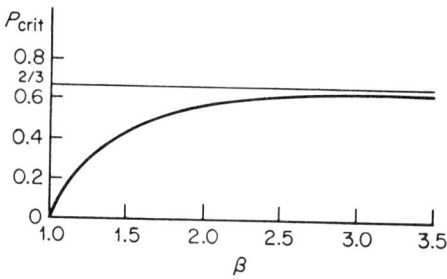

FIGURE 8.2.1 Variation of critical pressure with β, pressure loading only.

the inner surface toward the outer surface. Note that as β approaches infinity P_{crit} approaches $\frac{2}{3}$, so that if the pressure is equal to $\frac{2}{3}$ the yield stress, yielding is sure to take place, no matter what the dimensions of the sphere.

Considering the case of temperature only, equation (8.2.8) gives

$$\left| \frac{\beta \tau_0}{2(\beta^3 - 1)} \left[\frac{3\beta^2}{\rho^3} - \frac{\beta^2 + \beta + 1}{\rho} \right] \right| = 1 \quad (8.2.11)$$

For this case yielding will also first start at the inner surface. However, if both pressure and temperature are present, yielding may start at any radius, depending on the relative values of P, τ_0, and β.

As an example let $\beta = 2$. Then for the case of internal pressure alone, equation (8.2.10) gives

$$P_{\text{crit}} = \frac{7}{12} = 0.583$$

For temperature alone, equation (8.2.11) gives

$$\tau_{0,\text{crit}} = 1.4$$

Sec. 8-3] Hollow Sphere. Spread of Plastic Zone. Pressure Loading Only

so that assuming $E = 31 \times 10^6$, $\sigma_0 = 31,000$ psi, $\mu = 0.3$, and $\alpha = 7.5 \times 10^{-6}$ per °F, the temperature difference between the inner and outer surfaces for yielding to start is 130°F. For both pressure and temperature acting, equation (8.2.8) gives, for $\beta = 2$,

$$|12P + (7\rho^2 - 12)\tau_0| = 7\rho^3$$

For a value of $\tau_0 = 0.4$, which corresponds to only a 37°F temperature difference T_0, yielding will first occur at $\rho = 1$ for a value of

$$P_{\text{crit}} = 0.75$$

compared to 0.583 for pressure alone. The effect of the temperature gradient in this case has been to retard the onset of yielding.

A complete discussion of the effects of temperature and pressure on yielding is given in reference [1].

So far, only the start of yielding has been considered. The spread of the plastic zone through the sphere is investigated next. The pressure problem and the temperature problem will be discussed separately in Sections 8.3, 8.4, and 8.5 under the assumption that the material is perfectly plastic. The general solution for strain-hardening materials under combined pressure and thermal gradient is presented in Section 8.6.

8-3 HOLLOW SPHERE. SPREAD OF PLASTIC ZONE. PRESSURE LOADING ONLY

When only internal pressure is acting, yielding will begin at the inner surface at a pressure given by equation (8.2.10); i.e.,

$$P_{\text{crit}} = \frac{2}{3} \frac{\beta^3 - 1}{\beta^3} \qquad (8.3.1)$$

As the pressure increases, the plastic zone will spread outward toward the outer surface. Let the radius to the end of the plastic zone be r_c. Since the material is assumed perfectly plastic, at every point in the plastic region the equivalent stress is equal to the yield stress and since for this case $\sigma_\theta > \sigma_r$,

$$S_\theta - S_r = 1 \qquad (8.3.2)$$

in the plastic region. Substituting into the equilibrium equation (8.1.1) gives

$$\frac{dS_r}{d\rho} = \frac{2}{\rho}$$

or
$$S_r = 2 \ln \rho + C$$

But at

$$\rho = 1 \quad S_r = -P$$

Therefore,
$$C = -P$$

and
$$\left. \begin{array}{l} S_r = 2 \ln \rho - P \\ S_\theta = (2 \ln \rho + 1) - P \end{array} \right\} \rho \leq \rho_c \qquad (8.3.3)$$

Equations (8.3.3) give the stresses in the plastic region. Note that no stress–strain relation was needed to obtain these stresses. The problem is therefore called *statically determinate*.

At the plastic zone boundary, i.e., at $\rho = \rho_c$, the radial stress is

$$S_{r,c} = 2 \ln \rho_c - P \qquad (8.3.4)$$

We can now consider the elastic part of the sphere as a new sphere with inner radius r_c and outer radius b, with an internal pressure given by equation (8.3.4). Since at this new inner radius the sphere is just at the yield point, equation (8.3.1) must apply with β replaced by $\beta_c = b/r_c$, and $-P_{\text{crit}}$ replaced by $S_{r,c}$. Hence

$$2 \ln \rho_c - P = -\frac{2}{3} \frac{\beta_c^3 - 1}{\beta_c^3}$$

or
$$P = 2 \ln \rho_c + \frac{2}{3}\left(1 - \frac{1}{\beta_c^3}\right)$$

$$= 2 \ln \frac{r_c}{a} + \frac{2}{3}\left(1 - \frac{r_c^3}{b^3}\right) \qquad (8.3.5)$$

Equation (8.3.5) gives the pressure required to cause the plastic zone to reach a radius r_c or, alternatively, for a given internal pressure P, equation (8.3.5) could be solved for the plastic zone radius r_c. A plot of the pressure versus the plastic zone radius is given in Figure 8.3.1 for $\beta = 2$.

When r_c becomes equal to b, the sphere is completely plastic. This will occur at a pressure [from (8.3.5)]

$$P = 2 \ln \beta \qquad (8.3.6)$$

Sec. 8–3] Hollow Sphere. Spread of Plastic Zone. Pressure Loading Only 143

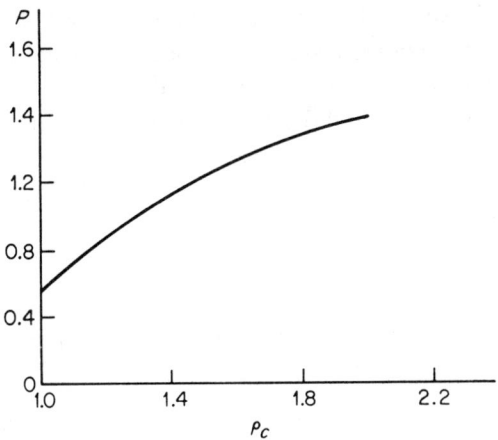

FIGURE 8.3.1 Plastic zone radius versus applied pressure, $\beta = 2$.

From (8.3.3) and (8.3.6) the stresses will be

$$S_r = 2 \ln \rho - 2 \ln \beta$$
$$= 2 \ln \frac{\rho}{\beta}$$
$$= 2 \ln \frac{r}{b} \qquad (8.3.7)$$

$$S_\theta = 1 + 2 \ln \frac{r}{b}$$

As a check, note that $S_\theta - S_r = 1$ for all r.

So far the stresses in the plastic part of the sphere have been computed. The stress distribution in the elastic part of the sphere can be readily obtained by considering the elastic portion of the sphere to be a new sphere with inner radius r_c, outer radius b, and with the pressure at the inner radius equal to the critical pressure for a sphere with these dimensions. The elastic solution, equations (8.2.5) (without temperature terms), can be used for this fictitious sphere replacing a by r_c and β by β_c. Thus

$$S_r = \frac{(r^3/r_c^3) - \beta_c^3}{(r^3/r_c^3)(\beta_c^3 - 1)} \left(\frac{2}{3} \frac{\beta_c^3 - 1}{\beta_c^3} \right)$$

or

$$\left. \begin{array}{l} S_r = \dfrac{2}{3} \dfrac{\rho^3 - \beta^3}{\rho^3 \beta_c^3} \\[1em] S_\theta = \dfrac{1}{3} \dfrac{2\rho^3 + \beta^3}{\rho^3 \beta_c^3} \end{array} \right\} \rho \geq \rho_c \qquad (8.3.8)$$

Equations (8.3.3) and (8.3.8) give the complete stress distribution in the sphere for given ratio of applied internal pressure to yield strength, with the plastic zone boundary ρ_c given by equation (8.3.5).

To calculate the strains and displacements in the sphere, the stress–strain relations and strain-displacement relations are used. For convenience we define, as was done for stresses in equations (8.2.4), "dimensionless" strains and displacements as follows:

$$\epsilon_r \equiv \frac{E\varepsilon_r}{\sigma_0} \quad \epsilon_\theta \equiv \frac{E\varepsilon_\theta}{\sigma_0} \quad U = \frac{Eu}{\sigma_0 a} \quad (8.3.9)$$

To compute the strains before yielding begins, equations (8.2.5) (with temperature terms deleted) are substituted into the stress–strain relations (8.1.4), resulting in

$$\epsilon_\theta = \frac{2(1 - 2\mu)\rho^3 + (1 + \mu)\beta^3}{2\rho^3(\beta^3 - 1)} P \quad (8.3.10)$$

and from the strain-displacement relation (8.1.2),

$$U = \left[(1 - 2\mu)\rho + \frac{(1 + \mu)\beta^3}{2\rho^2}\right] \frac{P}{\beta^3 - 1} \quad (8.3.11)$$

When yielding starts P is equal P_{crit} given by equation (8.2.10), and the displacement at the inner surface, $\rho = 1$, is

$$U_a = \frac{2}{3}\frac{1 - 2\mu}{\beta^3} + \frac{1}{3}(1 + \mu) \quad (8.3.12)$$

Note that if incompressibility had been assumed in the elastic region, $\mu = \frac{1}{2}$, the first term on the right of equation (8.3.12) disappears. For a Poisson's ratio of 0.3 and β of 2, the error in the displacement of the inner surface at the beginning of yield would be about 7 per cent. At the outer surface the error is 38 per cent.

As yielding progresses to some radius r_c, we can consider as before a new sphere with inner radius r_c, outer radius b, and critical pressure

$$P_{\text{crit}} = \frac{2}{3}\frac{\beta_c^3 - 1}{\beta_c^3}$$

Sec. 8-4] Hollow Sphere. Residual Stresses. Pressure Loading   145

acting at the inner radius r_c. From (8.3.11) the displacement at $r \geq r_c$ is obtained by replacing β by β_c and ρ by r/r_c, resulting in

$$U = \frac{2}{3\beta_c^3}\left[(1 - 2\mu)\rho + \frac{1 + \mu}{2}\frac{\beta^3}{\rho^2}\right] \qquad (8.3.13)$$

When the plastic zone reaches the outer radius b, $\rho = \beta$, $\beta_c = 1$, and

$$U_b = (1 - \mu)\beta \qquad (8.3.14)$$

The error in assuming $\mu = \tfrac{1}{2}$ is about 30 per cent, for $\mu = 0.3$.

In all the previous computations it was assumed that the dimensions of the sphere do not change as the pressure increases. This is, of course, not true, since the sphere grows with increase in internal pressure, the inner radius becoming $a + u_a$ and the outer radius $b + u_b$. A rigorous analysis would therefore have to take into account the change of dimensions of the shell. This can become particularly important for large strains.

8-4 HOLLOW SPHERE. RESIDUAL STRESSES. PRESSURE LOADING

If the pressure is removed from the sphere discussed in Section 8.3 after plastic flow has occurred over part of the sphere, residual stresses will result. To find the residual stresses it is necessary to superpose on the stress system due to the internal pressure p and temperature T a completely elastic stress system due to a pressure $-p$ and temperature $-T$. This will be correct as long as yielding in reverse does not occur; i.e., the residual stresses are not large enough to produce yielding. To see this, consider two stress systems satisfying the following two sets of equations:

$$\frac{d\sigma_r'}{dr} + 2\frac{\sigma_r' - \sigma_\theta'}{r} = 0$$

$$\frac{d\varepsilon_\theta}{dr} = \frac{\varepsilon_r' - \varepsilon_\theta'}{r}$$

$$\varepsilon_r' = \frac{1}{E}(\sigma_r' - 2\mu\sigma_\theta') + \alpha T + \varepsilon_r^P \qquad (8.4.1)$$

$$\varepsilon_\theta' = \frac{1}{E}[(1 - \mu)\sigma_\theta' - \mu\sigma_r'] + \alpha T + \varepsilon_\theta^P$$

$$\sigma_r'(a) = -p$$

$$\sigma_r'(b) = 0$$

and

$$\frac{d\sigma_r''}{dr} + 2\frac{\sigma_r'' - \sigma_\theta''}{r} = 0$$

$$\frac{d\varepsilon_r''}{dr} = \frac{\varepsilon_r'' - \varepsilon_\theta''}{r}$$

$$\varepsilon_r'' = \frac{1}{E}(\sigma_r'' - 2\mu\sigma_\theta'') - \alpha T \tag{8.4.2}$$

$$\varepsilon_\theta'' = \frac{1}{E}[(1 - \mu)\sigma_\theta'' - \mu\sigma_r''] - \alpha T$$

$$\sigma_r''(a) = p$$
$$\sigma_r''(b) = 0$$

The primed system corresponds to the system of stresses in the sphere with temperature T and internal pressure P. The double-primed system corresponds to the stresses in a sphere with temperature $-T$ and internal pressure $-P$. If the two systems are added together, there is obtained a system of stresses $\sigma_r = \sigma_r' + \sigma_r''$, etc., satisfying the following equations:

$$\frac{d\sigma_r}{dr} + 2\frac{\sigma_r - \sigma_\theta}{r} = 0$$

$$\frac{d\varepsilon_\theta}{dr} = \frac{\varepsilon_r - \varepsilon_\theta}{r}$$

$$\varepsilon_r = \frac{1}{E}(\sigma_r - 2\mu\sigma_\theta) + \varepsilon_r^P \tag{8.4.3}$$

$$\varepsilon_\theta = \frac{1}{E}[(1 - \mu)\sigma_\theta - \mu\sigma_r] + \varepsilon_\theta^P$$

$$\sigma_r(a) = 0$$
$$\sigma_r(b) = 0$$

Thus the resultant system corresponds to the unloaded sphere with permanent plastic strains due to the first system. If plastic flow occurs during the unloading, the elastic double-primed system can no longer be added to the original system, but it is necessary to solve another plastic flow problem for the new plastic strains.

For the case of pressure loading only, the elastic stresses due to a pressure equal to $-P$ are, from (8.2.5),

$$S_r'' = -\frac{\rho^3 - \beta^3}{\rho^3(\beta^3 - 1)}P \qquad S_\theta'' = -\frac{2\rho^3 + \beta^3}{2\rho^3(\beta^3 - 1)}P \tag{8.4.4}$$

Sec. 8–4] Hollow Sphere. Residual Stresses. Pressure Loading

Adding to the stresses given by (8.3.3) and by (8.3.8) gives for the residual stresses

$$S_r^r = 2\ln\rho - P\left(1 + \frac{\rho^3 - \beta^3}{\rho^3(\beta^3 - 1)}\right)$$

$$= 2\ln\rho - \frac{2}{3}\frac{P}{P_{\text{crit}}}\left(1 - \frac{1}{\rho^3}\right)$$

or

$$\left.\begin{aligned}S_r^r &= \frac{2}{3}\left[3\ln\rho - \frac{P}{P_{\text{crit}}}\left(1 - \frac{1}{\rho^3}\right)\right]\\ S_\theta^r &= \frac{2}{3}\left[\frac{3}{2} + 3\ln\rho - \frac{P}{P_{\text{crit}}}\left(1 + \frac{1}{2\rho^3}\right)\right]\end{aligned}\right\}\rho \le \rho_c \quad (8.4.5)$$

$$\left.\begin{aligned}S_r^r &= \frac{2}{3}\left[\left(\frac{P}{P_{\text{crit}}} - \rho_c^3\right)\left(\frac{1}{\rho^3} - \frac{1}{\beta^3}\right)\right]\\ S_\theta^r &= -\frac{2}{3}\left[\left(\frac{P}{P_{\text{crit}}} - \rho_c^3\right)\left(\frac{1}{2\rho^3} + \frac{1}{\beta^3}\right)\right]\end{aligned}\right\}\rho \ge \rho_c \quad (8.4.6)$$

The superscript r is used in the above equations to indicate residual stresses. When $\rho = 1$ (at the inner surface), $S_r^r = 0$, as expected, and

$$S_\theta^r = 1 - \frac{P}{P_{\text{crit}}} \quad (8.4.7)$$

and since $P \ge P_{\text{crit}}$, a residual compressive stress results. Upon reapplication of a pressure less than or equal to the original maximum, only elastic strains will occur. The shell has thus been strengthened by the initial pressurization. If the material work hardens, an even greater strengthening can be achieved.

In the above derivation it has been assumed that no plastic flow takes place during the unloading; i.e., there is no yielding in compression due to the residual stresses. If such yielding occurs, then not only is our assumption that the unloading is elastic violated, but the situation may be dangerous with regard to the safety of the sphere. The maximum value of applied pressure P such that if the sphere is unloaded there will be no reversed plastic flow is called the *shakedown pressure*, P_s. This pressure can be found as follows. For reversed yielding the yield criterion can be written

$$S_r^r - S_\theta^r = 1 \quad (8.4.8)$$

The maximum residual stress will occur at $\rho = 1$, where $S_r^r = 0$. From (8.4.7) it therefore follows that

$$P_s = 2P_{\text{crit}} \quad (8.4.9)$$

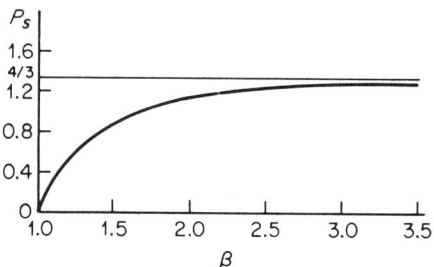

FIGURE 8.4.1 Variation of shakedown pressure with thickness ratio for hollow sphere with internal pressure.

As long as the applied pressure is less than twice the critical pressure, the residual stresses will be elastic. Making use of equation (8.3.1) the shakedown pressure can be written directly as a function of the thickness ratio β:

$$P_s = \frac{4}{3} \frac{\beta^3 - 1}{\beta^3} \qquad (8.4.10)$$

Figure 8.4.1 shows the shakedown pressure as a function of β.

8–5 HOLLOW SPHERE. THERMAL LOADING ONLY

For the case of a temperature gradient as given by equation (8.2.7), yielding will occur as given by equation (8.2.11). Then

$$\frac{\beta \tau_0}{2(\beta^3 - 1)} \left| \frac{3\beta^2}{\rho^3} - \frac{\beta^2 + \beta + 1}{\rho} \right| = 1 \qquad (8.5.1)$$

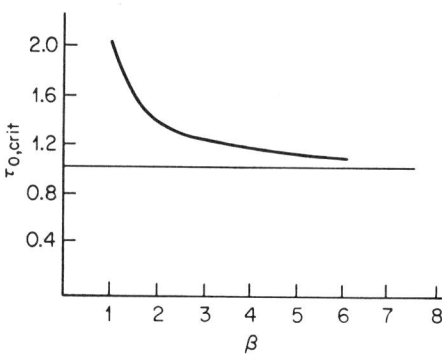

FIGURE 8.5.1 $\tau_{0,\text{crit}}$ as a function of β.

Sec. 8-5] Hollow Sphere. Thermal Loading Only

Yielding will first occur at $r = a$, ($\rho = 1$), and the critical temperature difference at which yielding will first start is given by

$$\tau_{0,\text{crit}} = \frac{2(\beta^3 - 1)}{\beta(2\beta^2 - \beta - 1)} = \frac{2(\beta^2 + \beta + 1)}{\beta(2\beta + 1)} \qquad (8.5.2)$$

A plot of $\tau_{0,\text{crit}}$ versus β is shown in Figure 8.5.1.

If τ_0 exceeds $\tau_{0,\text{crit}}$, the plastic zone will spread outward to some radius r_c. Within this zone, i.e., for $r \leq r_c$, the yield criterion $|S_\theta - S_r| = 1$ will apply. But since in this region the tangential stress will be a large compressive stress and the radial stress will be a small compressive stress, the yield criterion can be written

$$S_\theta - S_r = -1 \qquad (8.5.3)$$

The equilibrium equation now becomes

$$\frac{dS_r}{d\rho} = -\frac{2}{\rho}$$

or
$$S_r = -2 \ln \rho + C$$

and since $S_r(1) = 0$, $C = 0$. Hence

$$\left.\begin{array}{l} S_r = -2 \ln \rho \\ S_\theta = -1 - 2 \ln \rho \end{array}\right\} \rho \leq \rho_c \qquad (8.5.4)$$

Note that the stresses in the plastic region are independent of the temperature. The radius of the plastic zone, r_c, of course, depends on the temperature.

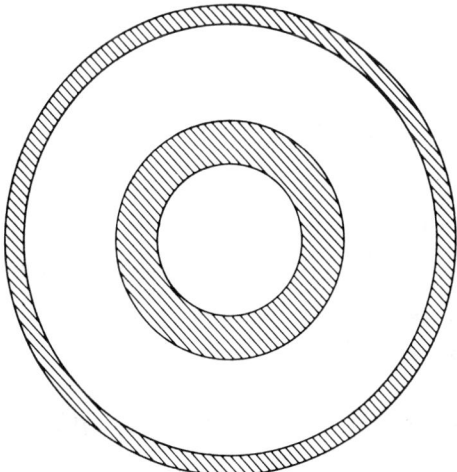

FIGURE 8.5.2 Two plastic zones due to temperature gradient.

As the temperature is further increased, a second plastic zone unconnected to the first may start at a new radius, depending on the value of β as shown in Figure 8.5.2. This is due to the fact that since there are no external forces acting on the sphere, the resultant force acting on any cross section must vanish. Thus the tangential stress varies from compression at the inner surface to tension at the outer surface. The inner surface will begin flowing plastically in compression, but if the temperature gradient is high enough, the outer surface will start flowing plastically in tension, thus producing two plastic zones, one in tension and one in compression. A detailed discussion is given in reference [1].

8–6 HOLLOW SPHERE OF STRAIN-HARDENING MATERIAL

We now consider the general case of a hollow sphere of strain-hardening material with both pressure and thermal loads. Equations (8.1.1) through (8.1.9) apply and in addition the assumption is made that the plastic strains are varying monotonically, so that equation (8.1.10) may be used. If this is not the case, equation (8.1.9) is used instead, and the calculation performed in steps or increments as described subsequently for more general types of problems. In addition, we use the dimensionless quantities defined by equations (8.2.4) and (8.3.9). The equilibrium compatibility and stress–strain relations are now written

$$\frac{dS_r}{d\rho} + \frac{2(S_r - S_\theta)}{\rho} = 0$$

$$\frac{d\epsilon_\theta}{d\rho} + \frac{\epsilon_\theta - \epsilon_r}{\rho} = 0 \qquad (8.6.1)$$

$$\epsilon_r = S_r - 2\mu S_\theta + (1 - \mu)\tau + \epsilon_r^P$$

$$\epsilon_\theta = (1 - \mu)S_\theta - \mu S_r + (1 - \mu)\tau + \epsilon_\theta^P$$

Substituting the last two of equations (8.6.1) into the second, combining with the first, and integrating results in the following equations:

$$S_r = -\frac{2}{\rho^3} \int_1^\rho \rho^2 \tau \, d\rho + \frac{1}{1-\mu} \int_1^\rho \frac{\epsilon_r^P}{\rho} \, d\rho + \left(1 - \frac{1}{\rho^3}\right) C_1 - P$$

$$S_\theta = -\tau - P + \frac{1}{\rho^3} \int_1^\rho \rho^2 \tau \, d\rho + \left(1 + \frac{1}{2\rho^3}\right) C_1 \qquad (8.6.2)$$

$$+ \frac{1}{2(1-\mu)} \epsilon_r^P + \frac{1}{1-\mu} \int_1^\rho \frac{\epsilon_r^P}{\rho} \, d\rho$$

Sec. 8-6] Hollow Sphere of Strain-Hardening Material

$$S = -\tau + \frac{3}{\rho^3}\int_1^\rho \rho^2 \tau \, d\rho + \frac{\epsilon_r^P}{2(1-\mu)} + \frac{3}{2\rho^3} C_1 \qquad (8.6.3)$$

$$C_1 = \frac{\beta^3}{\beta^3 - 1}\left[\frac{2}{\beta^3}\int_1^\beta \rho^2 \tau \, d\rho - \frac{1}{1-\mu}\int_1^\beta \frac{\epsilon_r^P}{\rho} d\rho + P\right] \qquad (8.6.4)$$

where

$$S \equiv S_\theta - S_r \qquad (8.6.5)$$

The boundary conditions used in deriving the above equations were

$$\begin{aligned} S_r(1) &= -P \\ S_r(\beta) &= 0 \end{aligned} \qquad (8.6.6)$$

For the elastic case equations (8.6.2) reduce to (8.2.5). For the case of a perfectly plastic material, the solution was given in the previous sections. We shall consider here only the case of a strain-hardening material.

To obtain a complete solution to the problem, it is necessary to determine the plastic strain distribution ϵ_r^P through the sphere. This will, of course, depend on the stress–strain curve of the material. The plastic strain distribution can be obtained in the following manner. The equivalent stress is related to the equivalent plastic strain through the stress–strain curve of the material. Thus

$$|S| = f(\epsilon_p)$$

or, for this case,

$$|S| = f(|\epsilon_r^P|)$$

where f is a known function representing the stress–strain curve. It therefore follows that

$$\begin{aligned} S &= \frac{S}{|S|} f(|\epsilon_r^P|) & |S| &\geq 1 \\ \epsilon_r^P &= 0 & |S| &\leq 1 \end{aligned} \qquad (8.6.7)$$

Also, from equation (8.6.3),

$$\epsilon_r^P = 2(1-\mu)\left(\tau - \frac{3}{\rho^3}\int_1^\rho \tau \rho^2 \, d\rho - \frac{3}{2\rho^3} C_1 + S\right) \qquad (8.6.8)$$

A complete solution can now be obtained by an iterative or successive approximation method. One chooses a distribution of ϵ_r^P (say zero). S is computed using equation (8.6.3), merely to determine its sign at the different radial positions. A first approximation to S is then obtained from equation (8.6.7), and a first approximation to the plastic strain distribution can be calculated from (8.6.8). A better value of S can then be computed from (8.6.7) and the next approximation for ϵ_r^P obtained from (8.6.8). If the process converges, we will thus obtain the proper values of ϵ_r^P and S such that (8.6.7) and (8.6.8) are satisfied simultaneously. The individual stresses S_r and S_θ can then readily be computed. Thus a complete solution is obtained in both the elastic and plastic regions. There is no need to treat the two regions separately as was done for the perfectly plastic material in previous sections. It should be noted, however, that equations (8.6.7) and (8.6.8) apply only for $|S| > 1$; for $|S| \leq 1$, ϵ_r^P is set equal to zero. The integrands ϵ_r^P/ρ appearing in the previous equations are therefore generally zero over part of the integration range. A similar technique is described in reference [9].

As a specific example, consider a sphere made of a material whose stress–strain curve is given by the following equation:

$$\sigma_e = 30{,}000 + 136{,}000 \epsilon_p^{1/2} \qquad \sigma_e \geq 30{,}000$$

Equation (8.6.7) now becomes

$$S = \frac{S}{|S|}(1 + 0.1434|\epsilon_r^P|^{1/2})$$

Results of calculations performed by the iterative procedure described are shown in Figures 8.6.1 and 8.6.2 for $\beta = 2$ and temperature distribution given by equation (8.2.7). In performing the calculations, the thickness of the sphere was divided into 40 equally spaced intervals and Simpson's rule

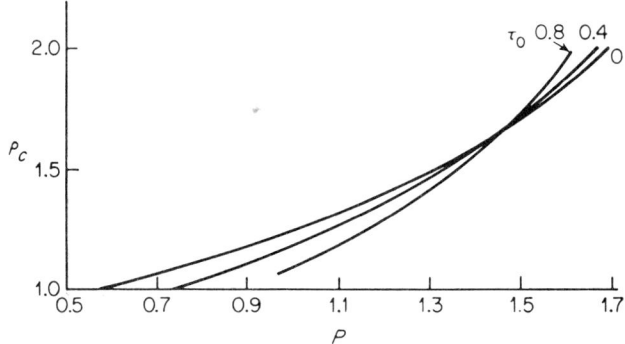

FIGURE 8.6.1 Variation of plastic zone radius with applied pressure for different temperature gradients:

$\beta = 2$, $\tau = \tau_0(\beta/\rho - 1)/(\beta - 1)$, $\sigma_e = 30{,}000 + 136{,}000\, \epsilon_p^{1/2}$.

Sec. 8-6] Hollow Sphere of Strain-Hardening Material

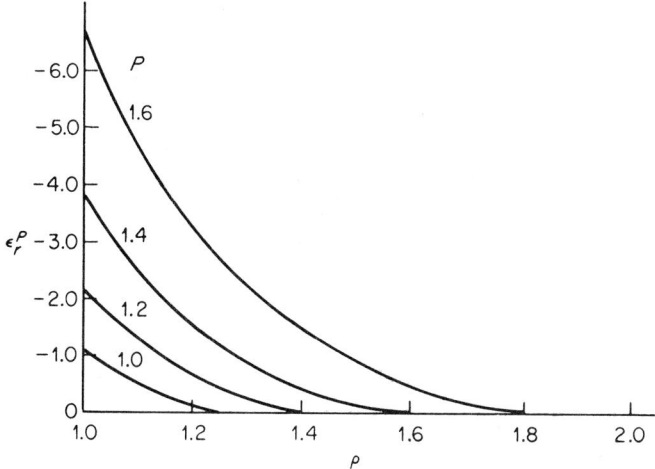

FIGURE 8.6.2 Variation of plastic strain with radius for various pressures: $\beta = 2$, $\tau_0 = 0$.

was used to perform the integrations. The cases shown are for illustrative purposes only. Any combination of geometry, loading, and material properties can be used and a rapid solution obtained. The time required to obtain a complete solution for a given loading condition, using a high-speed digital computer, is on the order of a few seconds.

This type of successive approximation method will be discussed at greater length in Chapter 9, where several numerical examples will be given. Right now it will be shown that if the material strain hardens linearly the solution can for some cases be obtained in closed form.

For linear strain hardening it follows from Figure 8.6.3 that

$$\epsilon_p = \frac{1-m}{m}(|S|-1) \tag{8.6.9}$$

where the *strain-hardening parameter m* is defined as the ratio of the slope of the strain-hardening part of the stress–strain curve to the elastic modulus. Then, from (8.1.10),

$$\epsilon_r^P = \frac{1-m}{m}(1-|S|)\frac{S}{|S|} = \frac{1-m}{m}\left(\frac{S}{|S|}-S\right) \tag{8.6.10}$$

Consider the case of pressure loading only. σ_θ will always be positive and σ_r will always be negative, so that $\sigma_\theta - \sigma_r > 0$. Therefore, $S/|S| = +1$ and

$$\epsilon_r^P = \frac{1-m}{m}(1-S) \tag{8.6.11}$$

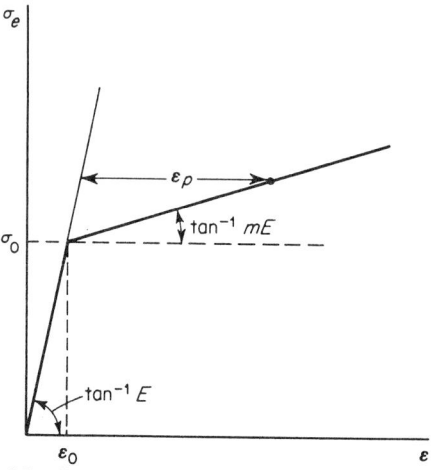
FIGURE 8.6.3 Stress–strain curve for linear strain hardening.

If the plastic zone extends to $\rho = \rho_c$, then, from (8.6.11), making use of the first of equations (8.6.1), it follows that

$$\int_1^\rho \frac{\epsilon_r^P}{\rho}\, d\rho = \begin{cases} \dfrac{1-m}{m}\ln\rho - \dfrac{1-m}{2m}(S_r + P) & \rho \leq \rho_c \\[2mm] \dfrac{1-m}{m}\ln\rho_c - \dfrac{1-m}{2m}(S_{r,c} + P) & \rho \geq \rho_c \end{cases} \quad (8.6.12)$$

where $S_{r,c}$ is the value of S_r at $\rho = \rho_c$. Substituting into (8.6.3) and (8.6.4) results in

$$C_1 = \frac{\beta^3}{\beta^3 - 1}\left[P - \frac{1-m}{(1-\mu)m}(\ln\rho_c - \tfrac{1}{2}S_{r,c} - \tfrac{1}{2}P)\right] \quad (8.6.13)$$

$$S = \frac{1-m}{2(1-\mu)m}(1 - S) + \frac{3}{2\rho^3}C_1 \quad (8.6.14)$$

and, since $S = 1$ when $\rho = \rho_c$, we have

$$\rho_c^3 = \tfrac{3}{2}C_1 \qquad C_1 = \tfrac{2}{3}\rho_c^3 \quad (8.6.15)$$

Substituting into (8.6.13) gives

$$\tfrac{2}{3}\rho_c^3 = \frac{\beta^3}{\beta^3 - 1}\left[P - \frac{1-m}{2(1-\mu)m}(2\ln\rho_c - S_{r,c} - P)\right] \quad (8.6.16)$$

At the onset of yield when $\rho_c = 1$, we have $S_{r,c} = -P$ and therefore

$$P_{\text{crit}} = \frac{2}{3}\frac{\beta^3 - 1}{\beta^3} \quad (8.6.17)$$

Sec. 8–6] Hollow Sphere of Strain-Hardening Material

which is the same result previously obtained for the perfectly plastic material. Obviously the onset of yield depends only on the yield stress. As P is increased, the plastic zone spreads to ρ_c and $-S_{r,c}$ can be considered to be the critical pressure acting on a sphere with inner radius ρ_c and outer radius $\beta_c = \beta/\rho_c$. Thus

$$S_{r,c} = -\frac{2}{3}\frac{\beta_c^3 - 1}{\beta_c^3} \qquad (8.6.18)$$

Hence

$$\frac{2}{3}\rho_c^3 = \frac{\beta^3}{\beta^3 - 1}\left[\left(1 + \frac{1-m}{2(1-\mu)m}\right)P - \frac{1-m}{(1-\mu)m}\ln\rho_c - \frac{1-m}{3(1-\mu)m}\frac{\beta_c^3 - 1}{\beta_c^3}\right] \qquad (8.6.19)$$

or

$$P = \frac{\frac{4}{3}(1-\mu)m[(\beta^3 - 1)/\beta^3]\rho_c^3 + 2(1-m)\ln\rho_c + \frac{2}{3}(1-m)(\beta_c^3 - 1)/\beta_c^3}{2m(1-\mu) + (1-m)} \qquad (8.6.20)$$

which relates the pressure P to the plastic zone radius ρ_c. Note that if $m = 0$,

$$P = 2\ln\rho_c + \frac{2}{3}\frac{\beta_c^3 - 1}{\beta_c^3}$$

which is the value previously obtained for the perfectly plastic material [equation (8.3.5)].

As an example, for $\beta = 2$, $m = 0.1$, and $\mu = 0.3$, the pressure required for yielding of the complete sphere, $\rho_c = \beta$, is 1.83, compared to 1.39 for a perfectly plastic material. It thus takes a 32 per cent higher pressure for the strain-hardening sphere to yield completely as compared to the perfectly plastic sphere.

To obtain the stresses we substitute into equations (8.6.2). Thus

$$\left.\begin{aligned}S_r &= -P + \left(1 - \frac{1}{\rho^3}\right)C_1 + \frac{1}{1-\mu}\left[\frac{1-m}{m}\ln\rho_c - \frac{1-m}{2m}(S_{r,c} + P)\right]\\ S_\theta &= -P + \left(1 + \frac{1}{2\rho^3}\right)C_1 + \frac{1}{1-\mu}\left[\frac{1-m}{m}\ln\rho_c - \frac{1-m}{2m}(S_{r,c} + P)\right]\end{aligned}\right\}\rho \geq \rho_c$$

(8.6.21)

$$\left.\begin{aligned}S_r &= -P + \frac{2(1-m)}{1-m+2(1-\mu)m}\ln\rho + \frac{\frac{4}{3}(1-\mu)m(\rho_c^3/\rho^3)(\rho^3 - 1)}{1-m+2(1-\mu)m}\\ S_\theta &= -P + \frac{2(1-m)}{1-m+2(1-\mu)m}\ln\rho + \frac{2}{3}\frac{(1-\mu)m(\rho_c^3/\rho^3)(2\rho^3 + 1)}{1-m+2(1-\mu)m}\end{aligned}\right\}\rho \leq \rho_c$$

(8.6.22)

156 Elastoplastic Problems of Spheres and Cylinders [Ch. 8]

Note that if $m = 0$ these reduce to the previously obtained values for the perfectly plastic material. Thus to obtain the complete stress distribution, p_c (or P) is obtained from (8.6.20), $S_{r,c}$ from (8.6.18), and then the stresses from (8.6.21) and (8.6.22).

8-7 PLASTIC FLOW IN THICK-WALLED TUBES

A considerable amount of work has been done on the problem of plastic flow in a thick-walled tube under internal pressure with and without temperature gradients. Solutions have been obtained, for example, in references [1] through [8]. These solutions differ in the yield criteria used and in the plastic stress–strain relations. Some solutions use the von Mises yield criterion and the associated flow rule [3]. Others use the Tresca criterion and its flow rule [7]. Reference [4] uses the Hencky total strain relations. In other papers complete incompressibility is assumed in both the elastic and plastic regions. Of the references cited, only [8] takes into account strain hardening of the material.

There are three cases that can be treated: (1) plane strain, $\varepsilon_z = 0$; (2) generalized plane strain, $\varepsilon_z = \text{constant} \neq 0$; and (3) tube with open ends, $P = 0$. We shall present a general solution for a strain-hardening material, including radial temperature gradients, which can take into account any of these cases.

The Tresca criterion and its associated flow rule will be used, since in this case it offers some simplifications. For this purpose it will be assumed that $\sigma_\theta > \sigma_z > \sigma_r$. It is shown by Koiter [7] that this is true for a large range of conditions.

We introduce the same dimensionless quantities as in the problem of the sphere; i.e.,

$$S_r = \frac{\sigma_r}{\sigma_0} \qquad S_\theta = \frac{\sigma_\theta}{\sigma_0} \qquad S_z = \frac{\sigma_z}{\sigma_0} \qquad P = \frac{p}{\sigma_0}$$

$$\tau = \frac{E\alpha T}{(1-\mu)\sigma_0} \qquad \gamma = \frac{E\alpha T_0}{(1+\mu)\sigma_0} \qquad \rho = \frac{r}{a} \qquad \beta = \frac{b}{a} \qquad (8.7.1)$$

$$\epsilon_r = \frac{\varepsilon_r}{\varepsilon_0} \qquad \epsilon_\theta = \frac{\varepsilon_\theta}{\varepsilon_0} \qquad \epsilon_z = \frac{\varepsilon_z}{\varepsilon_0}$$

$$S = S_\theta - S_r$$

where a and b are the internal and external radii, σ_0 is the yield stress, and ε_0 is the yield strain.

Sec. 8-7] Plastic Flow in Thick-Walled Tubes

Equations (8.1.11), (8.1.12), and (8.1.13) are now written

$$\frac{dS_r}{d\rho} = \frac{S_\theta - S_r}{\rho}$$

$$\frac{d\epsilon_\theta}{d\rho} = \frac{\epsilon_r - \epsilon_\theta}{\rho} \quad (8.7.2)$$

$$\epsilon_r = S_r - \mu(S_\theta + S_z) + (1 - \mu)\tau + \epsilon_r^P$$
$$\epsilon_\theta = S_\theta - \mu(S_r + S_z) + (1 - \mu)\tau + \epsilon_\theta^P \quad (8.7.3)$$
$$\epsilon_z = S_z - \mu(S_r + S_\theta) + (1 - \mu)\tau + \epsilon_z^P$$

If the Tresca criterion and its associated flow rule are used, then, assuming $S_\theta > S_z > S_r$,

$$\epsilon_z^P = 0$$
$$\epsilon_r^P = -\epsilon_\theta^P \quad (8.7.4)$$
and $\quad S = S_\theta - S_r = 1 \quad$ at yielding

For boundary conditions it is assumed that

$$S_r(a) = -P$$
$$S_r(b) = 0 \quad (8.7.5)$$

and the conditions at the end of the tube are determined by case 1, 2, or 3 above. For plane strain $\epsilon_z = 0$. For generalized plane strain, ϵ_z is a constant which can be determined from the end loads on the tube. Thus let the axial force acting on the tube be F. Define

$$F^* \equiv \frac{F}{2\pi a^2 \sigma_0}$$

Then it readily follows from the condition

$$F^* = \int_1^\beta S_z \rho \, d\rho$$

and the third of equations (8.7.3) that

$$\epsilon_z = \frac{2}{\beta^2 - 1}\left[F^* - \mu P + (1 - \mu)\int_1^\beta \tau \rho \, d\rho\right] \quad (8.7.6)$$

If the axial force is due to internal pressure only, then $F^* = P/2$ and

$$\epsilon_z = \frac{1}{\beta^2 - 1}\left[(1 - 2\mu)P + 2(1 - \mu)\int_1^\rho \tau\rho\, d\rho\right] \qquad (8.7.7)$$

For a tube with open ends, $P = 0$ and

$$\epsilon_z = \frac{2(1 - \mu)}{\beta^2 - 1}\int_1^\beta \tau\rho\, d\rho \qquad (8.7.8)$$

In any case ϵ_z is a known constant. From the last of equations (8.7.3),

$$S_z = \epsilon_z + \mu(S_r + S_\theta) - (1 - \mu)\tau \qquad (8.7.9)$$

Substituting this relation into the first two of equations (8.7.3), making use of the equilibrium and compatibility equations (8.7.2), and integrating results after some algebraic manipulations in the following solution:

$$S_r = -\frac{P}{\rho^2} - \frac{1}{\rho^2}\int_1^\rho \tau\rho\, d\rho + \frac{1}{1 - \mu^2}\int_1^\rho \frac{\epsilon_r^P}{\rho}\, d\rho + \left(1 - \frac{1}{\rho^2}\right)C_1$$

$$S_\theta = \frac{P}{\rho^2} - \tau + \frac{1}{\rho^2}\int_1^\rho \tau\rho\, d\rho + \frac{1}{1 - \mu^2}\left(\epsilon_r^P + \int_1^\rho \frac{\epsilon_r^P}{\rho}\, d\rho\right) + \left(1 + \frac{1}{\rho^2}\right)C_1 \qquad (8.7.10)$$

$$S = \frac{2P}{\rho^2} - \tau + \frac{2}{\rho^2}\int_1^\rho \tau\rho\, d\rho + \frac{1}{1 - \mu^2}\epsilon_r^P + \frac{2}{\rho^2}C_1 \qquad (8.7.11)$$

$$C_1 = \frac{1}{\beta^2 - 1}\left(P + \int_1^\beta \tau\rho\, d\rho - \frac{\beta^2}{1 - \mu^2}\int_1^\beta \frac{\epsilon_r^P}{\rho}\, d\rho\right) \qquad (8.7.12)$$

Yielding will begin at $\rho = 1$ when $S = 1$, so that the critical pressure will be

$$P_{\text{crit}} = \tfrac{1}{2}(1 + \tau) - C_1$$

and, from (8.7.12),

$$C_1 = \frac{1}{\beta^2 - 1}\left(\int_1^\beta \tau\rho\, d\rho + P_{\text{crit}}\right)$$

Therefore,

$$P_{\text{crit}} = \frac{\beta^2 - 1}{2\beta^2}\left(1 + \tau - \frac{2}{\beta^2 - 1}\int_1^\beta \tau\rho\, d\rho\right) \qquad (8.7.13)$$

For $P \le P_{\text{crit}}$, we have the elastic solution, which agrees with the classical elastic solution. For $P \ge P_{\text{crit}}$ a plastic zone will spread out to some radius ρ_c. The solution for general strain hardening can be obtained by an iterative or successive approximation method, as indicated for the sphere.

Sec. 8-7] Plastic Flow in Thick-Walled Tubes

Let the stress–strain curve be given by a relation of the form

$$|S| = f(\epsilon_p) \tag{8.7.14}$$

To relate ϵ_r^P to ϵ_p, the two methods indicated in Section 7.6 may be used. If the definition

$$d\epsilon_p = \sqrt{\tfrac{2}{3} d\epsilon_{ij}^P \, d\epsilon_{ij}^P}$$

is used, then from (8.7.4),

$$d\epsilon_p = \frac{2}{\sqrt{3}} |d\epsilon_r^P| \tag{8.7.15}$$

On the other hand, if the definition (7.6.20) is used, then it follows that

$$d\epsilon_p = |d\epsilon_r^P| \tag{8.7.16}$$

The two definitions differ by the familiar constant $2/\sqrt{3}$ and either one can be used. Since the definition based on the plastic work increment appears to be more consistent with the Tresca criterion, we shall use it, and assume for the case under consideration that (8.7.14) may be written

$$\begin{aligned} |S| &= f(|\epsilon_r^P|) & |S| &\geq 1 \\ \epsilon_r^P &= 0 & |S| &\leq 1 \end{aligned} \tag{8.7.17}$$

To use the successive approximation method, it is preferable, as was done the for case of the sphere, to rewrite equations (8.7.11) and (8.7.17) as follows:

$$\epsilon_r^P = (1 - \mu^2)\left(S - \frac{2P}{\rho^2} + \tau - \frac{2}{\rho^2}\int_1^\rho \tau\rho \, d\rho - \frac{2}{\rho^2} C_1\right) \tag{8.7.18}$$

$$\left.\begin{aligned} S &= f(|\epsilon_r^P|) \operatorname{sgn} S & |S| &\geq 1 \\ \epsilon_r^P &= 0 & |S| &\leq 1 \end{aligned}\right\} \tag{8.7.19}$$

An initial distribution of ϵ_r^P (such as zero) is assumed. The signs of S throughout the cross section of the tube are then determined from (8.7.11), and the actual values of S are calculated from (8.7.19). A better approximation can now be obtained for the ϵ_r^P using equation (8.7.18). The process is repeated until convergence is obtained.

For the cases of the sphere and the tube heretofore discussed, the successive approximation method has been found to converge fairly rapidly using the techniques described. However, this may not always be true. A general discussion of the convergence of the successive approximation method is given in Chapter 9.

160   Elastoplastic Problems of Spheres and Cylinders   [Ch. 8

For the case of linear strain hardening, a solution can be obtained in closed form for the above problem. As for the sphere problem, equation (8.6.11) is used:

$$\varepsilon_r^P = \frac{1-m}{m}(1-S)$$

$$\int_1^\beta \frac{\varepsilon_r^P}{\rho} d\rho = \frac{1-m}{m}(\ln \rho_c - S_{r,c} - P) \tag{8.7.20}$$

Therefore,

$$C_1 = \frac{1}{\beta^2 - 1}\left[\int_1^\beta \tau\rho \, d\rho + P - \frac{\beta^2(1-m)}{m(1-\mu^2)}(\ln \rho_c - S_{r,c} - P)\right] \tag{8.7.21}$$

Also, from (8.7.11),

$$S = -\tau + \frac{2}{\rho^2}\int_1^\rho \tau\rho \, d\rho + \frac{2}{\rho^2}(P + C_1) + \frac{1-m}{m(1-\mu^2)}(1-S) \quad \rho \leq \rho_c \tag{8.7.22}$$

When $\rho = \rho_c$, $S = 1$. Therefore, from (8.7.22),

$$C_1 = \frac{\rho_c^2}{2}\left[1 + \tau(\rho_c) - \frac{2}{\rho_c^2}\int_1^{\rho_c} \tau\rho \, d\rho\right] - P \tag{8.7.23}$$

Substituting into the expression for C_1, (8.7.21), we get

$$\frac{\rho_c^2}{2}\left[1 + \tau(\rho_c) - \frac{2}{\rho_c^2}\int_1^{\rho_c} \tau\rho \, d\rho\right]$$
$$= \frac{\beta^2}{\beta^2 - 1}\left[P + \frac{1}{\beta^2}\int_1^\beta \tau\rho \, d\rho - \frac{1-m}{m(1-\mu^2)}(\ln \rho_c - S_{r,c} - P)\right] \tag{8.7.24}$$

As a check, at the onset of yield, $\rho_c = 1$, $S_{r,c} = -P$, and (8.7.24) reduces to (8.7.13) for the critical pressure.

At $\rho = \rho_c$ we can consider a new tube with inner radius ρ_c and outer radius β_c with $S_{r,c}$ equal to $-P_{\text{crit}}$. Thus

$$S_{r,c} = -\frac{\beta_c^2 - 1}{2\beta_c^2}\left[1 + \tau(\rho_c) - \frac{2}{\beta_c^2 - 1}\int_{\rho_c}^{\beta_c} \tau\rho \, d\rho\right] \tag{8.7.25}$$

Solving (8.7.24) for P gives

$$P = \frac{\beta^2 - 1}{\beta^2}\frac{(1-\mu^2)m}{1-\mu^2 m}\left[\frac{\rho_c^2}{2}\left(1 + \tau(\rho_c) - \frac{2}{\rho_c^2}\int_1^{\rho_c} \tau\rho \, d\rho\right) - \frac{1}{\beta^2 - 1}\int_1^\beta \tau\rho \, d\rho\right]$$
$$+ \frac{1-m}{1-\mu^2 m}(\ln \rho_c - S_{r,c}) \tag{8.7.26}$$

Sec. 8-7] Plastic Flow in Thick-Walled Tubes

Equation (8.7.26) gives the relationship between the plastic zone radius ρ_c and the applied pressure P, for a given temperature distribution τ. For a perfectly plastic material this reduces to

$$P = \ln \rho_c - S_{r,c}$$

$$= \ln \rho_c + \frac{\beta_c^2 - 1}{2\beta_c^2}\left[1 + \tau(\rho_c) - \frac{2}{\beta_c^2 - 1}\int_{\rho_c}^{\beta_c} \tau\rho\, d\rho\right] \quad (8.7.27)$$

To obtain the stresses we now substitute into equations (8.7.10). Thus, since

$$\int_1^\rho \frac{\epsilon_r^p}{\rho}\, d\rho = \frac{1-m}{m}(\ln \rho - S_r - P) \qquad \rho \le \rho_c$$

$$\int_1^\rho \frac{\epsilon_r^p}{\rho}\, d\rho = \frac{1-m}{m}(\ln \rho_c - S_{r,c} - P) \qquad \rho \ge \rho_c$$

then, from (8.7.10),

$$\left.\begin{array}{l} S_r = \dfrac{m(1-\mu^2)}{1-\mu^2 m}\left(C_1\dfrac{\rho^2-1}{\rho^2} - \dfrac{1}{\rho^2}\int_1^\rho \rho\tau\, d\rho - \dfrac{P}{\rho^2}\right) \\[6pt] \qquad + \dfrac{1-m}{1-\mu^2 m}(\ln \rho - P) \\[10pt] S_\theta = \dfrac{m(1-\mu^2)}{1-\mu^2 m}\left(C_1\dfrac{\rho^2+1}{\rho^2} + \dfrac{1}{\rho^2}\int_1^\rho \rho\tau\, d\rho + \dfrac{P}{\rho^2} - \tau\right) \\[6pt] \qquad + \dfrac{1-m}{1-\mu^2 m}(1 + \ln \rho - P) \end{array}\right\} \rho \le \rho_c$$

and (8.7.28)

$$\left.\begin{array}{l} S_r = -\dfrac{1}{\rho^2}\int_1^\rho \tau\rho\, d\rho + \dfrac{1-m}{(1-\mu^2)m}(\ln \rho_c - S_{r,c} - P) \\[6pt] \qquad - \dfrac{P}{\rho^2} + \left(1 - \dfrac{1}{\rho^2}\right)C_1 \\[10pt] S_\theta = -\tau + \dfrac{1}{\rho^2}\int_1^\rho \tau\rho\, d\rho + \dfrac{1-m}{(1-\mu^2)m}(\ln \rho_c - S_{r,c} - P) \\[6pt] \qquad + \dfrac{P}{\rho^2} + \left(1 + \dfrac{1}{\rho^2}\right)C_1 \end{array}\right\} \rho \ge \rho_c$$

To obtain the complete stress distribution, we compute P or ρ_c from (8.7.26), C_1 from (8.7.23), $S_{r,c}$ from (8.7.25), and the stresses from (8.7.28).

Problems

1. Show that for the sphere with radial symmetry, the von Mises yield criterion becomes

$$|\sigma_\theta - \sigma_r| = \sigma_0$$

and the Prandtl–Reuss equations reduce to

$$d\varepsilon_r^P = -2d\varepsilon_\theta^P = d\varepsilon_p \, \text{sgn}\,(\sigma_r - \sigma_\theta)$$

2. Explain why one would expect the Tresca and von Mises yield criteria to coincide for the case of a sphere with radial symmetry.
3. Derive equations (8.2.1) and (8.2.3).
4. Obtain the equations for all the strains and displacements in the sphere before yielding begins for pressure loading only, for thermal loading only, and for the case when both thermal and pressure loading exist.
5. Show that the steady-state temperature distribution in a sphere of inner radius a and outer radius b is equal to

$$T = \frac{T_0 a}{b - a}\left(\frac{b}{r} - 1\right)$$

if the inner and outer surfaces are kept at temperatures of T_0 and zero, respectively.

6. Using equations (8.3.3) and (8.3.8), show that the stresses are continuous across the elastoplastic boundary.
7. Compute the displacements and strains in a sphere with pressure loading only, for $r \leq r_c$. Assume a perfectly plastic material and that the dimensions remain fixed. Determine the error in the displacements of the inner radius for the fully plastic case if μ is assumed to equal 0.5 instead of 0.3.
8. Show that for a hollow sphere with a temperature distribution given by equation (8.2.7), the tangential stress is compressive and the radial stress is near zero in the region adjacent to the inner circumference, so that the yield criterion in this region can be written

$$S_\theta - S_r = -1$$

9. Starting with equations (8.6.1), derive equations (8.6.2) and (8.6.4) using boundary conditions (8.6.6).
10. Derive equation (8.6.9).
11. Derive equations (8.6.12).
12. Derive equation (8.7.6).
13. Derive equations (8.7.10) through (8.7.12).
14. Show that the definition (7.6.20) for the equivalent plastic strain increment leads to equation (8.7.16) for the case of a tube with the Tresca criterion and associated flow rule, if $\sigma_\theta > \sigma_z > \sigma_r$.

General References

15. Plot the plastic zone as a function of the applied pressure for a sphere with linear strain hardening. Assume $\beta = 2$, $m = 0.1$, and $\mu = 0.3$. Compare the results with those for a perfectly plastic material.
16. Repeat Problem 15 for a tube.
17. Perform complete numerical analysis of the problem of the sphere, $a \leq r \leq b$, $T(a) = T_0$, and $T(b) = 0$. Assume $E = 30 \times 10^6$, $\mu = 0.3$, $\alpha = 10^{-5}$, $\sigma_e = 30{,}000 + 136{,}000\,(\varepsilon_p + 10^{-4})$ for $\sigma_e > 30{,}000$, and E, α, and μ are independent of temperature.

References

1. W. Johnson and P. B. Mellor, *Plasticity for Mechanical Engineers*, Van Nostrand, Princeton, N.J., 1962.
2. R. Hill, *The Mathematical Theory of Plasticity*, Oxford Univ. Press, London, 1950.
3. P. G. Hodge and G. N. White, A Quantitative Comparison of Flow and Deformation Theories of Plasticity, *J. Appl. Mech.*, **17**, 1950, pp. 180–184.
4. D. N. de G. Allen and D. G. Sopwith, The Stresses and Strains in a Partially Plastic Thick Tube Under Internal Pressure and End-Load, *Proc. Roy. Soc. (London)*, **A205**, 1951, pp. 69–83.
5. M. C. Steele, Partially Plastic Thick-Walled Cylinder Theory, *J. Appl. Mech.*, **19**, 1952, pp. 133–140.
6. R. Hill, E. H. Lee, and S. J. Tupper, The Theory of Combined Plastic and Elastic Deformation with Particular Reference to a Thick Tube Under Internal Pressure, *Proc. Roy. Soc. (London)*, **A191**, 1947, pp. 278–303.
7. W. T. Koiter, On Partially Plastic Thick-Walled Tubes, *Biezeno Anniversary Volume in Applied Mechanics*, N. V. De Technische Uitgeverij H. Stam, Haarlem, 1953, pp. 232–251.
8. D. R. Bland, Elastoplastic Thick-Walled Tubes of Work-Hardening Material Subject to Internal and External Pressures and to Temperature Gradients, *J. Mech. Phys. Solids*, **4**, pp. 209–229.
9. I. S. Tuba, Elastic–Plastic Analysis for Hollow Spherical Media Under Uniform Radial Loading, *J. Franklin Inst.*, **280**, 1965, pp. 343–355.

General References

Hill, R., *The Mathematical Theory of Plasticity*, Oxford Univ. Press, London, 1950.

Hoffman, O., and G. Sachs, *Introduction to the Theory of Plasticity for Engineers*, McGraw-Hill, New York, 1953.

Johnson, W., and P. M. Mellor, *Plasticity for Mechanical Engineers*, Van Nostrand, Princeton, N.J., 1962.

CHAPTER 9

THE METHOD OF SUCCESSIVE ELASTIC SOLUTIONS

9-1 GENERAL DESCRIPTION OF THE METHOD

In Chapter 8 it was indicated how the sphere and tube problems can be solved for arbitrary strain hardening by a successive-approximation method. This method is nothing more than the extension of Picard's method (see reference [1]) of successive approximations to nonlinear equations. The method was apparently first used in plastic flow problems by Ilyushin [2] in his treatment of a thin shell. Ilyushin refers to it as the *method of successive elastic solutions*, since each iteration involves essentially the solution of an elastic problem.

Before the advent of modern high-speed computing machinery, this method could be used only for relatively simple problems. However, with current widespread availability and use of digital computers it now becomes possible to solve simply and quickly many problems whose *elastic solution can be obtained by numerical methods*.

Before proceeding to describe this method for the general elastoplastic problem, we shall first give an illustration of the method of successive approximations for a simple differential equation [3]. Consider the equation

$$\frac{dy}{dx} - y = 0 \qquad y(0) = 1 \qquad (9.1.1)$$

Sec. 9-1] General Description of the Method

The solution is known to be $y = e^x$. To find the solution by Picard's method, we proceed as follows. Integrate (9.1.1) to give

$$y = 1 + \int_0^x y \, dx \qquad (9.1.2)$$

Assume as a first approximation for y,

$$y^{(1)} = 1$$

Substitute this value for y on the right side of (9.1.2) and calculate a second approximation for y:

$$y^{(2)} = 1 + \int_0^x y^{(1)} \, dx = 1 + x$$

Substitute the second approximation for y and calculate the third approximation:

$$y^{(3)} = 1 + \int_0^x y^{(2)} \, dx = 1 + x + \frac{x^2}{2}$$

Continue in this way to get

$$y^{(4)} = 1 + \int_0^x y^{(3)} \, dx = 1 + x + \frac{x^2}{2!} + \frac{x^3}{3!}$$

$$y^{(n+1)} = 1 + \int_0^x y^{(n)} \, dx = 1 + x + \frac{x^2}{2!} + \cdots + \frac{x^n}{n!}$$

As n gets larger and larger, it is seen that the infinite series for e^x is approached. The exact solution can thus be approached as closely as desired by taking more and more approximations.

This technique can be directly extended to the general elastoplastic problem in the following manner. For convenience the pertinent equations given in previous chapters will be repeated here. The equilibrium and compatibility equations are independent of the plasticity relations and are given by equations (3.2.2) and (4.7.2); i.e.,

$$\frac{\partial \sigma_x}{\partial x} + \frac{\partial \tau_{xy}}{\partial y} + \frac{\partial \tau_{xz}}{\partial z} = -F_x$$

$$\frac{\partial \tau_{xy}}{\partial x} + \frac{\partial \sigma_y}{\partial y} + \frac{\partial \tau_{yz}}{\partial z} = -F_y \qquad (9.1.3)$$

$$\frac{\partial \tau_{xz}}{\partial x} + \frac{\partial \tau_{yz}}{\partial y} + \frac{\partial \sigma_z}{\partial z} = -F_z$$

and

$$\frac{\partial^2 \varepsilon_x}{\partial y^2} + \frac{\partial^2 \varepsilon_y}{\partial x^2} = 2 \frac{\partial^2 \varepsilon_{xy}}{\partial x \, \partial y}$$

$$\frac{\partial^2 \varepsilon_y}{\partial z^2} + \frac{\partial^2 \varepsilon_z}{\partial y^2} = 2 \frac{\partial^2 \varepsilon_{yz}}{\partial y \, \partial z}$$

$$\frac{\partial^2 \varepsilon_z}{\partial x^2} + \frac{\partial^2 \varepsilon_x}{\partial z^2} = 2 \frac{\partial^2 \varepsilon_{xz}}{\partial z \, \partial x}$$

$$\frac{\partial}{\partial x}\left(-\frac{\partial \varepsilon_{yz}}{\partial x} + \frac{\partial \varepsilon_{zx}}{\partial y} + \frac{\partial \varepsilon_{xy}}{\partial z}\right) = \frac{\partial^2 \varepsilon_x}{\partial y \, \partial z} \quad (9.1.4)$$

$$\frac{\partial}{\partial y}\left(-\frac{\partial \varepsilon_{zx}}{\partial y} + \frac{\partial \varepsilon_{xy}}{\partial z} + \frac{\partial \varepsilon_{yz}}{\partial x}\right) = \frac{\partial^2 \varepsilon_y}{\partial z \, \partial x}$$

$$\frac{\partial}{\partial z}\left(-\frac{\partial \varepsilon_{xy}}{\partial z} + \frac{\partial \varepsilon_{yz}}{\partial x} + \frac{\partial \varepsilon_{zx}}{\partial y}\right) = \frac{\partial^2 \varepsilon_z}{\partial x \, \partial y}$$

The stress–strain relations depend on the plasticity theory used and we can write

$$\varepsilon_x = \frac{1}{E}[\sigma_x - \mu(\sigma_y + \sigma_z)] + \alpha T + \varepsilon_x^P + \Delta \varepsilon_x^P$$

$$\varepsilon_y = \frac{1}{E}[\sigma_y - \mu(\sigma_z + \sigma_x)] + \alpha T + \varepsilon_y^P + \Delta \varepsilon_y^P$$

$$\varepsilon_z = \frac{1}{E}[\sigma_z - \mu(\sigma_x + \sigma_y)] + \alpha T + \varepsilon_z^P + \Delta \varepsilon_z^P$$

$$\varepsilon_{xy} = \frac{1}{2G} \tau_{xy} + \varepsilon_{xy}^P + \Delta \varepsilon_{xy}^P \quad (9.1.5)$$

$$\varepsilon_{yz} = \frac{1}{2G} \tau_{yz} + \varepsilon_{yz}^P + \Delta \varepsilon_{yz}^P$$

$$\varepsilon_{zx} = \frac{1}{2G} \tau_{zx} + \varepsilon_{zx}^P + \Delta \varepsilon_{zx}^P$$

where ε_x^P, ε_y^P, etc., are the total accumulated plastic strains up to, but not including, the current increment of loading, $\Delta \varepsilon_x^P$, $\Delta \varepsilon_y^P$, $\Delta \varepsilon_z^P$, etc., are the plastic strain increments due to the current increment of loading. The plastic strain increments are related to the stresses through the yield criterion and

Sec. 9-1] General Description of the Method

the associated flow rule. For definiteness we shall consider the Prandtl–Reuss relations, but any other set of relations can be used equally well. Thus

$$\Delta \varepsilon_x^P = \frac{\Delta \varepsilon_p}{2\sigma_e}(2\sigma_x - \sigma_y - \sigma_z)$$

$$\Delta \varepsilon_y^P = \frac{\Delta \varepsilon_p}{2\sigma_e}(2\sigma_y - \sigma_z - \sigma_x)$$

$$\Delta \varepsilon_z^P = -\Delta \varepsilon_x^P - \Delta \varepsilon_y^P$$

$$\Delta \varepsilon_{xy}^P = \frac{3}{2}\frac{d\varepsilon_p}{\sigma_e}\tau_{xy} \quad (9.1.6)$$

$$\Delta \varepsilon_{yz}^P = \frac{3}{2}\frac{d\varepsilon_p}{\sigma_e}\tau_{yz}$$

$$\Delta \varepsilon_{zx}^P = \frac{3}{2}\frac{d\varepsilon_p}{\sigma_e}\tau_{zx}$$

where

$$\Delta \varepsilon_p = \frac{\sqrt{2}}{3}\sqrt{(\Delta \varepsilon_x^P - \Delta \varepsilon_y^P)^2 + (\Delta \varepsilon_y^P - \Delta \varepsilon_z^P)^2 + (\Delta \varepsilon_z^P - \Delta \varepsilon_x^P)^2 + 6[(\Delta \varepsilon_{xy}^P)^2 + (\Delta \varepsilon_{yz}^P)^2 + (\Delta \varepsilon_{zx}^P)^2]}$$
(9.1.7)

$$\sigma_e = \frac{1}{\sqrt{2}}\sqrt{(\sigma_x - \sigma_y)^2 + (\sigma_y - \sigma_z)^2 + (\sigma_z - \sigma_x)^2 + 6(\tau_{xy}^2 + \tau_{yz}^2 + \tau_{zx}^2)}$$

Alternatively, the plastic strain increments can be related to the modified total strains as described in Section 7.9; i.e.,

$$\Delta \varepsilon_x^P = \frac{\Delta \varepsilon_p}{3\varepsilon_{et}}(2\varepsilon_x' - \varepsilon_y' - \varepsilon_z')$$

$$\Delta \varepsilon_y^P = \frac{\Delta \varepsilon_p}{3\varepsilon_{et}}(2\varepsilon_y' - \varepsilon_z' - \varepsilon_x')$$

$$\Delta \varepsilon_z^P = -\Delta \varepsilon_x^P - \Delta \varepsilon_y^P$$

$$\Delta \varepsilon_{xy}^P = \frac{\Delta \varepsilon_p}{\varepsilon_{et}}\varepsilon_{xy}' \quad (9.1.8)$$

$$\Delta \varepsilon_{yz}^P = \frac{\Delta \varepsilon_p}{\varepsilon_{et}}\varepsilon_{yz}'$$

$$\Delta \varepsilon_{zx}^P = \frac{\Delta \varepsilon_p}{\varepsilon_{et}}\varepsilon_{zx}'$$

where

$$\varepsilon'_x = \varepsilon_x - \varepsilon^P_x \quad \varepsilon'_y = \varepsilon_y - \varepsilon^P_y \quad \text{etc.}$$

$$\varepsilon'_{xy} = \varepsilon_{xy} - \varepsilon^P_{xy} \quad \text{etc.} \tag{9.1.9}$$

$$\varepsilon_{et} = \frac{\sqrt{2}}{3} \sqrt{(\varepsilon'_x - \varepsilon'_y)^2 + (\varepsilon'_y - \varepsilon'_z)^2 + (\varepsilon'_z - \varepsilon'_x)^2 + 6[(\varepsilon'_{xy})^2 + (\varepsilon'_{yz})^2 + (\varepsilon'_{zx})^2]}$$

Finally, the equivalent plastic strain increment $\Delta\varepsilon_p$ is related to the equivalent stress σ_e or the equivalent total strain ε_{et} through the stress–strain curve. In addition, of course, the boundary conditions must always be satisfied. For complete generality, the full three-dimensional equations have been written out above. In practice only one- and two-dimensional problems can usually be solved, which is equally true for elasticity problems.

The method of successive approximations now proceeds as follows. The loading path is divided into a number of increments. For the first increment of load, a distribution is assumed for the plastic strain increments $\Delta\varepsilon^P_x$, $\Delta\varepsilon^P_y$, etc. The total plastic strains ε^P_x, etc., are zero. The set of equations (9.1.3), (9.1.4), and (9.1.5) are now solved as for any elasticity problem and a first approximation obtained for the stresses and total strains. At the same time, using the assumed values of the plastic strain increments, an equivalent plastic strain increment $\Delta\varepsilon_p$ is computed by the first of equations (9.1.7). From the stress–strain curve the corresponding value of σ_e can be determined. A new approximation can now be obtained for the individual plastic strain increments using the Prandtl–Reuss relations (9.1.6).

Using these new plastic strain increments, equations (9.1.3) through (9.1.5) are solved again as a new elastic problem. A second and presumably better approximation is obtained for the stresses and total strains. At the same time, using these last values of the plastic strain increments a new approximation is computed for the equivalent plastic strain increment $\Delta\varepsilon_p$ by the first of equations (9.1.7). Using this value of $\Delta\varepsilon_p$, a new value is obtained for σ_e from the stress–strain curve. New approximations are now obtained for the plastic strain increments $\Delta\varepsilon^P_x$, etc., using the Prandtl–Reuss relations (9.1.6). The process is continued until convergence is obtained; i.e., the differences between two successive sets of strain increments are less than some prescribed values. The calculation scheme is illustrated by the flow diagram of Figure 9.1.1.

In this manner the solution is obtained for the first increment of loading. For the next increment of load, an exactly similar calculation is made except that ε^P_x, ε^P_y, etc., are no longer zero but are equal to the known values of $\Delta\varepsilon^P_x$, $\Delta\varepsilon^P$, etc., obtained for the first increment of loading. The complete stress

Sec. 9-1] General Description of the Method

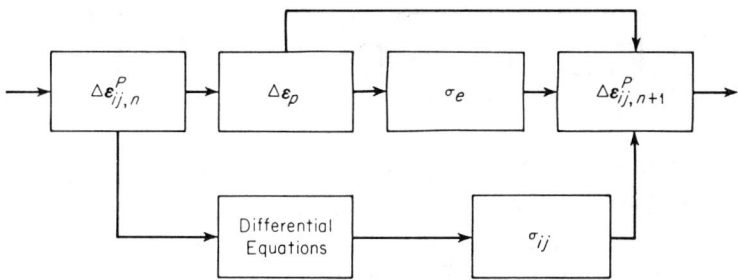

FIGURE 9.1.1 Block diagram for computing plastic strain increments by successive approximations.

and strain history can thus be obtained for any loading and unloading path. If a deformation type of theory of plasticity is used, only one loading step is required for the calculation.

The question of course which immediately arises is: What about convergence? Will this method always converge and, if not, under what conditions will it converge? A rigorous discussion of the convergence problem for the complicated set of nonlinear equations representing the elastoplastic problem is beyond the scope of this book. Our discussion will therefore be based on experience, together with some qualitative observations.

From the previous description of the method and Figure 9.1.1, it is seen that the equivalent plastic strain increment $\Delta \varepsilon_p$ is computed from the individual plastic strains by means of the first of equations (9.1.7), and the equivalent stress σ_e is then determined from $\Delta \varepsilon_p$ and the stress–strain curve. This is illustrated by path $OABC$ of Figure 9.1.2. On the other hand, it would seem more direct to compute σ_e from the second of equations (9.1.7) and then to determine $\Delta \varepsilon_p$ from σ_e and the stress–strain curve as shown by the flow diagram of Figure 9.1.3 and path $OCBA$ of Figure 9.1.2. However, experience

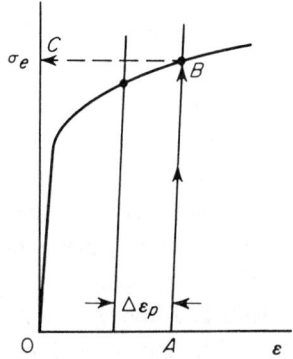

FIGURE 9.1.2 Determination of σ_e from $\Delta \varepsilon_p$.

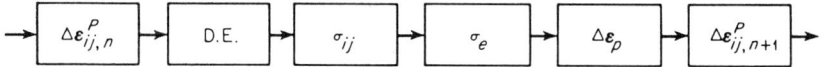

FIGURE 9.1.3   Block diagram for nonconvergent scheme.

has shown that this latter scheme will generally not converge, whereas the former method will *converge provided the loading increment is made sufficiently small*. The reason for this can be seen qualitatively from Figure 9.1.2. Since the stress–strain curve is very flat in the plastic region, it follows that a small error in $\Delta\varepsilon_p$ will produce a smaller error in σ_e, but a small error in σ_e will produce a much larger error in $\Delta\varepsilon_p$, and so the second method described will not converge. The reason for the suggested order of the computations for the sphere and tube of Chapter 8 [equations (8.6.8) and (8.7.18)] now becomes apparent.

It was mentioned above that the first method described will generally converge if the increment is made small enough. The question of what increment size is sufficiently small for a given problem can usually be determined only by trial and error; i.e., one picks an increment size and if the process diverges the increment size is reduced. As a rough rule of thumb it has been found that if $\Delta\varepsilon_p$ is less than about 0.3 per cent, convergence will *usually* take place, whereas if it is greater than 0.3 per cent, the process may diverge. The reason for this divergence can again be seen in a qualitative way from the Prandtl–Reuss relations. Thus consider the equations (9.1.6),

$$\Delta\varepsilon_x^P = \frac{\Delta\varepsilon_p}{2\sigma_e}(2\sigma_x - \sigma_y - \sigma_z) \quad \text{etc.}$$

In using this equation the value of $\Delta\varepsilon_p$ is computed from the previous values of $\Delta\varepsilon_x^P$, $\Delta\varepsilon_y^P$, etc., using the first of equations (9.1.7). The value of σ_e is then obtained from the stress–strain curve, using this computed value of $\Delta\varepsilon_p$, as shown by path *OABC* of Figure 9.1.2. The stresses appearing in this equation are obtained from the solution of equations (9.1.3) through (9.1.5), using the previous values of the plastic strain increments, as indicated by Figure 9.1.1. If we compute σ_e by the second of equations (9.1.7) using these stresses, the value of σ_e will not in general agree with the value of σ_e obtained from the stress–strain curve, until convergence is obtained. There is thus an inconsistency in the Prandtl–Reuss equations (9.1.6), this inconsistency diminishing as convergence is approached. If the increments of strain are too large, this initial inconsistency will produce divergence.

In the second method described above and shown in Figure 9.1.3, this type of inconsistency is avoided, since σ_e is computed from the calculated stresses and $\Delta\varepsilon_p$ is then obtained from the stress–strain curve. However, as

Sec. 9-1] General Description of the Method

pointed out above, the second method diverges because of the flatness of the stress–strain curve.

Both these difficulties can, however, be avoided by using the plastic strain–total strain equations described in Section 7.9 and given by equations (9.1.8) and (9.1.9). In using these equations the total strains are obtained from the solution of equations (9.1.3) through (9.1.5) with assumed or previously calculated approximate values of $\Delta\varepsilon_x^P$, $\Delta\varepsilon_y^P$, $\Delta\varepsilon_z^P$, etc. The modified or shifted total strains ε'_x, ε'_y, ε'_z, etc., are then computed from the first of equations (9.1.9) and the equivalent total strain from the last of equations (9.1.9). The equivalent plastic strain increment is then determined from ε_{et} and the stress–strain curve using the relation (7.9.16); i.e.,

$$\Delta\varepsilon_p = \frac{\varepsilon_{et} - \frac{2}{3}[(1 + \mu)/E]\sigma_{e,i-1}}{1 + \frac{2}{3}[(1 + \mu)/E](d\sigma_e/d\varepsilon_p)_{i-1}} \tag{9.1.10}$$

where $\sigma_{e,i-1}$ is the value of σ_e at the end of the previous increment of loading, and, if we are dealing with the first increment, $\sigma_{e,i-1}$ is equal to the yield stress. The new values of the plastic strain increments are then computed from equations (9.1.8), and the process is continued until convergence is obtained. A flow diagram of the calculation scheme is shown in Figure 9.1.4.

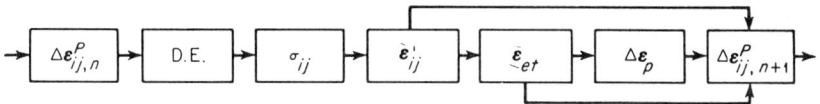

FIGURE 9.1.4 Block diagram for rapidly convergent scheme.

In this case, the values of ε'_x, ε'_y, ε'_z, etc., appearing in equations (9.1.8) are consistent with the value of ε_{et} appearing in these equations. Also, examination of (9.1.10) shows that since the denominator of the right side is approximately unity, a plot of $\Delta\varepsilon_p$ versus ε_{et} will have a slope of approximately unity, so that a small error in ε_{et} will produce the same order-of-magnitude error in $\Delta\varepsilon_p$, without any magnification.

As a result, it has been found from experience that this last method is convergent for even large size increments and converges more rapidly than the first method described when that method converges. The plastic strain–total strain method is therefore generally recommended for use wherever possible.

We shall now present a series of examples of the use of the successive approximation method. These examples are taken from references [4] and [5] and also unpublished results. In all the examples of this chapter the von Mises criterion and the Prandtl–Reuss equations will be used.

9-2 THIN FLAT PLATE

As a first example, consider the uniaxial case of a thin infinite plate of width $2c$ with a temperature distribution $T(y)$ across the width, as shown in Figure 9.2.1. For a thin infinite strip, as shown, the only nonzero stress is

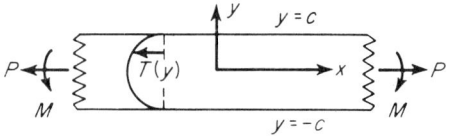

FIGURE 9.2.1 Thin infinite strip with temperature distribution across the width.

assumed to be $\sigma_x(y)$ and the stress–strain relation can be written

$$\varepsilon_x = \frac{1}{E}\sigma_x + \alpha T + \varepsilon_x^P \qquad (9.2.1)$$

Introduce the dimensionless quantities

$$S = \frac{\sigma_x}{\sigma_0} \quad \epsilon_x = \frac{\varepsilon_x}{\varepsilon_0} \quad \tau = \frac{\alpha T}{\varepsilon_0} \quad \epsilon_x^P = \frac{\varepsilon_x^P}{\varepsilon_0} \quad \eta = \frac{y}{c} \quad H = \frac{E}{E_0} \qquad (9.2.2)$$

where σ_0 is the yield stress at some reference temperatures T_0, $\varepsilon_0 = \sigma_0/E_0$ the yield strain at the reference temperature, and E_0 the modulus of elasticity at the reference temperature. Note that the modulus and yield stress may be functions of temperature. Equation (9.2.1) may now be written

$$\epsilon_x = \frac{S}{H} + \tau + \epsilon_p \qquad (9.2.3)$$

where $\epsilon_x^P = \epsilon_p$ for the uniaxial case under consideration, with ϵ_p taken positive for tension and negative for compression.

The assumed stress distribution, i.e., $\sigma_x = \sigma_x(y)$, all other stresses being zero, satisfies the equilibrium equations identically. Also, all the compatibility equations are identically satisfied, except one, i.e.,

$$\frac{\partial^2 \epsilon_x}{\partial \eta^2} = 0 \qquad (9.2.4)$$

which gives

$$\epsilon_x = a + b\eta \qquad (9.2.5)$$

Sec. 9-2] Thin Flat Plate

Equation (9.2.5) expresses the usual assumption of simple beam theory—that plane sections remain plane.

To determine the constants a and b, the boundary conditions are used. If it is assumed that at the ends of the strip, an axial force P and a bending moment M are applied, then it follows that for equilibrium

$$\int_{-1}^{1} S \, d\eta = \frac{P}{\sigma_0 c h} \equiv P^*$$

$$\int_{-1}^{1} S\eta \, d\eta = \frac{M}{\sigma_0 c^2 h} \equiv M^*$$

(9.2.6)

where h is the thickness of the plate. Combining equations (9.2.5), (9.2.3), and (9.2.6) enables one to obtain the constants a and b:

$$a = A_1 \int_{-1}^{1} H(\tau + \epsilon_p) d\eta - A_2 \int_{-1}^{1} H(\tau + \epsilon_p) \eta \, d\eta + A_1 P^* - A_2 M^*$$

$$b = -A_2 \int_{-1}^{1} H(\tau + \epsilon_p) d\eta + A_3 \int_{-1}^{1} H(\tau + \epsilon_p) \eta \, d\eta - A_2 P^* + A_3 M^*$$

(9.2.7)

where

$$A_1 = \frac{\int_{-1}^{1} H\eta^2 \, d\eta}{\int_{-1}^{1} H \, d\eta \int_{-1}^{1} H\eta^2 \, d\eta - \left(\int_{-1}^{1} H\eta \, d\eta\right)^2}$$

$$A_2 = \frac{\int_{-1}^{1} H\eta \, d\eta}{\int_{-1}^{1} H \, d\eta \int_{-1}^{1} H\eta^2 \, d\eta - \left(\int_{-1}^{1} H\eta \, d\eta\right)^2}$$

(9.2.8)

$$A_3 = \frac{\int_{-1}^{1} H \, d\eta}{\int_{-1}^{1} H \, d\eta \int_{-1}^{1} H\eta^2 \, d\eta - \left(\int_{-1}^{1} H\eta \, d\eta\right)^2}$$

Substitution of (9.2.7) into (9.2.5) now gives

$$\epsilon_x = (A_1 - A_2 \eta) \left[\int_{-1}^{1} H(\tau + \epsilon_p) d\eta + P^*\right]$$

$$- (A_2 - A_3 \eta) \left[\int_{-1}^{1} H(\tau + \epsilon_p) \eta \, d\eta + M^*\right]$$

(9.2.9)

or, for the mechanical strain, defined by $\epsilon \equiv \epsilon_x - \tau$,

$$\epsilon = (A_1 - A_2\eta) \left(\int_{-1}^{1} H\tau \, d\eta + P^* \right)$$

$$- (A_2 - A_3\eta) \left(\int_{-1}^{1} H\tau\eta \, d\eta + M^* \right)$$

$$+ (A_1 - A_2\eta) \int_{-1}^{1} H\epsilon_p \, d\eta - (A_2 - A_3\eta) \int_{-1}^{1} H\epsilon_p\eta \, d\eta$$

$$- \tau \tag{9.2.10}$$

If E if a constant, independent of temperature, then $H = 1$, $A_1 = \frac{1}{2}$, $A_2 = 0$, $A_3 = \frac{3}{2}$, and

$$\epsilon = -\tau + \frac{1}{2} \left(\int_{-1}^{1} \tau \, d\eta + P^* \right) + \frac{3}{2} \eta \left(\int_{-1}^{1} \tau\eta \, d\eta + M^* \right)$$

$$+ \frac{1}{2} \int_{-1}^{1} \epsilon_p \, d\eta + \frac{3}{2} \eta \int_{-1}^{1} \epsilon_p\eta \, d\eta \tag{9.2.11}$$

It is also to be noted that if the temperature distribution is symmetric (an even function of y), then $\int_{-1}^{1} \tau\eta \, d\eta = 0$. Furthermore, if there are no external forces and moments and the temperature distribution is symmetric, equation (9.2.11) reduces to the simple form

$$\epsilon = -\tau + \int_{0}^{1} \tau \, d\eta + \int_{0}^{1} \epsilon_p \, d\eta \tag{9.2.12}$$

In equations (9.2.10) through (9.2.12) the plastic strain ratio ϵ_p will, in general, be a nonlinear function of the total mechanical strain ratio ϵ. They will be related to each other through the stress–strain curve of the material, as shown in Figure 9.2.2. The mechanical strain ε is the abscissa, the equivalent stress σ_e is the ordinate, and the plastic part of the strain is ε_p. If the stress–strain curve of the material is temperature dependent in the range of temperatures being considered, a different stress–strain curve, similar to Figure 9.2.2, must be used for each point of the plate. This does not appreciably complicate the analysis.

Equations (9.2.10) through (9.2.12) therefore represent integral equations for the solution of the elastoplastic strains during the loading of a thermally stressed plate made of a work-hardening material with temperature-dependent properties. For the general case of nonlinear strain hardening, these equations are nonlinear and can be solved by the successive-approximation method described in Section 9.1, as will be shown. If the strain hardening is linear, these equations are linear Fredholm equations of the second kind and the solution can often be obtained in closed form.

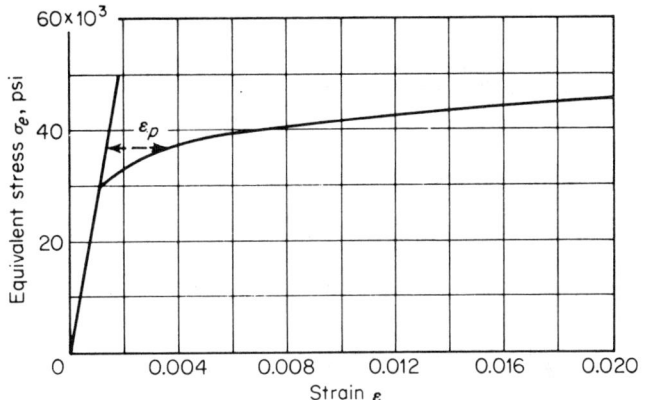

FIGURE 9.2.2 Typical stress–strain curve.

Nonlinear Strain Hardening

The solution for the case of nonlinear strain hardening can readily be obtained by successive approximations as follows:

1. Assume ϵ_p to be zero everywhere.
2. Calculate ϵ from equations (9.2.10), (9.2.11), or (9.2.12). This gives the elastic solution.
3. For each value of ϵ read ϵ_p from the stress–strain curve. A plot of ϵ_p versus ϵ, which can be obtained from the stress–strain curve, would be convenient for this purpose. If the material properties are temperature-dependent, a different curve is used at every station.
4. Using equation (9.2.10), (9.2.11), or (9.2.12) calculate better approximations to ϵ. The integrals appearing in these equations are evaluated numerically.
5. Repeat steps 3 and 4 until convergence is obtained. It will be found that the foregoing process converges fairly rapidly, a half a dozen iterations generally being sufficient for practical purposes.

To illustrate the procedure, consider the following example: Let the temperature be given by

$$T = 600 \, (y^2 - \tfrac{1}{3})$$
$$E = 28 \times 10^6$$
$$c = 1$$
$$\alpha = 9.5 \times 10^{-6}$$
$$\sigma_0 = 28{,}000$$

TABLE 9.2.1 Plastic Strain Calculation for Flat Plate

n	η	ε_p	ε Eq. (9.2.12) $\times 10^{-3}$	ε_p Fig. 9.2.2 $\times 10^{-3}$	ε Eq. (9.2.12) $\times 10^{-3}$	ε_p Fig. 9.2.2 $\times 10^{-3}$	ε Eq. (9.2.12) $\times 10^{-3}$	ε_p Fig. 9.2.2 $\times 10^{-3}$	ε Eq. (9.2.12) $\times 10^{-3}$	ε_p Fig. 9.2.2 $\times 10^{-3}$
1	0	0	1.9	0.7	1.7	0.5	1.6	0.4	1.6	0.4
2	0.1	0	1.8	0.6	1.6	0.4	1.5	0.4	1.5	0.4
3	0.2	0	1.7	0.5	1.5	0.4	1.4	0.3	1.4	0.3
4	0.3	0	1.4	0.3	1.2	0.2	1.1	0.1	1.1	0.1
5	0.4	0	1.0	0	0.8	0	0.7	0	0.7	0
6	0.5	0	0.5	0	0.3	0	0.2	0	0.2	0
7	0.6	0	-0.2	0	-0.4	0	-0.4	0	-0.5	0
8	0.7	0	-0.9	0	-1.1	-0.1	-1.2	-0.2	-1.2	-0.2
9	0.8	0	-1.7	-0.5	-1.9	-0.7	-2.0	-0.8	-2.0	-0.8
10	0.9	0	-2.7	-1.5	-2.9	-1.6	-3.0	-1.7	-3.0	-1.7
11	1.0	0	-3.8	-2.5	-4.0	-2.7	-4.0	-2.7	-4.1	-2.8

$\tau = 5.7(\eta^2 - \tfrac{1}{3})$. Stress–strain curve of Figure 9.2.2.

Sec. 9-2] Thin Flat Plate

and let P and M be zero. Then

$$\tau = 5.7(\eta^2 - \tfrac{1}{3}) \tag{9.2.13}$$

Assume a stress–strain curve independent of temperature, as shown in Figure 9.2.2. We can then use equation (9.2.12), which becomes

$$\epsilon = -5.7(\eta^2 - \tfrac{1}{3}) + \int_0^1 \epsilon_p \, d\eta \tag{9.2.14}$$

By using Figure 9.2.2 and the procedure outlined above, Table 9.2.1 was constructed. Eleven equally spaced stations were used and the integral was evaluated by the trapezoidal rule. If greater accuracy is desired, a greater number of stations and a more accurate integration formula can be used. However, the accuracy of Table 9.2.1 is sufficient for engineering purposes, as can be seen in Figures 9.2.3 and 9.2.4, where the results of Table 9.2.1 as well as the results obtained by using 51 stations and Simpson's rule are

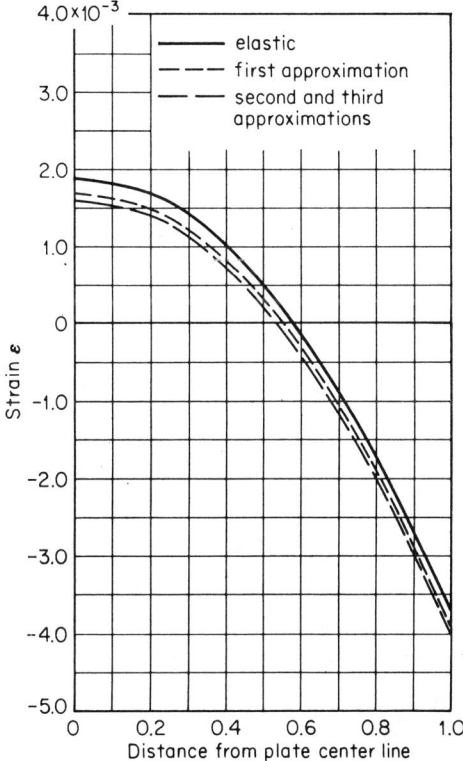

FIGURE 9.2.3 Strain distribution in infinite strip.

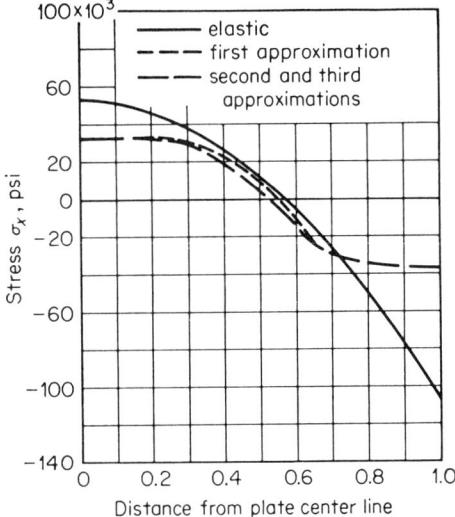

FIGURE 9.2.4 Stress distribution in infinite strip.

plotted. Table 9.2.1 was constructed in approximately $\frac{1}{2}$ hour using a desk calculator. ϵ_p at each station was computed from the equation of the stress–strain curve with ϵ_p, of course, being set equal to zero at those stations where $|\epsilon| \leq 1$.

Linear Strain Hardening

For linear strain hardening the solution can often be obtained in closed form. We shall consider only the simple type of temperature distribution of the previous example. A more general formulation is given in reference [5]. Thus let $\tau = \tau_0(\eta^2 - \frac{1}{3})$. τ_0 will be called the *loading parameter*. For linear strain hardening it can readily be shown that (assuming constant material properties)

$$\epsilon_p = (1 - m)(\epsilon \mp 1) \tag{9.2.15}$$

where m, the strain-hardening parameter, is the ratio of the tangent modulus to the elastic modulus. The minus sign is used when ϵ is positive (tension), and the plus sign when ϵ is negative (compression). Furthermore, since $|\epsilon|$ must be greater than 1 for plastic flow to occur, $\epsilon_p = 0$ if $|\epsilon| \leq 1$. Substituting into (9.2.12) gives

$$\epsilon = -\tau_0(\eta^2 - \tfrac{1}{3}) + (1 - m)\int_0^1 (\epsilon \mp 1)K(\eta, \xi)d\xi \tag{9.2.16}$$

Sec. 9-2] Thin Flat Plate

where

$$K(\eta, \xi) = 0 \qquad |\epsilon| \leq 1$$
$$K(\eta, \xi) = 1 \qquad |\epsilon| > 1$$

If $|\epsilon| < 1$ everywhere, no plastic flow will occur and the first term on the right of (9.2.16) gives the elastic solution. The maximum value of $|\epsilon|$ will obviously occur at $\eta = 1$ and yielding will therefore begin at the edge, $\eta = 1$, when $\epsilon = -1$, i.e., in compression. The value of τ_0 for yielding to begin is then

$$\tau_{0,\text{crit}} = \tfrac{3}{2}$$

If $\tau_0 < \tfrac{3}{2}$, no yielding will occur. As the loading parameter τ_0 is increased beyond $\tfrac{3}{2}$, a compressive plastic zone will spread inward from the edge. The center of the plate, $\eta = 0$, will meanwhile be in a state of tension, the tensile stress increasing as τ_0 increases. Eventually at some value of τ_0 designated by $\tau_{0,c}$ a tensile plastic zone will begin spreading from the center toward the compressive zone spreading from the edge.

Let η_1 be the value of η at the edge of the tensile plastic zone extending from the center, and let η_2 be the value of η at the edge of the compressive zone spreading from the edge; i.e., for $0 < \eta < \eta_1$ there is a tensile plastic zone with $\epsilon \geq 1$, and for $\eta_2 \leq \eta \leq 1$ there is a compressive plastic zone with $\epsilon \leq -1$. As long as $\tau_0 \leq \tau_{0,c}$, $\eta_1 = 0$, and equation (9.2.16) becomes

$$\begin{aligned}\epsilon &= -\tau_0(\eta^2 - \tfrac{1}{3}) + (1-m)\int_{\eta_2}^{1}(\epsilon+1)d\xi \\ &= -\tau_0(\eta^2 - \tfrac{1}{3}) + (1-m)(1-\eta_2) + (1-m)\int_{\eta_2}^{1}\epsilon\,d\xi \qquad \tau_0 \leq \tau_{0,c}\end{aligned}$$
(9.2.17)

Let us now apply the method of successive approximations to equation (9.2.17). Thus let

$$\epsilon^{(1)} = -\tau_0(\eta^2 - \tfrac{1}{3}) + (1-m)(1-\eta_2)$$

$$\begin{aligned}\epsilon^{(2)} &= -\tau_0(\eta^2 - \tfrac{1}{3}) + (1-m)(1-\eta_2) + (1-m)\int_{\eta_2}^{1}\epsilon^{(1)}\,d\xi \\ &= -\tau_0(\eta^2 - \tfrac{1}{3}) + (1-m)(1-\eta_2) + (1-m) \\ &\quad \times \left[\frac{\tau_0}{3}(\eta_2^3 - \eta_2) + (1-m)(1-\eta_2)^2\right]\end{aligned}$$

$$\begin{aligned}\epsilon^{(3)} &= -\tau_0(\eta^2 - \tfrac{1}{3}) + (1-m)(1-\eta_2) + (1-m)^2(1-\eta_2)^2 \\ &\quad + (1-m)^3(1-\eta_2)^3 + (1-m)\frac{\tau_0}{3}(\eta_2^3 - \eta_2) \\ &\quad \times [1 + (1-m)(1-\eta_2)]\end{aligned}$$

$$\epsilon^{(4)} = -\tau_0(\eta^2 - \tfrac{1}{3}) + (1-m)(1-\eta_2) + (1-m)^2(1-\eta_2)^2$$
$$+ (1-m)^3(1-\eta_2)^3 + (1-m)^4(1-\eta_2)^4$$
$$+ (1-m)\frac{\tau_0}{3}(\eta_2^3 - \eta_2)[1 + (1-m)(1-\eta_2)$$
$$+ (1-m)^2(1-\eta_2)^2]$$

or

$$\epsilon^{(n+1)} = -\tau_0(\eta^2 - \tfrac{1}{3}) + (1-m)(1-\eta_2)\sum_{i=0}^{n}(1-m)^i(1-\eta_2)^i$$
$$+ (1-m)\frac{\tau_0}{3}(\eta_2^3 - \eta_2)\sum_{i=0}^{n-1}(1-m)^i(1-\eta_2)^i$$

and

$$\lim_{n\to\infty}\epsilon^{(n+1)} = \epsilon = -\tau_0(\eta^2 - \tfrac{1}{3}) + (1-m)\frac{1-\eta_2 + (\tau_0/3)(\eta_2^3 - \eta_2)}{1 - (1-m)(1-\eta_2)}$$
$$\tau_0 \leq \tau_{0,c} \qquad (9.2.18)$$

For $\tau_0 > \tau_{0,c}$, a tensile plastic zone will extend from $\eta = 0$ to $\eta = \eta_1$. Equation (9.2.16) now becomes

$$\epsilon = -\tau_0(\eta^2 - \tfrac{1}{3}) + (1-m)\int_0^{\eta_1}(\epsilon - 1)d\xi + (1-m)\int_{\eta_2}^{1}(\epsilon + 1)d\xi$$

or $\epsilon = -\tau_0(\eta^2 - \tfrac{1}{3}) + (1-m)(1-\eta_2-\eta_1) + (1-m)\left(\int_0^{\eta_1} + \int_{\eta_2}^{1}\epsilon\, d\xi\right)$

(9.2.19)

and a similar successive approximation as above leads to

$$\epsilon = -\tau_0(\eta^2 - \tfrac{1}{3}) + (1-m)\frac{1 - \eta_1 - \eta_2 + \tfrac{1}{3}\tau_0(\eta_2^3 - \eta_2 - \eta_1^3 + \eta_1)}{1 - (1-m)(1 + \eta_1 - \eta_2)}$$

(9.2.20)

The above derivations of equations (9.2.18) and (9.2.20) are actually not necessary, since the integral equations (9.2.19) and (9.2.17) are Fredholm equations of the second kind with degenerate kernels and the solutions can therefore be obtained directly by standard methods (see, for example, reference [6]). However, the successive approximation method was used in obtaining the solution to show how this method leads to the exact solution for this more complicated case.

Equations (9.2.18) and (9.2.20) give the complete solution to the problem if η_1, η_2, and $\tau_{0,c}$ were known. These can be determined as follows. If $\tau_0 \leq \tau_{0,c}$, there is one region of plastic flow extending from η_2 to 1. At

Sec. 9-2] Thin Flat Plate

$\eta = \eta_2$, ϵ must equal -1. Therefore, substituting $\eta = \eta_2$, $\epsilon = -1$, into equation (9.2.18) a relationship is obtained between τ_0, η_2, and m:

$$\tfrac{2}{3}(1 - m)\eta_2^3 + m(\eta_2^2 - \tfrac{1}{3}) = \frac{1}{\tau_0} \tag{9.2.21}$$

Thus for given values of m and τ_0, η_2 can be determined or, more simply, for given values of m and η_2, τ_0 can be calculated directly. For a perfectly plastic material, $m = 0$, and

$$\eta_2 = \left(\frac{3}{2\tau_0}\right)^{1/3} \tag{9.2.22}$$

Once η_2 is known, the complete strain distribution is obtained from (9.2.18), the plastic strains and the stresses from (9.2.15) and (9.2.3).

If τ_0 is greater than $\tau_{0,c}$, a tensile plastic zone will spread from the center to η_1. To determine the values of η_1 and η_2 in this case, one first sets $\eta = \eta_1$ and $\epsilon = 1$ in (9.2.20) and then $\eta = \eta_2$ and $\epsilon = -1$ in (9.2.20). This results in the following two simultaneous equations for determining η_1 and η_2:

$$\begin{aligned} 1 &= -\tau_0(\eta_1^2 + \tfrac{1}{3}) + C \\ -1 &= -\tau_0(\eta_2^2 + \tfrac{1}{3}) + C \end{aligned} \tag{9.2.23}$$

where

$$C = (1 - m)\frac{1 - \eta_1 - \eta_2 + \tfrac{1}{3}\tau_0(\eta_2^3 - \eta_2 - \eta_1^3 + \eta_1)}{1 - (1 - m)(1 + \eta_1 - \eta_2)}$$

If the second equation is subtracted from the first, there results

$$2 = \tau_0(\eta_2^2 - \eta_1^2) \tag{9.2.24}$$

The value of $\tau_{0,c}$ at which the center begins to flow plastically can now be found. For this condition, $\eta_1 = 0$ and (9.2.24) gives

$$\eta_2 = \sqrt{\frac{2}{\tau_{0,c}}} \tag{9.2.25}$$

Substituting this value for η_2 together with $\eta_1 = 0$ into either of equations (9.2.23) results in

$$\tau_{0,c}^3 + 6\frac{1 - 2m}{m}\tau_{0,c} + 9\left(\frac{1 - 2m}{m}\right)^2 \tau_{0,c} - 32\left(\frac{1 - m}{m}\right)^2 = 0 \tag{9.2.26}$$

Only one of the roots of (9.2.26) will be physically meaningful. For a perfectly plastic material, $m = 0$, and

$$\tau_{0,c} = \frac{32}{9} = 3.555 \tag{9.2.27}$$

Figure 9.2.5 shows the variation of $\tau_{0,c}$ with m. The value of $\tau_{0,c}$ is rather insensitive to the strain-hardening parameter m.

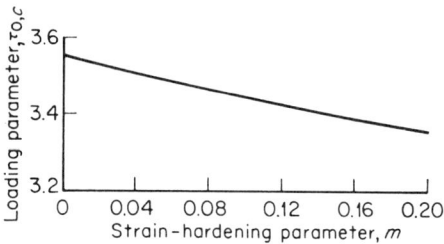

FIGURE 9.2.5 Variation of $\tau_{0,c}$ with m.

If τ_0 is greater than $\tau_{0,c}$, two regions of plastic flow will exist. The corresponding values of η_1 and η_2 are found by solving simultaneously (9.2.24) and one of (9.2.23). Thus eliminating η_1 from the first of (9.2.23) by means of (9.2.24) results in the following quintic for η_2:

$$\eta_2^5 + \left(\frac{3m}{4(1-m)} + 2\frac{1-m}{m\tau_0}\right)\eta_2^4 + \frac{1}{3m}\left(m + \frac{3}{\tau_0}\right)\eta_2^3$$
$$- \left[\frac{m + 3/\tau_0}{2(1-m)} + \frac{4(1-m)}{m\tau_0^2}\right]\eta_2^2 + \frac{(m + 3/\tau_0)^2 + 32(1-m)^2/\tau_0^3}{12m(1-m)} = 0$$
(9.2.28)

A plot of η_1 and η_2 showing the growth of the plastic regions is shown in Figure 9.2.6.

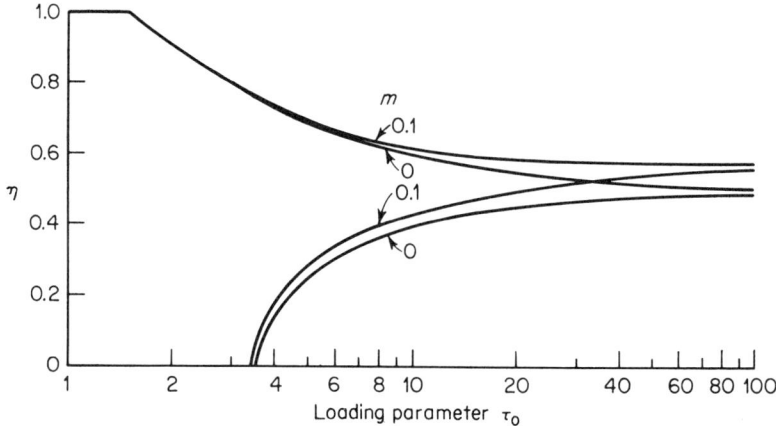

FIGURE 9.2.6 Growth of plastic region with loading parameter τ_0.

9-3 THIN CIRCULAR SHELL

The next problem we shall consider is that of a thin circular shell with an axial temperature gradient. We shall use the deformation theory of plasticity for this problem and assume a one-step thermal load, although this is not essential.

The equilibrium equations for the shell are given in reference [7]. These are

$$N_x = 0$$
$$\frac{R}{l^2}\frac{d^2 M_x}{dx^2} + N_\theta = 0 \qquad (9.3.1)$$

where

$$N_x = \int_{-h/2}^{h/2} \sigma_x \, dz$$
$$N_\theta = \int_{-h/2}^{h/2} \sigma_\theta \, dz \qquad (9.3.2)$$
$$M_x = \int_{-h/2}^{h/2} \sigma_x z \, dz$$

where R is the mean radius of the shell and l is the characteristic length, equal to

$$l = \sqrt[4]{\frac{R^2 h^2}{3(1 - \mu^2)}} \qquad (9.3.3)$$

h is the thickness, x is the dimensionless axial coordinate, the actual coordinate divided by l, and z is the radial coordinate measured from the middle surface, positive inward.

The stress–strain relations are

$$\sigma_x = \frac{E}{1 - \mu^2}[\varepsilon_x - \alpha T - \varepsilon_x^P + \mu(\varepsilon_\theta - \alpha T - \varepsilon_\theta^P)]$$
$$\sigma_\theta = \frac{E}{1 - \mu^2}[\varepsilon_\theta - \alpha T - \varepsilon_\theta^P + \mu(\varepsilon_x - \alpha T - \varepsilon_x^P)] \qquad (9.3.4)$$

and σ_z is assumed to be zero for a thin shell.

The strain-displacement relations are

$$\varepsilon_x = \frac{1}{l}\frac{du}{dx} - \frac{d^2 w}{l^2 \, dx^2} z$$
$$\varepsilon_\theta = -\frac{w}{R} \qquad (9.3.5)$$

where u is the axial displacement of a point on the middle surface of the shell and w is the radial displacement, positive inward. Substituting (9.3.5) into (9.3.4) gives

$$\sigma_x = \frac{E}{1-\mu^2}\left[\frac{du}{l\,dx} - \frac{d^2w}{l^2\,dx^2}z - \alpha T - \varepsilon_x^P + \mu\left(-\frac{w}{R} - \varepsilon_\theta^P - \alpha T\right)\right]$$
$$\sigma_\theta = \frac{E}{1-\mu^2}\left[-\frac{w}{R} - \alpha T - \varepsilon_\theta^P + \mu\left(\frac{du}{l\,dx} - \frac{d^2w}{l^2\,dx^2}z - \alpha T - \varepsilon_x^P\right)\right]$$
(9.3.6)

From the first of (9.3.1), using (9.3.2),

$$\frac{du}{l\,dx} = \mu\frac{w}{R} + (1+\mu)\alpha T + \frac{1}{h}\int_{-h/2}^{h/2}(\varepsilon_x^P + \mu\varepsilon_\theta^P)\,dz \qquad (9.3.7)$$

and from (9.3.2), making use of (9.3.7),

$$N_\theta = -Eh\left(\frac{w}{R} + \frac{1}{h}\int_{-h/2}^{h/2}\varepsilon_\theta^P\,dz + \alpha T\right)$$
$$M_x = -\frac{Eh^3}{12(1-\mu^2)}\frac{d^2w}{l^2\,dx^2} - \frac{E}{1-\mu^2}\int_{-h/2}^{h/2}(\varepsilon_x^P + \mu\varepsilon_\theta^P)z\,dz$$
(9.3.8)

Also eliminating du/dx from (9.3.5) by means of (9.3.7) gives

$$\varepsilon_x = \mu\frac{w}{R} + (1+\mu)\alpha T + \frac{1}{h}\int_{-h/2}^{h/2}(\varepsilon_x^P + \mu\varepsilon_\theta^P)\,dz - \frac{1}{l^2}\frac{d^2w}{dx^2}z \qquad (9.3.9)$$

and, since $\sigma_z = 0$, the third stress–strain relation gives

$$\varepsilon_z = -\frac{\mu}{E}(\sigma_x + \sigma_\theta) + \varepsilon_z^P + \alpha T$$

and using (9.3.4) results in

$$\varepsilon_z = -\frac{\mu}{1-\mu}(\varepsilon_x + \varepsilon_\theta) + \frac{1+\mu}{1-\mu}\alpha T - \frac{1-2\mu}{1-\mu}(\varepsilon_x^P + \varepsilon_\theta^P) \qquad (9.3.10)$$

Finally from the second of (9.3.1), making use of (9.3.8), there results

$$\frac{d^4w}{dx^4} + 4w = -4R\alpha T - \frac{d^2P}{dx^2} - Q \qquad (9.3.11)$$

where
$$P = \frac{12l^2}{h^3} \int_{-h/2}^{h/2} (\varepsilon_x^P + \mu\varepsilon_\theta^P) z \, dz$$

$$Q = \frac{4R}{h} \int_{-h/2}^{h/2} \varepsilon_\theta^P \, dz \qquad (9.3.12)$$

Equation (9.3.11) is recognizable as the equation for the deflections of a beam on an elastic foundation, the right side representing the load. If P and Q are zero, the problem is elastic. If P and Q are not zero, the problem is an elastoplastic problem which can be solved by successive approximations.

We note first that the elastic solution can be obtained in closed form. The homogeneous solution can be written

$$w_H = C_1 \cos x \cosh x + C_2 \cos x \sinh x$$
$$+ C_3 \sin x \cosh x + C_4 \sin x \sinh x \qquad (9.3.13)$$

A particular solution can be obtained by the use of the Green's function. Without going into the details, it can be verified by differentiation that a particular solution is given by

$$w_P = -4R \int_0^x \alpha T(\xi) G(x - \xi) d\xi - \int_0^x P(\xi) \frac{d^2 G(x - \xi)}{dx^2} d\xi$$
$$- \int_0^x Q(\xi) G(x - \xi) d\xi \qquad (9.3.14)$$

where $G(x - \xi)$, the Green's function, is

$$G(x - \xi) = \tfrac{1}{4}[\sin(x - \xi) \cosh(x - \xi) - \cos(x - \xi) \sinh(x - \xi)] \qquad (9.3.15)$$

The complete solution is then

$$w = w_H + w_P \qquad (9.3.16)$$

where the constants C_1 to C_4 are determined from the boundary conditions.

The method of solution will be illustrated for a single step load using the plastic strain–total strain relations; i.e.,

$$\varepsilon_x^P = \frac{\varepsilon_p}{3\varepsilon_{et}} (2\varepsilon_x - \varepsilon_\theta - \varepsilon_z)$$

$$\varepsilon_\theta^P = \frac{\varepsilon_p}{3\varepsilon_{et}} (2\varepsilon_\theta - \varepsilon_x - \varepsilon_z) \qquad (9.3.17)$$

$$\varepsilon_{et} = \frac{\sqrt{2}}{3} [(\varepsilon_x - \varepsilon_\theta)^2 + (\varepsilon_\theta - \varepsilon_z)^2 + (\varepsilon_z - \varepsilon_x)^2]^{1/2}$$

To facilitate the solution, the stress-strain curve can be cross-plotted to obtain a strain-strain curve as shown in Figure 9.3.1.

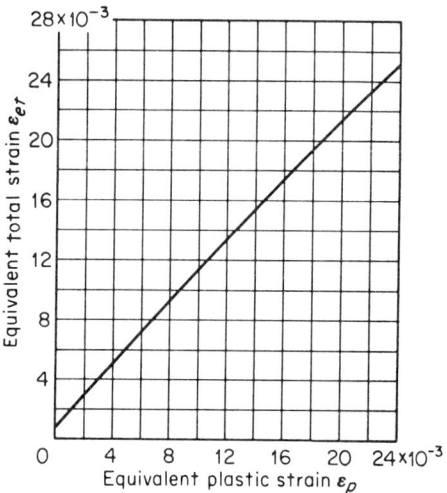

FIGURE 9.3.1   Equivalent total strain versus equivalent plastic strain.

The solution now proceeds as follows. Assume as a zeroth approximation that ε_x^p and ε_θ^p are zero everywhere. w can then be computed from (9.3.13) through (9.3.16), and the total strains ε_x, ε_θ, and ε_z from (9.3.5), (9.3.9), and (9.3.10). ε_{et} is computed from the last of (9.3.17) and ε_p determined from Figure 9.3.1. First approximations for the plastic strains are now computed from the first two equations of (9.3.17). P and Q are then computed from (9.3.12) and w from (9.3.13) through (9.3.16). The process is repeated until convergence is obtained. For every iteration, new values of C_1, C_2, C_3, and C_4 must be determined from the boundary conditions. Once convergence has been obtained, the stresses can be computed from (9.3.4).

As a specific example consider a thin circular shell with the following geometric and physical properties:

$$L = 48 \text{ in.} \quad R = 12 \text{ in.} \quad h = 2 \text{ in.} \quad \mu = 0.3 \quad E = 28 \times 10^6$$
$$\alpha = 9.5 \times 10^{-6}$$

Then from (9.3.3), $l = 3.81$ in. and $0 \le x \le 12.6$. Assume a temperature distribution given by $T = 2.21x^2$ (corresponding to a 350°F rise from one end of the shell to the other end), and boundary conditions (fixed ends)

$$w(0) = w'(0) = w(12.6) = w'(12.6) = 0 \qquad (9.3.18)$$

The functions P and Q become

$$P = 21.8 \int_{-1}^{1} (\varepsilon_x^P + 0.3\varepsilon_\theta^P)z\, dz$$

$$Q = 24 \int_{-1}^{1} \varepsilon_\theta^P\, dz \qquad (9.3.19)$$

From the first two boundary conditions,

$$C_1 = 0 \qquad C_2 = -C_3$$

Also

$$4R \int_0^x \alpha T(\xi)G(x - \xi)d\xi = 0.000252x^2 + 0.000252 \sin x \sinh x$$

and the equation for w becomes

$$\begin{aligned} w = &\, C_2 (\cos x \sinh x - \sin x \cosh x) \\ &+ C_4 \sin x \sinh x \\ &- 0.000252x^2 - I_1(x) - I_2(x) \end{aligned} \qquad (9.3.20)$$

where

$$I_1(x) = \int_0^x Q(\xi)G(x - \xi)d\xi$$

$$I_2(x) = \int_0^x P(\xi)\frac{d^2G(x - \xi)}{dx^2}d\xi \qquad (9.3.21)$$

Twenty-one stations were taken along the length of the shell and six stations through the thickness. For the zeroth approximation, it is assumed that ε_x^P and ε_θ^P and hence I_1 and I_2 are zero. The function $w(x)$ is calculated from (9.3.20) with the constants C_2 and C_4 determined from the last two boundary conditions of equation (9.3.18). The strains ε_x, ε_θ, and ε_z are then computed from equations (9.3.5), (9.3.9), and (9.3.10). First approximations to ε_x^P and ε_θ^P are obtained by computing ε_{et} from the third of equations (9.3.17), reading ε_p from Figure 9.3.1 and calculating ε_x^P and ε_θ^P from the first two of equations (9.3.17). With these values of ε_x^P and ε_θ^P, P and Q are computed from equations (9.3.19), I_1 and I_2 from (9.3.21), and w from (9.3.20) with new values computed for C_2 and C_4. New values are then computed for ε_x, ε_θ, and ε_z and the process continued until convergence is obtained.

Wherever derivatives of w are needed, the following relations are useful:

$$\frac{dI_1}{dx} = \int_0^x Q(\xi) \frac{dG(x - \xi)}{dx} d\xi$$

$$\frac{dI_2}{dx} = \int_0^x P(\xi) \frac{d^3G(x - \xi)}{dx^3} d\xi$$

$$\frac{d^2I_1}{dx^2} = \int_0^x Q(\xi) \frac{d^2G(x - \xi)}{dx^2} d\xi \qquad (9.3.22)$$

$$\frac{d^2I_2}{dx^2} = P(x) - 4\int_0^x P(\xi)G(x - \xi)d\xi$$

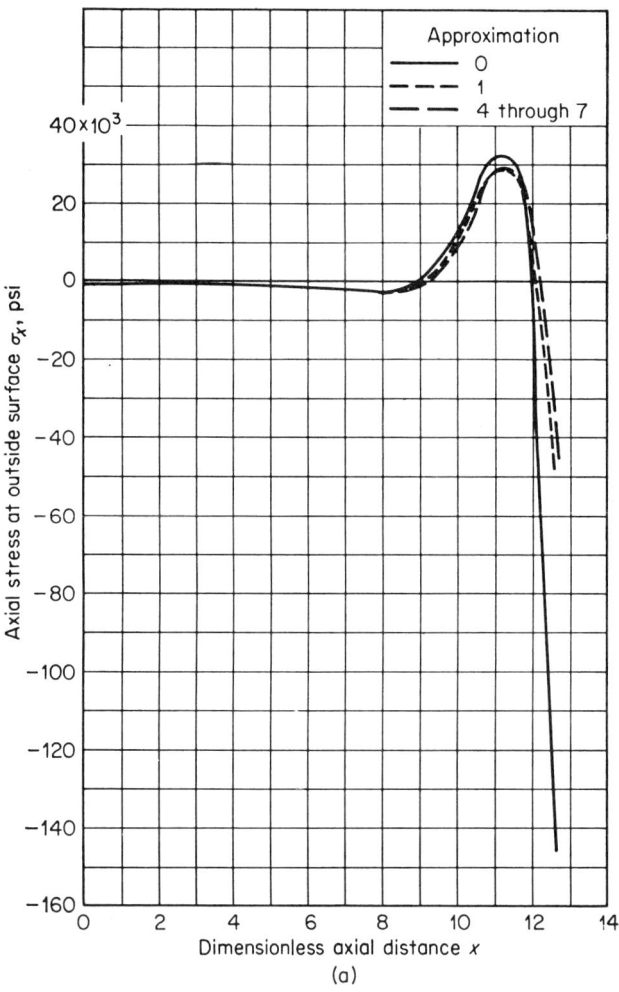

FIGURE 9.3.2 Variations of stresses and strains in shell.

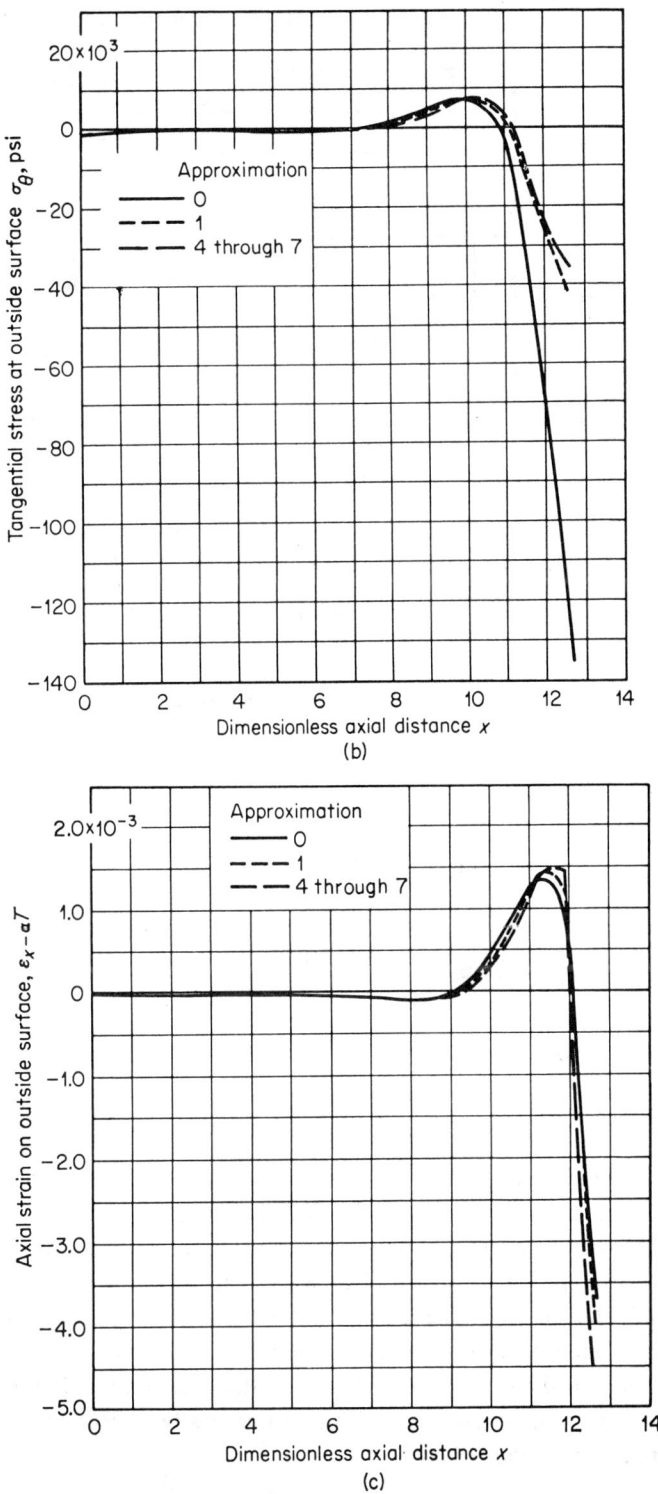

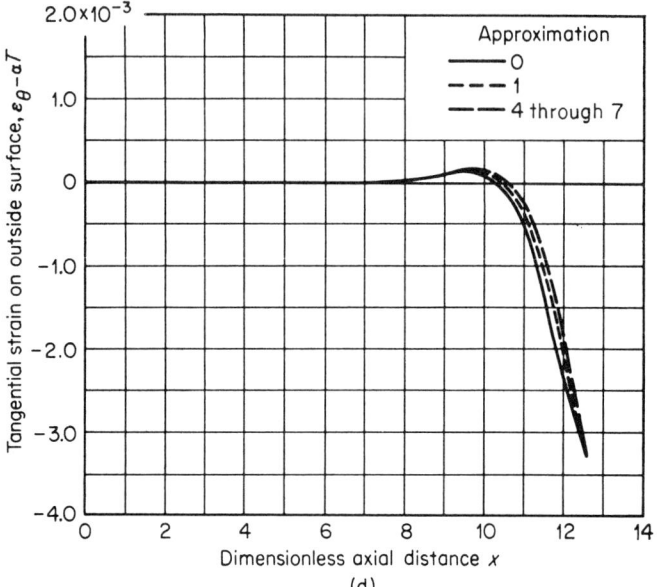

FIGURE 9.3.2 (*continued*)

and

$$\frac{dG}{dx} = \tfrac{1}{2} \sin(x - \xi) \sinh(x - \xi)$$

$$\frac{d^2G}{dx^2} = \tfrac{1}{2} [\sin(x - \xi) \cosh(x - \xi) - \cos(x - \xi) \sinh(x - \xi) \quad (9.3.23)$$

$$\frac{d^3G}{dx^3} = \cos(x - \xi) \cosh(x - \xi)$$

All the integrals in the previous equations were evaluated using the trapezoidal rule. The results for this problem are shown in Figure 9.3.2. An abbreviated calculation setup for one of the successive approximations at the last two stations is given in Table 9.3.1. As can be seen from the figures, the differences between the seventh and the fourth approximations are very small. From an engineering viewpoint the first approximation is actually sufficient. This is probably due to the fact that the total strains do not change much, because of the plastic flow. This will generally be true when the only loads are thermal loads.

Sec. 9-2] Thin Circular Shell

TABLE 9.3.1 Calculation of Plastic Strains in Thin Shells

Station	x	z	αT	ε_x^P prev. approx.	ε_θ^P prev. approx.	I_1 prev. approx.	I_2 prev. approx.	w (9.3.20)
				$\times 10^{-3}$	$\times 10^{-3}$	$\times 10^{-6}$	$\times 10^{-6}$	
20	11.97	-1.0	-3.008	0.550	-0.800	-24.14	-581.1	-0.01372
		-0.6667		0.480	-0.770			
		-0.3333		0.420	-0.740			
		0		0.365	-0.725			
		0.3333		0.310	-0.710			
		0.6667		0.260	-0.700			
		1.0		0.215	-0.695			
21	12.60	-1.0	-3.333	-3.20	-2.500	-1113	-1932	0
		-0.6667		-1.57	-2.270			
		-0.3333		-0.175	-2.075			
		0		0.940	-2.030			
		0.3333		2.170	-2.300			
		0.6667		3.800	-2.750			
		1.0		5.500	-2.850			

Station	$\varepsilon_x - \alpha T$ (9.3.9)	$\varepsilon_\theta - \alpha T$ (9.3.5)	ε_x^P (9.3.17)	ε_θ^P (9.3.17)	Q (9.3.19)	P	I_1 (9.3.21)	I_2	w (9.3.20)
	$\times 10^{-3}$	$\times 10^{-3}$	$\times 10^{-3}$	$\times 10^{-3}$			$\times 10^{-6}$	$\times 10^{-6}$	
20	1.314	-1.865	0.600	-0.780	-0.03295	-0.003013	-22.56	-545.6	-0.01414
	1.110		0.505	-0.745					
	0.9100		0.420	-0.705					
	0.7089		0.340	-0.670					
	0.5100		0.270	-0.650					
	0.3100		0.200	-0.640					
	0.1000		0.135	-0.650					
21	-4.414	-3.333	-3.350	-2.500	-0.1130	0.06130	-1043	-2195	0
	-2.490		-1.625	-2.280					
	-0.5700		-0.175	-2.075					
	1.349		0.985	-2.030					
	3.270		2.285	-2.325					
	5.190		4.000	-2.750					
	7.110		5.800	-2.850					

Plastic Strain Charts

Instead of using equations (9.3.17) and the strain-strain curve of Figure 9.3.1, it is sometimes more convenient to use a plastic strain chart, as shown in Figure 9.3.3. The ordinate and abscissa are the total mechanical strains, and the curves shown are two intersecting families representing curves of constant ε_θ^P and constant ε_x^P, respectively. Thus for a given pair of values of ε_x and ε_θ, the values of ε_x^P and ε_θ^P can be read directly from this chart. Interpolation between curves may be necessary. This procedure avoids the

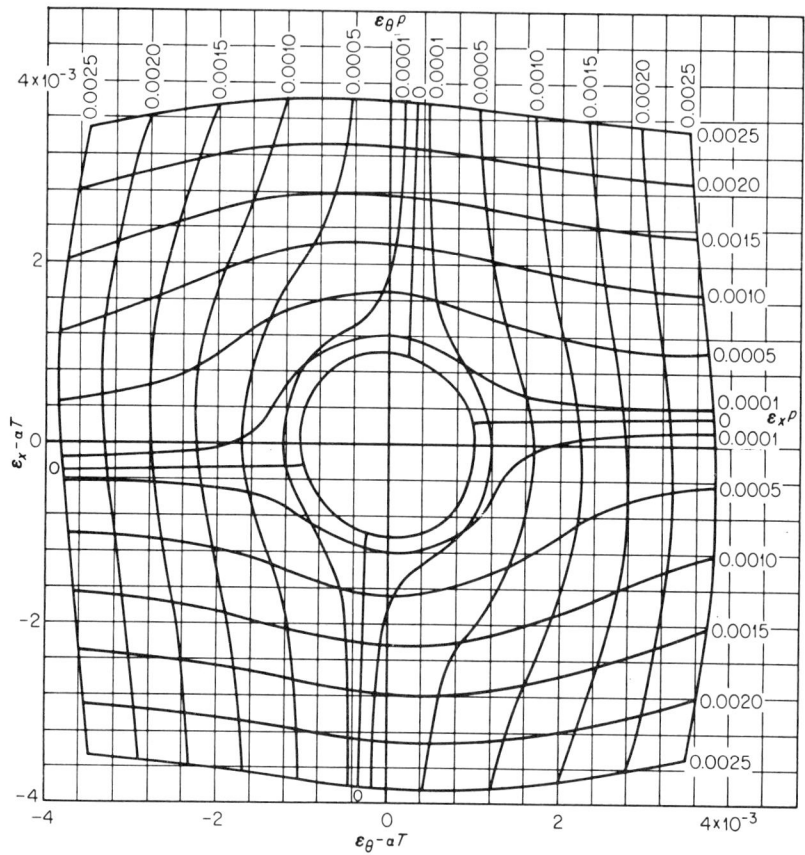

FIGURE 9.3.3 Plastic strain chart.

necessity of computing ε_z and ε_{et}, reading ε_p from the strain–strain curve (Figure 9.3.1), and finally computing ε_x^P and ε_θ^P. However, constructing such a chart involves a considerable amount of labor, and a different chart must be constructed for each stress–strain curve used. However, if many similar computations are to be made using the same stress–strain curve, such a chart can be very time saving, particularly if a digital computer is not being used.

To obtain this chart, the stress–strain relations with $\sigma_z = 0$ are written

$$\varepsilon_x - \alpha T = \frac{1}{E}(\sigma_x - \mu\sigma_\theta) + \varepsilon_x^P$$

$$\varepsilon_\theta - \alpha T = \frac{1}{E}(\sigma_\theta - \mu\sigma_x) + \varepsilon_\theta^P$$

(9.3.24)

Sec. 9–4] Long Solid Cylinder

Also, the Prandtl–Reuss relations solved for the stresses are

$$\sigma_x = \frac{2}{3} \frac{\sigma_e}{\varepsilon_p} (2\varepsilon_x^P + \varepsilon_\theta^P)$$

$$\sigma_\theta = \frac{2}{3} \frac{\sigma_e}{\varepsilon_p} (2\varepsilon_\theta^P + \varepsilon_x^P)$$

(9.3.25)

Substituting (9.3.25) into (9.3.24) gives

$$\varepsilon_x - \alpha T = \left(1 + \frac{2(2-\mu)}{3E} \frac{\sigma_e}{\varepsilon_p}\right) \varepsilon_x^P + \frac{2(1-2\mu)}{3E} \frac{\sigma_e}{\varepsilon_p} \varepsilon_\theta^P$$

$$\varepsilon_\theta - \alpha T = \left(1 + \frac{2(2-\mu)}{3E} \frac{\sigma_e}{\varepsilon_p}\right) \varepsilon_\theta^P + \frac{2(1-2\mu)}{3E} \frac{\sigma_e}{\varepsilon_p} \varepsilon_x^P$$

(9.3.26)

With the above equations and a given stress–strain curve, a two-parameter family of curves can be plotted giving the total strains for any pair of plastic strains ε_x^P and ε_θ^P. Thus

1. An arbitrary convenient value is chosen for ε_θ^P.
2. A series of values are chosen for ε_x^P. For each of these values, (a) ε_p is computed from the first of (9.1.7) (without the Δ's), (b) σ_e is then read from the stress–strain curve, and (c) $\varepsilon_x - \alpha T$ and $\varepsilon_\theta - \alpha T$ are computed from (9.3.26). This gives one curve of the family.
3. To obtain the other curves, the process is repeated for new values of ε_θ^P. The limiting curve of zero plastic strain is an ellipse about the origin as shown. Any point inside this ellipse corresponds to zero plastic strain.

9–4 LONG SOLID CYLINDER

Several problems involving rotational symmetry will next be considered. We start with the problem of a long solid circular cylinder with a radial temperature distribution. The cylinder, having a radius R, is assumed in a state of generalized plane strain; i.e., the axial strain ε_z is assumed to be a constant. The ends of the cylinder are assumed to be unloaded; hence the solution will not be valid close to the ends.

The equilibrium, compatibility, and stress–strain relations for this problem can be written

$$\frac{d\sigma_r}{dr} + \frac{\sigma_r - \sigma_\theta}{r} = 0 \qquad (9.4.1)$$

$$\frac{d\varepsilon_\theta}{dr} + \frac{\varepsilon_\theta - \varepsilon_r}{r} = 0 \qquad (9.4.2)$$

194 The Method of Successive Elastic Solutions [Ch. 9

$$\sigma_r = \lambda(\theta - 3\alpha T) + 2G(\varepsilon_r - \alpha T - \varepsilon_r^P)$$
$$\sigma_\theta = \lambda(\theta - 3\alpha T) + 2G(\varepsilon_\theta - \alpha T - \varepsilon_r^P) \tag{9.4.3}$$
$$\sigma_z = \lambda(\theta - 3\alpha T) + 2G(\varepsilon_z - \alpha T + \varepsilon_r^P + \varepsilon_\theta^P)$$

where

$$\lambda = \frac{\mu E}{(1+\mu)(1-2\mu)} \qquad G = \frac{E}{2(1+\mu)} \qquad \theta = \varepsilon_r + \varepsilon_\theta + \varepsilon_z$$

To save writing, ε_r^P and ε_θ^P are here defined as the total accumulated plastic strains, including those for the current increment of load; i.e.,

$$\varepsilon_r^P = \sum_{k=1}^{i-1} \Delta\varepsilon_{r,k}^P + \Delta\varepsilon_{r,i}^P \tag{9.4.3a}$$

$$\varepsilon_\theta^P = \sum_{k=1}^{i-1} \Delta\varepsilon_{\theta,k}^P + \Delta\varepsilon_{\theta,i}^P$$

where the summations are assumed to have already been computed and $\Delta\varepsilon_{\theta,i}^P$ and $\Delta\varepsilon_{r,i}^P$ are to be determined. Assuming E, μ, and ε_z constant, and substituting (9.4.3) into (9.4.1) and eliminating ε_r by the use of (9.4.2), a differential equation is obtained for ε_θ, which upon integration results in

$$\varepsilon_\theta = \frac{1+\mu}{1-\mu}\frac{1}{r^2}\int_0^r \alpha T r\, dr + \frac{1-2\mu}{1-\mu}\frac{1}{r^2}\int_0^r \varepsilon_r^P r\, dr$$
$$+ \frac{1-2\mu}{1-\mu}\frac{1}{r^2}\int_0^r r\int_0^r \frac{\varepsilon_r^P - \varepsilon_\theta^P}{r}\, dr\, dr + C_1 + \frac{C_2}{r^2} \tag{9.4.4}$$

For a solid cylinder C_2 must vanish. Now substituting equation (9.4.4) into the compatibility relation (9.4.2) and solving for ε_r gives

$$\varepsilon_r = -\varepsilon_\theta + \frac{1+\mu}{1-\mu}\alpha T + \frac{1-2\mu}{1-\mu}\varepsilon_r^P + \frac{1-2\mu}{1-\mu}\int_0^r \frac{\varepsilon_r^P - \varepsilon_\theta^P}{r}\, dr + 2C_1 \tag{9.4.5}$$

To determine ε_z use is made of the fact that ε_z is a constant and that for unloaded ends

$$\int_0^R \sigma_z r\, dr = 0 \tag{9.4.6}$$

Substituting into the third of the stress–strain relations (9.4.3) and solving for ε_z results in

$$\varepsilon_z = \frac{2}{R^2}\left[\int_0^R \alpha T r\, dr - \int_0^R (\varepsilon_r^P + \varepsilon_\theta^P) r\, dr\right] \tag{9.4.7}$$

Sec. 9–4] Long Solid Cylinder

The constant C_1 is determined from the boundary condition $\sigma_r(R) = 0$. Substituting the first of equations (9.4.3) into this condition gives for C_1:

$$C_1 = \frac{1-3\mu}{1-\mu}\frac{1}{R^2}\int_0^R \alpha Tr\, dr + \frac{1-2\mu}{2(1-\mu)}\int_0^R \frac{\varepsilon_r^P - \varepsilon_\theta^P}{r}\, dr$$

$$+ \frac{1}{2(1-\mu)}\frac{1}{R^2}\int_0^R (\varepsilon_r^P + \varepsilon_\theta^P)r\, dr \quad (9.4.8)$$

To summarize, the total strains are given by

$$\varepsilon_\theta = \frac{1+\mu}{1-\mu}\frac{1}{r^2}\int_0^r \alpha Tr\, dr + \frac{1-2\mu}{2(1-\mu)}\frac{1}{r^2}\int_0^r (\varepsilon_r^P + \varepsilon_\theta^P)r\, dr$$

$$+ \frac{1-2\mu}{2(1-\mu)}\int_0^r \frac{\varepsilon_r^P - \varepsilon_\theta^P}{r}\, dr + C_1$$

$$\varepsilon_r = -\varepsilon_\theta + \frac{1+\mu}{1-\mu}\alpha T + \frac{1-2\mu}{1-\mu}\varepsilon_r^P + \frac{1-2\mu}{1-\mu}\int_0^r \frac{\varepsilon_r^P - \varepsilon_\theta^P}{r}\, dr + 2C_1$$

$$\varepsilon_z = \frac{2}{R^2}\left[\int_0^R \alpha Tr\, dr - \int_0^R (\varepsilon_r^P + \varepsilon_\theta^P)r\, dr\right]$$

(9.4.9)

where the double integral in (9.4.4) has been integrated by parts.

Equations (9.4.9) can now be solved by successive approximations as for the previous examples. The zeroth approximation to the total strains is obtained from equations (9.4.9) and (9.4.8) by assuming $\Delta\varepsilon_r^P$ and $\Delta\varepsilon_\theta^P$ zero. The total equivalent strain is computed from equations (9.1.9) and the equivalent plastic strain increment determined from the stress–strain curve and equation (9.1.10). First approximations are then calculated for $\Delta\varepsilon_r^P$ and $\Delta\varepsilon_\theta^P$ from the first two equations of (9.1.8) (with x replaced by r and y by θ).

TABLE 9.4.1 Calculation of Plastic Strains in Long Solid Cylinder

n	r	αT	ε_r^P prev. approx.	ε_θ^P prev. approx.	ε_r Eq. (9.4.9)	ε_θ Eq. (9.4.9)	ε_z Eq. (9.4.9)	ε_{et} Eq. (9.1.9)	ε_p Fig. 9.3.1	ε_r^P Eq. (9.1.8)	ε_θ^P Eq. (9.1.8)
			$\times 10^{-3}$		$\times 10^{-3}$	$\times 10^{-3}$	$\times 10^{-3}$	$\times 10^{-3}$	$\times 10^{-3}$	$\times 10^{-3}$	$\times 10^{-3}$
1	0	9.50	0	0	9.457	9.457	8.912	0.363	0	0	0
2	0.75	9.50	0	0	9.462	9.452		0.363	0	0	0
3	0.80	9.29	0	0	9.085	9.439		0.310	0	0	0
4	0.85	8.81	0	0	8.240	9.393		0.668	0	0	0
5	0.90	7.68	0	0	6.236	9.299		1.926	0.91	−0.907	0.545
6	0.95	7.06	0	0	5.274	9.109		2.494	1.40	−1.397	0.754
7	1.000	6.83	0	0	5.045	8.911		2.578	1.48	−1.477	0.738

These values of $\Delta\varepsilon_r^P$ and $\Delta\varepsilon_\theta^P$ are substituted into equations (9.4.8) and (9.4.9) and new approximations obtained for ε_r, ε_θ, and ε_z. The process is repeated as many times as necessary to obtain the desired degree of convergence. After

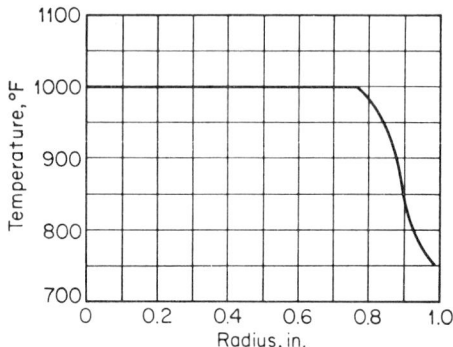

FIGURE 9.4.1 Temperature distribution in long solid cylinder.

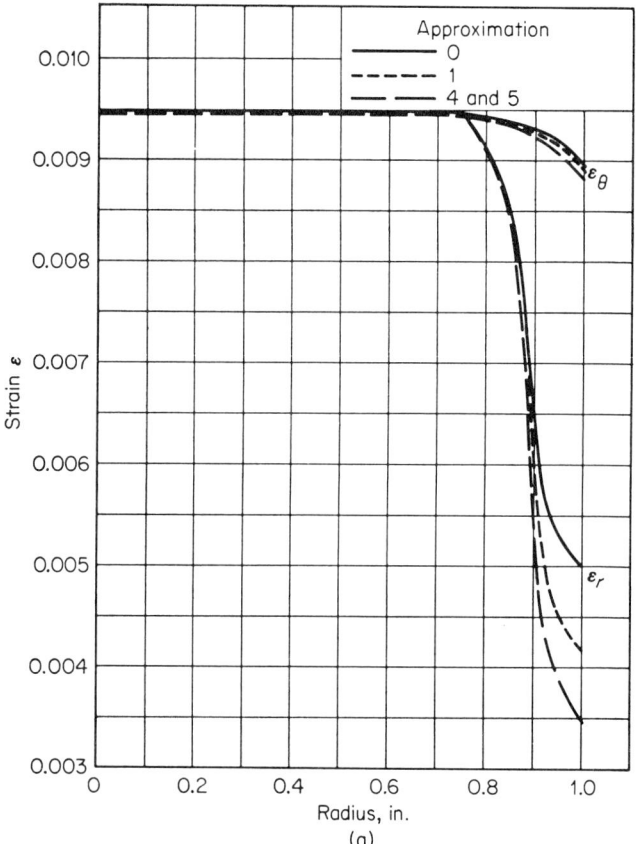

FIGURE 9.4.2 Strain and stress distributions in long solid cylinder.

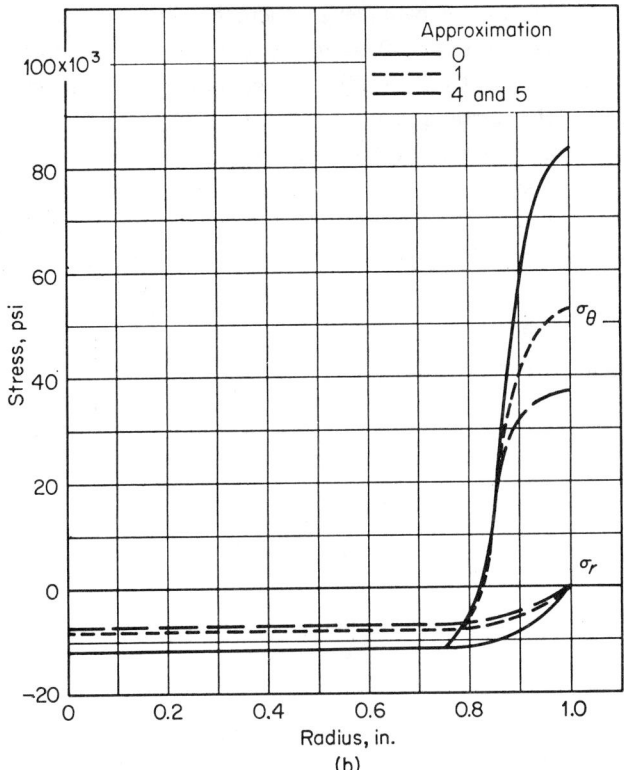

FIGURE 9.4.2 (*continued*)

convergence, the stresses can be computed from (9.4.3). The load is then incremented and the process repeated.

The above calculations have been carried out for a 1-in.-radius cylinder with a temperature distribution as shown in Figure 9.4.1, using the strain–strain curve of Figure 9.3.1. The computations are shown in Table 9.4.1 for one iteration and the results are plotted in Figure 9.4.2. Very little difference is found for this problem between the deformation theory and the incremental theory assuming the temperature distribution applied in several steps.

9–5 ROTATING DISK WITH TEMPERATURE GRADIENT

The previous problem was one of plane strain. We now consider a problem of plane stress—a rotating disk with a temperature gradient. Two methods will be presented for solving this problem. Both use successive approximations. The first converts the problem to solving a set of integral equations in the strains similar to the cylinder problem of the previous section. The second method converts the differential equations to finite-difference form and solves the resulting finite-difference equations. The integral-equation formulation has the advantages of being conceptually simpler and of including the boundary conditions automatically. On the other hand, it is difficult by this approach to take into account variations of material properties with temperature as well as variations in disk thickness. These can readily be taken into account by the finite-difference method.

Integral-Equation Formulation

We start as usual with the equilibrium and compatibility equations. These can be written

$$\frac{d}{dr}(hr\sigma_r) - h\sigma_\theta + \rho\omega^2 hr^2 = 0 \quad (9.5.1)$$

$$\frac{d\varepsilon_\theta}{dr} + \frac{\varepsilon_\theta - \varepsilon_r}{r} = 0 \quad (9.5.2)$$

where h is the disk thickness, ρ the density, and ω the rotational speed. For the plane stress problem $\sigma_z = 0$, and the stress–strain relations can be written [see equations (9.3.4)]

$$\sigma_r = \frac{E}{1 - \mu^2}[\varepsilon_r + \mu\varepsilon_\theta - (\varepsilon_r^P + \mu\varepsilon_\theta^P) - (1 + \mu)\alpha T]$$

$$\sigma_\theta = \frac{E}{1 - \mu^2}[\varepsilon_\theta + \mu\varepsilon_r - (\varepsilon_\theta^P + \mu\varepsilon_r^P) - (1 + \mu)\alpha T] \quad (9.5.3)$$

The terms ε_r^P and ε_θ^P are here defined, as for the cylinder, by equations (9.4.3a).

Proceeding as for the case of the cylinder in Section 9.4, equations (9.5.3) are substituted into (9.5.1) and ε_r eliminated by the use of (9.5.2), to give a differential equation in ε_θ. Assuming that h, E, and μ are constant, this differential equation is integrated to give

$$\varepsilon_\theta = -\frac{1 - \mu^2}{E}\frac{\rho\omega^2 r^2}{8} + \frac{1 + \mu}{r^2}\int_0^r \alpha Tr\,dr + \frac{1 - \mu}{2}\int_0^r \frac{\varepsilon_r^P - \varepsilon_\theta^P}{r}\,dr$$

$$+ \frac{1 + \mu}{2r^2}\int_0^r (\varepsilon_r^P + \varepsilon_\theta^P)r\,dr + \frac{C_1}{2} + \frac{C_2}{r^2} \quad (9.5.4)$$

TABLE 9.5.1 Calculation of Plastic Strains in Rotating Disk.

n	r	αT	ε_r^P prev. approx.	ε_θ^P prev. approx.	ε_r Eq. (9.5.5)	ε_θ Eq. (9.5.4)	ε_z Eq. (9.5.7)	ε_{et} Eq. (9.1.9)	ε_p Fig. 9.3.1	ε_r^P Eq. (9.1.8)	ε_θ^P Eq. (9.1.8)
		$\times 10^{-3}$	$\times 10^{-3}$	$\times 10^{-3}$	$\times 10^{-3}$	$\times 10^{-3}$	$\times 10^{-3}$	$\times 10^{-3}$	$\times 10^{-3}$	$\times 10^{-3}$	$\times 10^{-3}$
1	0	0.95	1.567	1.567	3.757	3.757	−3.247	4.669	3.441	1.720	1.720
2	0.5	1.045	1.515	1.517	3.780	3.784	−3.033	4.543	3.317	1.657	1.660
3	1.0	1.33	1.459	1.351	3.983	3.845	−2.490	4.270	3.048	1.573	1.474
4	1.5	1.71	1.376	1.144	4.250	3.942	−1.776	3.919	2.721	1.466	1.252
5	2.0	2.47	1.256	0.7269	4.862	4.104	−0.3885	3.277	2.124	1.298	0.8067
6	2.5	3.80	1.089	0.1111	6.022	4.381	1.913	2.389	1.304	1.047	0.1508
7	3.0	5.415	1.070	−0.4288	7.614	4.797	4.371	2.035	0.9972	0.9899	−0.3908
8	3.5	7.315	1.410	−1.175	9.773	5.362	6.964	2.578	1.469	1.371	−1.142
9	4.0	9.5	1.904	−2.298	12.28	6.076	10.00	3.623	2.446	1.908	−2.279
10	4.5	11.88	2.290	−3.637	14.83	6.911	13.51	4.896	3.664	2.303	−3.620
11	5.0	14.25	2.498	−4.996	17.18	7.824	17.18	6.235	4.982	2.491	−4.982

The disk is assumed to be solid; hence C_2 must vanish. To obtain ε_r, equation (9.5.4) is substituted into the compatibility equation (9.5.2), resulting in

$$\varepsilon_r = -\varepsilon_\theta - \frac{1-\mu^2}{E}\frac{\rho\omega^2 r^2}{2} + (1+\mu)\alpha T + \varepsilon_r^P + \mu\varepsilon_\theta^P$$

$$+ (1-\mu)\int_0^r \frac{\varepsilon_r^P - \varepsilon_\theta^P}{r} dr + C_1 \quad (9.5.5)$$

ε_z is now obtained by substituting equations (9.5.3) and $\sigma_z = 0$ into the stress–strain relation

$$\varepsilon_z = \frac{1}{E}[\sigma_z - \mu(\sigma_r + \sigma_\theta)] + \alpha T - \varepsilon_r^P - \varepsilon_\theta^P \quad (9.5.6)$$

to give

$$\varepsilon_z = -\frac{\mu}{1-\mu}(\varepsilon_r + \varepsilon_\theta) - \frac{1-2\mu}{1-\mu}(\varepsilon_r^P + \varepsilon_\theta^P) + \frac{1+\mu}{1-\mu}\alpha T \quad (9.5.7)$$

Finally, the constant C_1 is evaluated from the known rim loading. If the rim stress due to the rim loading is $\sigma_r(R)$, the first of equations (9.5.3) gives (R being the radius of the disk)

$$\sigma_r(R) = \frac{E}{1-\mu^2}[\varepsilon_r + \mu\varepsilon_\theta - (\varepsilon_r^P + \mu\varepsilon_\theta^P) - (1+\mu)\alpha T]_{r=R} \quad (9.5.8)$$

and substituting equations (9.5.4) and (9.5.5) into (9.5.8) gives for C_1:

$$C_1 = 2(1+\mu)\left[\frac{\sigma_r(R)}{E} + \frac{3+\mu}{8E}\rho\omega^2 R^2 + \frac{1}{R^2}\int_0^R \alpha T r\, dr\right.$$

$$\left. - \frac{1}{2}\int_0^R \frac{\varepsilon_r^P - \varepsilon_\theta^P}{r} dr + \frac{1}{2R^2}\int_0^R (\varepsilon_r^P + \varepsilon_\theta^P) r\, dr\right] \quad (9.5.9)$$

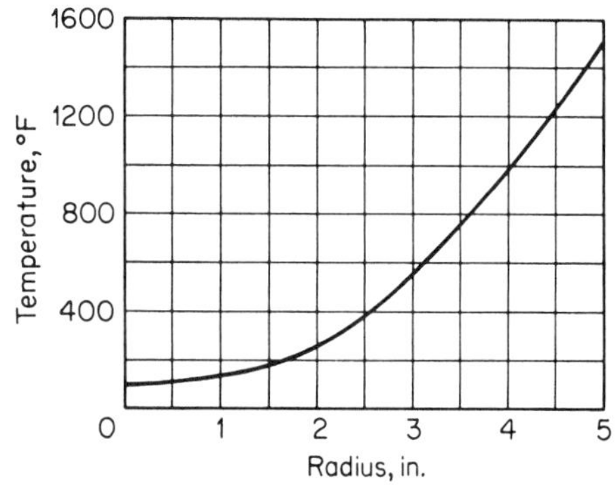

FIGURE 9.5.1 Temperature distribution in rotating disk.

Sec. 9–5] Rotating Disk with Temperature Gradient 201

The solution is now obtained by successive approximations exactly as in the previous example for a solid cylinder. By starting with assumed values of $\Delta \varepsilon_r^p$ and $\Delta \varepsilon_\theta^p$ equal to zero, ε_θ, ε_r, and ε_z are computed from (9.5.4), (9.5.5), and (9.5.7), ε_{et} from the last of equations (9.1.9), $\Delta \varepsilon_p$ from the stress–strain curve and equation (9.1.10), and $\Delta \varepsilon_r^p$ and $\Delta \varepsilon_\theta^p$ from the first two of equations (9.1.8), with x replaced by r and y by θ. New values of ε_r, ε_θ, and ε_z are computed from (9.5.4), (9.5.6), and (9.5.7), and the process is repeated.

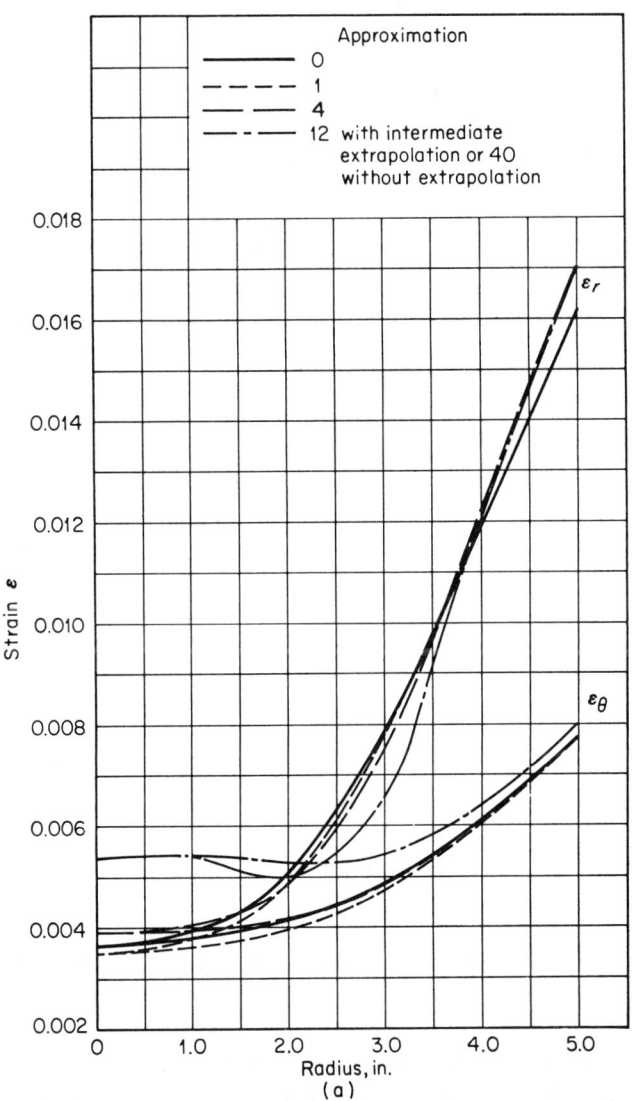

Figure 9.5.2 Strain and stress distributions in rotating disk.

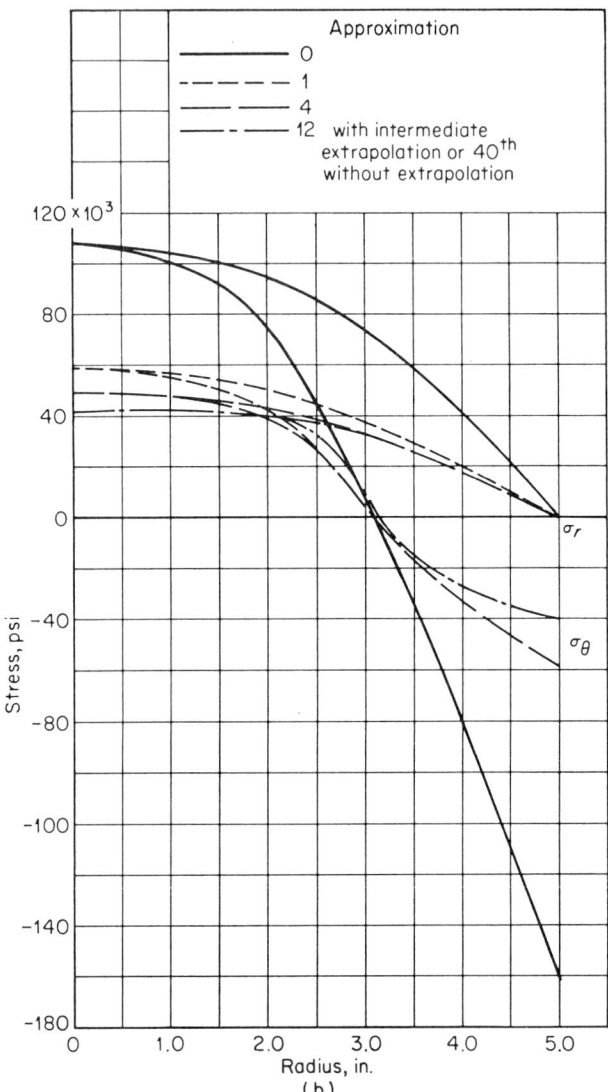

FIGURE 9.5.2 (*continued*)

As an example, the solution obtained in this manner for a 10-in.-diameter parallel-sided disk is shown in Figure 9.5.2. The value of $\rho\omega^2$ was taken as 1,500 and a temperature was assumed as shown in Figure 9.5.1. The strain–strain curve of Figure 9.3.1 was used. Again little difference was found between deformation theory and incremental theory. The computations for one iteration are shown in Table 9.5.1. As seen from Figure 9.5.2, a straight-

Sec. 9-5] Rotating Disk with Temperature Gradient   203

forward application of this method requires about 40 iterations for accurate results. The reason a greater number of iterations is required here compared to the cylinder problem of Section 9.4 is that there are external loads acting besides the thermal loads and the strains are larger. However, the convergence can be greatly increased by performing three or four iterations, taking the differences between successive iterations for the various strains, and extrapolating to a zero difference, as shown in Figure 9.5.3. A straight line is

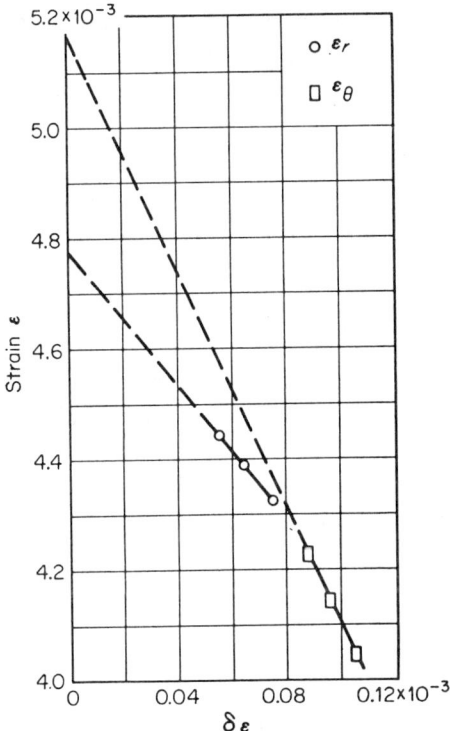

FIGURE 9.5.3   Variation of strain with change of strain.

drawn as shown and the intercept at zero $\delta \varepsilon$ is obtained. This furnishes a new starting estimate. Three or four more successive approximations are carried out, and another, similar, extrapolation is made. This technique reduced the number of successive approximations for this problem from 40 to 12. However, it should be noted that if a high-speed digital computor is used, the time per iteration is on the order of 0.1 sec and the number of iterations required is of secondary importance.

Finite-Difference Formulation

If it is desired to take into account variations of E, μ, or h along the disk, then it is necessary to use the finite-difference method [8]. This formulation is very general and can take into account not only initial variations in the disk thickness, but also the changes in thickness and radius during the loading process, if the disk should grow and change shape. In addition, we shall also consider the case of a disk with a central hole.

We start again with the equilibrium and compatibility equations (9.5.1) and (9.5.2). It is convenient this time to solve the problem in terms of stresses rather than strains, since the boundary conditions are generally given in terms of stresses. The stress–strain relations are written as

$$\varepsilon_r = \frac{1}{E}(\sigma_r - \mu\sigma_\theta) + \alpha T + \varepsilon_r^P + \Delta\varepsilon_r^P$$

$$\varepsilon_\theta = \frac{1}{E}(\sigma_\theta - \mu\sigma_r) + \alpha T + \varepsilon_\theta^P + \Delta\varepsilon_\theta^P \qquad (9.5.10)$$

$$\varepsilon_z = -\frac{\mu}{E}(\sigma_r + \sigma_\theta) + \alpha T - \varepsilon_r^P - \varepsilon_\theta^P - \Delta\varepsilon_r^P - \Delta\varepsilon_\theta^P$$

where ε_r^P and ε_θ^P represent the total plastic strains up to the current increment of loading, and $\Delta\varepsilon_r^P$ and $\Delta\varepsilon_\theta^P$ are the increments of plastic strain due to the current increment of loading.

Substituting (9.5.10) into the compatibility equation (9.5.2) gives the compatibility equation in terms of stresses:

$$\frac{d}{dr}\left[\frac{\sigma_\theta}{E} - \frac{\mu\sigma_r}{E} + \alpha T + \varepsilon_\theta^P + \Delta\varepsilon_\theta^P\right] = \frac{1+\mu}{E}\frac{\sigma_r - \sigma_\theta}{r} + \frac{\varepsilon_r^P - \varepsilon_\theta^P}{r} + \frac{\Delta\varepsilon_r^P - \Delta\varepsilon_\theta^P}{r}$$
(9.5.11)

Equations (9.5.1) and (9.5.11) are two equations for the two stresses σ_r and σ_θ. We proceed by putting them into finite-difference form as follows. Let the disk radius be divided into N intervals (not necessarily equal). There are thus $N + 1$ stations, the first station being at the center for a solid disk, or at the inner radius for a hollow disk. The last station is at the outer radius. Equations (9.5.1) and (9.5.11) are written in finite-difference form at the midpoints of these intervals. Thus at the midpoint of the $(i - 1)$st interval,

$$\frac{d}{dr}(hr\sigma_r)_{i-(1/2)} = \frac{h_i r_i \sigma_{r,i} - h_{i-1} r_{i-1} \sigma_{r,i-1}}{r_i - r_{i-1}}$$

$$(h\sigma_\theta)_{i-(1/2)} = \frac{h_{i-1}\sigma_{\theta,i-1} + h_i\sigma_{\theta,i}}{2}$$
(9.5.12)

Sec. 9-5] Rotating Disk with Temperature Gradient

In this manner equations (9.5.1) and (9.5.11) can be written

$$C_i\sigma_{r,i} - D_i\sigma_{\theta,i} = F_i\sigma_{r,i-1} + G_i\sigma_{\theta,i-1} - H_i$$
$$C_i'\sigma_{r,i} - D_i'\sigma_{\theta,i} = F_i'\sigma_{r,i-1} - G_i'\sigma_{\theta,i-1} + H_i' - P_i' \quad (9.5.13)$$

where

$$C_i = r_i h_i \qquad D_i = \tfrac{1}{2}(r_i - r_{i-1})h_i \qquad F_i = C_{i-1} \qquad G_i = \frac{h_{i-1}}{h_i}D_i$$

$$C_i' = \frac{\mu_i}{E_i} + a_i \qquad F_i' = \frac{\mu_{i-1}}{E_{i-1}} - b_i \qquad D_i' = \frac{1}{E_i} + a_i \qquad G_i' = \frac{1}{E_{i-1}} - b_i$$

$$H_i = \frac{\omega^2}{2}(r_i - r_{i-1})(\rho_i h_i r_i^2 + \rho_{i-1}h_{i-1}r_{i-1}^2) \qquad H_i' = \alpha_i T_i - \alpha_{i-1}T_{i-1}$$

$$P_i' = c_i \left[\varepsilon_{r,i}^P + \Delta\varepsilon_{r,i}^P + \frac{r_i}{r_{i-1}}(\varepsilon_{r,i-1}^P + \Delta\varepsilon_{r,i-1}^P) \right.$$

$$\left. - \left(1 + \frac{1}{c_i}\right)(\varepsilon_{\theta,i}^P + \Delta\varepsilon_{\theta,i}^P) + \left(\frac{1}{c_i} - \frac{r_i}{r_{i-1}}\right)(\varepsilon_{\theta,i-1}^P + \Delta\varepsilon_{\theta,i-1}^P) \right]$$

$$c_i = \frac{r_i - r_{i-1}}{2r_i} \qquad a_i = \frac{1 + \mu_i}{E_i}c_i \qquad b_i = \frac{E_i r_i}{E_{i-1}r_{i-1}}a_i \qquad (9.5.14)$$

It is to be noted that if the disk dimensions do not change appreciably with loading, all these coefficients except P_i' depend only on the initial geometry, material properties, and operating conditions of the disk and are evaluated once and for all. Only P_i' is a function of the plastic flow. Also for a solid disk, r_{i-1} can equal zero and so some of the primed coefficients become infinite. This can be avoided by assuming for the first radial station some small nonzero number, rather than zero.

Equations (9.5.13) give the stresses at the ith station in terms of the stresses at the $(i - 1)$st station. Solving for the stresses at the ith station gives

$$\sigma_{r,i} = l_{11,i}\sigma_{r,i-1} + l_{12,i}\sigma_{\theta,i-1} + m_{1,i}$$
$$\sigma_{\theta,i} = l_{21,i}\sigma_{r,i-1} + l_{22,i}\sigma_{\theta,i-1} + m_{2,i} \quad (9.5.15)$$

where

$$l_{11,i} = \frac{D_i'F_i - D_iF_i'}{C_iD_i' - C_i'D_i} \qquad l_{12,i} = \frac{D_i'G_i + D_iG_i'}{C_iD_i' - C_i'D_i}$$

$$l_{21,i} = \frac{C_i'F_i - C_iF_i'}{C_iD_i' - C_i'D_i} \qquad l_{22,i} = \frac{C_i'G_i + C_iG_i'}{C_iD_i' - C_i'D_i} \quad (9.5.16)$$

$$m_{1,i} = \frac{D_iP_i' - H_iD_i' - H_i'D_i}{C_iD_i' - C_i'D_i} \qquad m_{2,i} = \frac{C_iP_i' - H_iC_i' - H_i'C_i}{C_iD_i' - C_i'D_i}$$

or, in matrix notation,

$$\begin{bmatrix} \sigma_{r,i} \\ \sigma_{\theta,i} \end{bmatrix} = \begin{bmatrix} l_{11,i} & l_{12,i} \\ l_{21,i} & l_{22,i} \end{bmatrix} \begin{bmatrix} \sigma_{r,i-1} \\ \sigma_{\theta,i-1} \end{bmatrix} + \begin{bmatrix} m_{1,i} \\ m_{2,i} \end{bmatrix}$$

or
$$\sigma_i = L_i \sigma_{i-1} + M_i \qquad (9.5.17)$$

where σ_i, σ_{i-1}, L_i, and M_i are the indicated matrices. Equation (9.5.17) represents a linear recurrence relation between the stresses at the ith station and the stresses at the $(i-1)$st station. Obviously by successive application of (9.5.17), the stresses at the ith station can be linearly related to the stresses at the first station. Let this linear relation be written

$$\sigma_i = A_i \sigma_1 + B_i \qquad (9.5.18)$$

where A_i and B_i are as yet unknown and σ_1 are the radial and tangential stresses at the first station. Substituting (9.5.18) into (9.5.17) gives

$$A_i \sigma_1 + B_i = L_i(A_{i-1}\sigma_1 + B_{i-1}) + M_i$$

or
$$(A_i - A_{i-1}L_i)\sigma_1 = L_i B_{i-1} - B_i + M_i \qquad (9.5.19)$$

Now σ_1 will depend on the boundary conditions and is completely arbitrary, whereas (9.5.19) must be true for all values of σ_1. It therefore follows that both sides of the equation must vanish identically. Hence

$$\begin{aligned} A_i &= A_{i-1}L_i \\ B_i &= L_i B_{i-1} + M_i \end{aligned} \qquad (9.5.20)$$

Also, for the second station, equation (9.5.18) gives

$$\sigma_2 = A_2 \sigma_1 + B_2$$

and equation (9.5.17) gives

$$\sigma_2 = L_2 \sigma_1 + M_2$$

Hence $\quad A_2 = L_2 \quad$ and $\quad B_2 = M_2 \qquad (9.5.21)$

Beginning therefore with A_2 and B_2 as given by (9.5.21), all the other A's and B's can be computed successively by the recurrence relations (9.5.20). For the last station [the $(N+1)$st], r_{N+1} equals R, the disk radius, and equation (9.5.18) becomes

$$\sigma_{N+1} = A_{N+1}\sigma_1 + B_{N+1}$$

or

$$\begin{bmatrix} \sigma_{r,N+1} \\ \sigma_{\theta,N+1} \end{bmatrix} = \begin{bmatrix} a_{11,N+1} & a_{12,N+1} \\ a_{21,N+1} & a_{22,N+1} \end{bmatrix} \begin{bmatrix} \sigma_{r,1} \\ \sigma_{\theta,1} \end{bmatrix} + \begin{bmatrix} b_{1,N+1} \\ b_{2,N+1} \end{bmatrix} \qquad (9.5.22)$$

Sec. 9-5] Rotating Disk with Temperature Gradient

Therefore,

$$\sigma_r(R) = \sigma_{r,N+1} = a_{11,N+1}\sigma_{r,1} + a_{12,N+1}\sigma_{\theta,1} + b_{1,N+1}$$

For a solid disk $\sigma_{r,1} = \sigma_{\theta,1}$; therefore,

$$\sigma_{r,1} = \sigma_{\theta,1} = \frac{\sigma_r(R) - b_{1,N+1}}{a_{11,N+1} + a_{12,N+1}} \tag{9.5.23}$$

For a disk with a central hole of radius R_0, with inner prescribed pressure $\sigma_{r,1} = \sigma_r(R_0)$, we get

$$\sigma_{\theta,1} = \frac{\sigma_r(R) - a_{11,N+1}\sigma_r(R_0) - b_{1,N+1}}{a_{12,N+1}} \tag{9.5.24}$$

Thus σ_1 is now known. The stresses at every station can then be directly computed by means of (9.5.18).

To summarize then, the L_i and M_i matrices are computed from (9.5.16) and (9.5.14), the A_i and B_i matrices from (9.5.20) and (9.5.21), and then the stresses are computed from (9.5.18), using either (9.5.23) or (9.5.24).

It is to be noted that this straightforward procedure takes into account with equal ease variations of E, h, ρ, or even μ along the radius of the disk. Furthermore, if the dimensions of the disk are changing during the plastic flow process, this can readily be taken into account. For if r is the current radius to a given point P, and r' was the radius to the point P before plastic flow took place, then approximately

$$r = r'(1 + \varepsilon_\theta^P) \tag{9.5.25}$$

and, similarly, if H was the underformed thickness at the point P and h is actual thickness after deformation, then approximately

$$h = \frac{H}{(1 + \varepsilon_r^P)(1 + \varepsilon_\theta^P)} \tag{9.5.26}$$

Thus at any stage of the plastic flow process the values of r_i and h_i appearing in equations (9.5.14) can be corrected by means of (9.5.25) and (9.5.26).

The finite-difference formulation presented will of course give directly and quickly the elastic solution for a disk of arbitrary profile with variable properties, if P_i' is set equal to zero in equations (9.5.16). For the plastic problem P_i' is not zero, and its values can be determined by successive approximations, as thoroughly described in the previous examples. We shall return to this formulation at a later time in discussing the creep of a rotating disk. Solution by this method of the disk problem of the previous example gave almost identical answers.

9-6 CIRCULAR HOLE IN UNIFORMLY STRESSED INFINITE PLATE

As a final example of a problem involving rotational symmetry, consider the case of a thin infinite plate uniformly stressed containing a circular hole of radius a. Solutions to this problem by iterative methods similar to that discussed herein are given in references [9] and [10]. The present solution is taken from reference [10].

The equilibrium and compatibility equations are given by (9.4.1) and (9.4.2); i.e.,

$$\frac{d\sigma_r}{dr} + \frac{\sigma_r - \sigma_\theta}{r} = 0$$
$$\frac{d\varepsilon_\theta}{dr} + \frac{\varepsilon_\theta - \varepsilon_r}{r} = 0 \quad (9.6.1)$$

The stress–strain relations for the case of plane stress are

$$\varepsilon_r = \frac{1}{E}(\sigma_r - \mu\sigma_\theta) + \varepsilon_r^P + \Delta\varepsilon_r^P$$
$$\varepsilon_\theta = \frac{1}{E}(\sigma_\theta - \mu\sigma_r) + \varepsilon_\theta^P + \Delta\varepsilon_\theta^P \quad (9.6.2)$$

Substituting (9.6.2) into the second of equations (9.6.1), the compatibility equation is obtained in terms of stresses. A *stress function* ϕ is now introduced defined by

$$\sigma_r = \frac{1}{r}\frac{d\phi}{dr} \qquad \sigma_\theta = \frac{d^2\phi}{dr^2} \quad (9.6.3)$$

The equilibrium equation is identically satisfied by the stress function, and the compatibility equation becomes

$$\frac{d}{dr}\left[\frac{1}{r}\frac{d}{dr}\left(r\frac{d\phi}{dr}\right)\right] = -g(r) \quad (9.6.4)$$

where $\quad g(r) = E\left[\frac{d}{dr}(\varepsilon_\theta^P + \Delta\varepsilon_\theta^P) + \frac{\varepsilon_\theta^P - \varepsilon_r^P + \Delta\varepsilon_\theta^P - \Delta\varepsilon_r^P}{r}\right] \quad (9.6.5)$

Sec. 9–6] Circular Hole in Uniformly Stressed Infinite Plate

Integrating results in

$$\phi = Ar^2 + B \ln r + C - \int_a^r \frac{1}{r} \int_a^r r \int_a^r g(r) dr \, dr \, dr \qquad (9.6.6)$$

The triple integral can be somewhat simplified by making use of (9.6.5) and integrating by parts. The constants A and B are determined from the boundary conditions

$$\sigma_r(a) = 0 \qquad \sigma_r(\infty) = \sigma_\infty \qquad (9.6.7)$$

where σ_∞ is the applied uniform stress at infinity. Note that $\sigma_\infty < \sigma_0$, where σ_0 is the yield stress. Then

$$A = \frac{\sigma_\infty}{2} \qquad B = -a^2 \sigma_\infty \qquad (9.6.8)$$

and the constant C can arbitrarily be set equal to zero without affecting the solution.

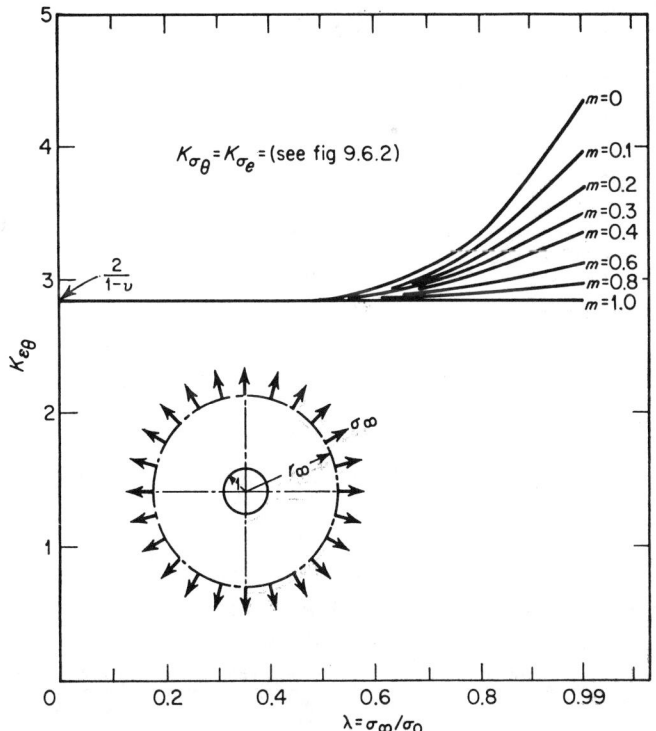

FIGURE 9.6.1 Tangential strain concentration factors at hole.

Detailed calculations were carried out by Tuba [10] using the successive-approximation technique with linear strain hardening for various values of the strain-hardening parameter m (ratio of slope of the strain-hardening part of curve to slope of the elastic part), and for various ratios of σ_∞ to the yield stress σ_0. His results, taken from reference [10], are shown in Figures 9.6.1 through 9.6.3. In these figures the stress and strain concentration factors due

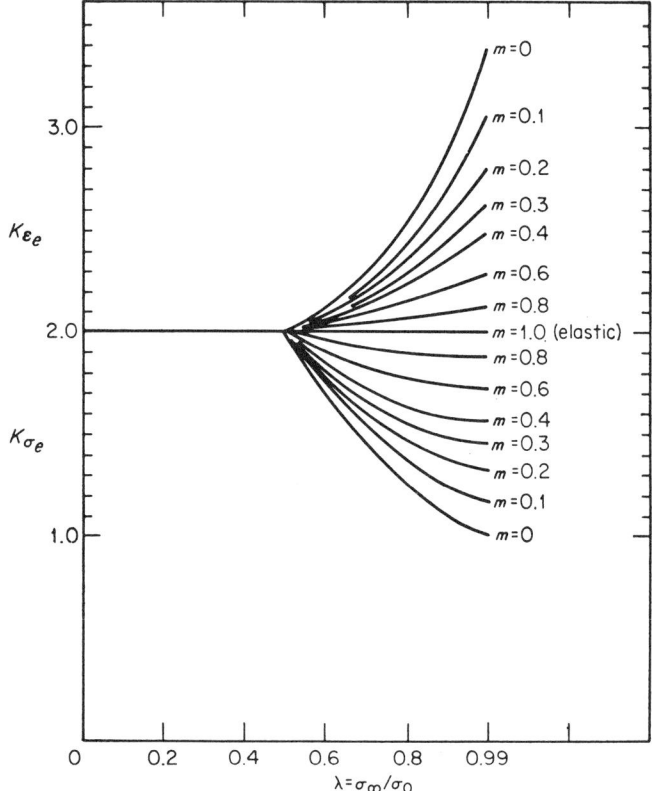

FIGURE 9.6.2 Equivalent stress and strain concentration factors at hole.

to the hole are plotted as functions of the dimensionless parameters m and λ. The various parameters are defined as follows:

$$m = \frac{\text{slope of plastic part of stress–strain curve}}{E}$$

$$\lambda = \frac{\sigma_\infty}{\sigma_0}$$

Sec. 9-6] Circular Hole in Uniformly Stressed Infinite Plate

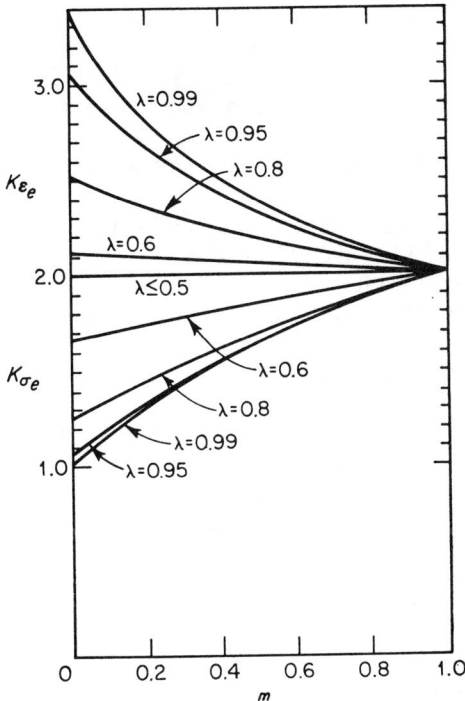

FIGURE 9.6.3 Equivalent stress and strain concentration factors at hole.

$$K_{\sigma_\theta} = \frac{\sigma_\theta(a)}{\sigma_\infty} = K_{\sigma_e}$$

$$K_{\varepsilon_\theta} = \frac{\varepsilon_\theta(a)}{\varepsilon_\theta(\infty)} = \frac{\varepsilon_\theta(a)}{[(1-\mu)/E]\sigma_\infty}$$

$$K_{\varepsilon_e} = \frac{\varepsilon_e(a)}{\varepsilon_e(\infty)} = \frac{\varepsilon_e(a)}{\frac{2}{3}[(1+\mu)/E]\sigma_\infty}$$

where
$$\varepsilon_e = \frac{\sqrt{2}}{3}\sqrt{(\varepsilon_r - \varepsilon_\theta)^2 + (\varepsilon_r - \varepsilon_z)^2 + (\varepsilon_\theta - \varepsilon_z)^2}$$

It is also noted in reference [10] that if the load is increased monotonically, the deformation and incremental theories give the same results for this problem. However, since in this case we have proportional loading at the hole where the stress concentration factors are determined, this is not too surprising.

In Chapter 10 the much more difficult two-dimensional plane elastoplastic problem is formulated and several examples of plate problems are solved.

Problems

1. Use the successive-approximation method to solve the differential equation

$$\frac{dy}{dx} + y = 0 \quad y(0) = 1$$

   Show that in the limit the exact solution is obtained.
2. Verify equations (9.2.7) and (9.2.8).
3. Derive equation (9.2.15).
4. Show that the method of successive approximations leads to equation (9.2.20).
5. Derive equation (9.4.4).
6. Derive equation (9.5.4).
7. Obtain the equations corresponding to (9.5.4) through (9.5.9) for a hollow disk with inner radius a and outer radius R.
8. Derive formulas (9.5.25) and (9.5.26).
9. Derive equations (9.6.4), (9.6.6), and (9.6.8).

References

1. E. L. Ince, *Ordinary Differential Equations*, Dover, New York, 1944.
2. A. A. Ilyushin, Some Problems in the Theory of Plastic Deformation, *RMB-12*, translation from *Prikl. Math. Mech.*, 7, 1943, pp. 245–272, by Grad. Div. Appl. Math., Brown Univ., 1946.
3. S. S. Manson, *Thermal Stress and Low Cycle Fatigue*, McGraw-Hill, New York, 1966, p. 100.
4. A. Mendelson and S. S. Manson, Practical Solution of Plastic Deformation Problems in the Elastic–Plastic Range, *NASA Tech. Rept. R-28*, 1959.
5. A. Mendelson and S. W. Spero, A General Solution for the Elastoplastic Thermal Stresses in a Strain-Hardening Plate with Arbitrary Material Properties, *J. Appl. Mech.*, 29, 1962, pp. 151–158.
6. S. G. Mikhlin, *Integral Equations*, Pergamon Press, London, 1957, p. 19.
7. C. T. Wang, *Applied Elasticity*, McGraw-Hill, New York, 1953.
8. A. Mendelson, M. H. Hirschberg, and S. S. Manson, A General Approach to the Practical Solution of Creep Problems, *Trans. ASME*, **81D**, 1959, pp. 585–598.
9. E. A. Davis, Extension of Iteration Method for Determining Strain Distributions to the Uniformly Stressed Plate with a Hole, *J. Appl. Mech.*, 30, 1963, pp. 210–214; discussions, *ibid.*, 31, 1964, pp. 362–364.
10. I. S. Tuba, Elastic–Plastic Stress and Strain Concentration Factors at a Circular Hole in a Uniformly Stressed Infinite Plate, *J. Appl. Mech.*, 32, 1965, pp. 710–711.

CHAPTER 10

THE PLANE ELASTOPLASTIC PROBLEM

10-1 GENERAL RELATIONS

By plane elastoplastic problems we mean the usual generalized plane strain or plane stress problems. Generalized plane strain is characterized by the stresses and strains being functions of x and y only and the strain in the z direction, ε_z, being equal to a constant. This is the type of problem encountered in long cylinders under certain loading conditions such as discussed in Sections 8.7 and 9.4. Generalized plane stress problems are encountered in thin plates and, in this case, the stresses and strains are taken as the average values through the thickness of the plate and σ_z is assumed to be zero. In both types of problems the shear stresses and strains in the z direction are assumed to be zero.

In all that follows, it is assumed that the material is homogeneous, isotropic, and strain hardens isotropically. The material properties such as modulus of elasticity, Poisson's ratio, and coefficient of linear thermal expansion are assumed independent of the temperature, and body forces are not considered. Rotational symmetry, as for the cylinder and disk problems of Chapter 9, is not assumed to exist. As before, the von Mises yield criterion and the associated flow rule will be used.

The equilibrium, compatibility, and stress–strain relations for the *plane stress* problem are

$$\frac{\partial \sigma_x}{\partial x} + \frac{\partial \tau_{xy}}{\partial y} = 0$$

$$\frac{\partial \sigma_y}{\partial y} + \frac{\partial \tau_{xy}}{\partial x} = 0$$

(10.1.1)

$$\frac{\partial^2 \varepsilon_x}{\partial y^2} + \frac{\partial^2 \varepsilon_y}{\partial x^2} - 2\frac{\partial^2 \varepsilon_{xy}}{\partial x\, \partial y} = 0 \qquad (10.1.2)$$

$$\varepsilon_x = \frac{1}{E}(\sigma_x - \mu\sigma_y) + \alpha T + \varepsilon_x^P + \Delta\varepsilon_x^P$$

$$\varepsilon_y = \frac{1}{E}(\sigma_y - \mu\sigma_x) + \alpha T + \varepsilon_y^P + \Delta\varepsilon_y^P$$

$$\varepsilon_z = \frac{-\mu}{E}(\sigma_x + \sigma_y) + \alpha T - \varepsilon_x^P - \varepsilon_y^P - \Delta\varepsilon_x^P - \Delta\varepsilon_y^P$$

(10.1.3)

$$\varepsilon_{xy} = \frac{1}{2G}\tau_{xy} + \varepsilon_{xy}^P + \Delta\varepsilon_{xy}^P$$

To these we append the Prandtl–Reuss relations,

$$\Delta\varepsilon_x^P = \frac{\Delta\varepsilon_p}{2\sigma_e}(2\sigma_x - \sigma_y)$$

$$\Delta\varepsilon_y^P = \frac{\Delta\varepsilon_p}{2\sigma_e}(2\sigma_y - \sigma_x) \qquad (10.1.4)$$

$$\Delta\varepsilon_{xy}^P = \frac{3}{2}\frac{\Delta\varepsilon_p}{\sigma_e}\tau_{xy}$$

or the corresponding plastic strain–total strain relations,

$$\Delta\varepsilon_x^P = \frac{\Delta\varepsilon_p}{3\varepsilon_{et}}(2\varepsilon_x' - \varepsilon_y' - \varepsilon_z')$$

$$\Delta\varepsilon_y^P = \frac{\Delta\varepsilon_p}{3\varepsilon_{et}}(2\varepsilon_y' - \varepsilon_z' - \varepsilon_x') \qquad (10.1.5)$$

$$\Delta\varepsilon_{xy}^P = \frac{\Delta\varepsilon_p}{\varepsilon_{et}}\varepsilon_{xy}'$$

where, as previously defined,

$$\varepsilon_x^P = \sum_{k=1}^{i-1} \Delta\varepsilon_{x,k}^P$$

Sec. 10-1] General Relations

is the plastic strain accumulated during the first $i - 1$ increments of load and $\Delta\varepsilon_x^P$ is the unknown plastic strain increment occurring during the ith or current increment of load. The subscript i has been deleted [see equations (7.10.2)] since no confusion can arise. Similar definitions hold for $\Delta\varepsilon_y^P$ and ε_{xy}^P. The definitions of some of the other quantities entering into the above equations will be also repeated here, for convenience. Thus

$$\varepsilon_x' = \varepsilon_x - \varepsilon_x^P \quad \text{etc.}$$

$$\Delta\varepsilon_p = \frac{2}{\sqrt{3}}[(\Delta\varepsilon_x^P)^2 + (\Delta\varepsilon_y^P)^2 + \Delta\varepsilon_x^P\Delta\varepsilon_y^P + (\Delta\varepsilon_{xy}^P)^2]^{1/2} \quad (10.1.6)$$

$$\varepsilon_{et} = \frac{\sqrt{2}}{3}[(\varepsilon_x' - \varepsilon_y')^2 + (\varepsilon_y' - \varepsilon_z')^2 + (\varepsilon_z' - \varepsilon_x')^2 + 6(\varepsilon_{xy}')^2]^{1/2}$$

$$\sigma_e = (\sigma_x^2 + \sigma_y^2 - \sigma_x\sigma_y + 3\tau_{xy}^2)^{1/2}$$

and $\Delta\varepsilon_p$ is related to σ_e through the uniaxial stress–strain curve as shown in Figure 7.9.1 or to ε_{et} by the uniaxial stress–strain curve and the relation

$$\Delta\varepsilon_p \cong \frac{\varepsilon_{et} - \frac{2}{3}[(1 + \mu)/E]\sigma_{e,i-1}}{1 + \frac{2}{3}[(1 + \mu)/E](d\sigma_e/d\varepsilon_p)_{i-1}} \quad (10.1.7)$$

All these relations have been previously described in detail.

For the case of *plane strain*, all the above relations remain unchanged except for equations (10.1.3) and (10.1.4) and the definition of σ_e. Since σ_z is no longer zero,

$$\varepsilon_x = \frac{1}{E}[\sigma_x - \mu(\sigma_y + \sigma_z)] + \alpha T + \varepsilon_x^P + \Delta\varepsilon_x^P$$

$$\varepsilon_y = \frac{1}{E}[\sigma_y - \mu(\sigma_z + \sigma_x)] + \alpha T + \varepsilon_y^P + \Delta\varepsilon_y^P \quad (10.1.8)$$

$$\varepsilon_z = \frac{1}{E}[\sigma_z - \mu(\sigma_x + \sigma_y)] + \alpha T - \varepsilon_x^P - \varepsilon_y^P - \Delta\varepsilon_x^P - \Delta\varepsilon_y^P$$

$$\varepsilon_{xy} = \frac{1 + \mu}{E}\tau_{xy} + \varepsilon_{xy}^P + \Delta\varepsilon_{xy}^P$$

where ε_z is a constant. Also

$$\sigma_e = \frac{1}{\sqrt{2}}[(\sigma_x - \sigma_y)^2 + (\sigma_y - \sigma_z)^2 + (\sigma_z - \sigma_x)^2 + 6\tau_{xy}^2]^{1/2} \quad (10.1.9)$$

and

$$\Delta\varepsilon_x^P = \frac{\Delta\varepsilon_p}{2\sigma_e}(2\sigma_x - \sigma_y - \sigma_z)$$

$$\Delta\varepsilon_y^P = \frac{\Delta\varepsilon_p}{2\sigma_e}(2\sigma_y - \sigma_z - \sigma_x) \qquad (10.1.10)$$

$$\Delta\varepsilon_{xy}^P = \frac{3}{2}\frac{\Delta\varepsilon_p}{\sigma_e}\tau_{xy}$$

We now proceed to introduce a stress function, as is common in plane problems of elasticity. First by substituting equations (10.1.3) or (10.1.8) into (10.1.2), the compatibility equation is obtained in terms of stresses. Making use of the equilibrium equation (10.1.1) to eliminate the shear stress and performing some algebraic manipulations results in

$$\nabla^2(\sigma_x + \sigma_y) = -\bar{E}\nabla^2(\alpha T) - [g(x, y) + \Delta g(x, y)] \qquad (10.1.11)$$

where $\nabla^2 \equiv (\partial^2/\partial x^2) + (\partial^2/\partial y^2)$ is the Laplacian operator in two dimensions and

$$\begin{aligned}\bar{E} &\equiv E & \text{for plane stress} \\ \bar{E} &\equiv \frac{E}{1-\mu} & \text{for plane strain}\end{aligned} \qquad (10.1.12)$$

$$\left.\begin{aligned}g(x, y) &\equiv E\left(\frac{\partial^2 \varepsilon_x^P}{\partial y^2} + \frac{\partial^2 \varepsilon_y^P}{\partial x^2} - 2\frac{\partial^2 \varepsilon_{xy}^P}{\partial x\,\partial y}\right) \\ \Delta g(x, y) &\equiv E\left(\frac{\partial^2(\Delta\varepsilon_x^P)}{\partial y^2} + \frac{\partial^2(\Delta\varepsilon_y^P)}{\partial x^2} - 2\frac{\partial^2(\Delta\varepsilon_{xy}^P)}{\partial x\,\partial y}\right)\end{aligned}\right\} \text{plane stress} \qquad (10.1.13)$$

$$\left.\begin{aligned}g(x, y) &\equiv \frac{E}{1-\mu^2}\left(\frac{\partial^2 \varepsilon_x^P}{\partial y^2} + \frac{\partial^2 \varepsilon_y^P}{\partial x^2} - 2\frac{\partial^2 \varepsilon_{xy}^P}{\partial x\,\partial y}\right) \\ &\quad - \frac{\mu E}{1-\mu^2}\nabla^2(\varepsilon_x^P + \varepsilon_y^P) \\ \Delta g(x, y) &\equiv \frac{E}{1-\mu^2}\left(\frac{\partial^2(\Delta\varepsilon_x^P)}{\partial y^2} + \frac{\partial^2(\Delta\varepsilon_y^P)}{\partial x^2} - 2\frac{\partial^2(\Delta\varepsilon_{xy}^P)}{\partial x\,\partial y}\right) \\ &\quad - \frac{\mu E}{1-\mu^2}\nabla^2(\Delta\varepsilon_x^P + \Delta\varepsilon_y^P)\end{aligned}\right\} \begin{aligned}&\text{plane strain} \\ &(10.1.14)\end{aligned}$$

Define a stress function ϕ by the relations

$$\frac{\partial^2 \phi}{\partial x^2} = \sigma_y \qquad \frac{\partial^2 \phi}{\partial y^2} = \sigma_x \qquad \frac{\partial^2 \phi}{\partial x\,\partial y} = -\tau_{xy} \qquad (10.1.15)$$

Sec. 10-1] General Relations

The equilibrium equations (10.1.1) are identically satisfied and the compatibility equation (10.1.11) becomes

$$\nabla^4 \phi = -\bar{E}\nabla^2(\alpha T) - (g + \Delta g) \tag{10.1.16}$$

where ∇^4 is the biharmonic operator:

$$\nabla^4 \equiv \frac{\partial^4}{\partial x^4} + 2\frac{\partial^4}{\partial x^2 \partial y^2} + \frac{\partial^4}{\partial y^4}$$

The boundary conditions to be satisfied by the stress function ϕ are (see, for example, reference [1])

$$\frac{\partial \phi}{\partial y} = \int^s T_x \, ds + C_1 \equiv f_1(s) + C_1$$
$$\frac{\partial \phi}{\partial x} = -\int^s T_y \, ds + C_2 \equiv f_2(s) + C_2 \tag{10.1.17}$$

where T_x and T_y are the x and y components, respectively, of the external force acting at a point of the boundary. The integrations are performed along the boundary from some arbitrary point. The constants C_1 and C_2 are arbitrary and can be set equal to zero without affecting the stresses. For a multiply connected region, however, the constants will be different for each of the contours and they can be arbitrarily chosen only on one contour. On the other contours they must be determined so that the displacements are single-valued. This greatly complicates even the elastic problem, and multiply connected regions will not be considered here.

Alternatively, by means of equations (10.1.17), the boundary conditions can be written in terms of ϕ and its normal derivative on the boundary [1],

$$\phi = \int^s \left(\frac{\partial \phi}{\partial x}\frac{dx}{ds} + \frac{\partial \phi}{\partial y}\frac{dy}{ds}\right) ds = f(s) + C_3$$
$$\frac{d\phi}{dn} = \frac{\partial \phi}{\partial x}\frac{dy}{ds} - \frac{\partial \phi}{\partial y}\frac{dx}{ds} = h(s) \tag{10.1.18}$$

so that if $\partial\phi(s)/\partial y$ and $\partial\phi(s)/\partial y$ are known from (10.1.17), $\partial\phi/dn$ and $\phi(s)$ can be computed from (10.1.18). For an unloaded boundary, $T_x = T_y = 0$ and consequently $f_1(s) = f_2(s) = 0$, so that $\partial\phi(s)/\partial y = C_1$ and $\partial\phi(s)/\partial x = C_2$, and, since the stresses depend only on the second derivatives of ϕ, we can arbitrarily subtract $C_1 y + C_2 x + C_3$ from ϕ (if the region is simply connected) so that

$$\frac{\partial \phi(s)}{\partial y} = \frac{\partial \phi(s)}{\partial x} = 0$$

or $\quad\dfrac{d\phi(s)}{dn} = \phi(s) = 0$ (10.1.19)

To solve the plane elastoplastic problem it is therefore necessary to solve the inhomogeneous biharmonic equation (10.1.16) subject to the boundary conditions (10.1.17) or (10.1.18) and the appropriate plasticity relations.

Note that the plane strain and plane stress problems differ primarily in the definition of the g function appearing in (10.1.11) and the inclusion of σ_z in the definition of σ_e and in the Prandtl–Reuss equations. The calculation procedures for the two problems therefore differ in only minor details, as will be shown subsequently.

Solutions to several specific plate problems will now be presented using the successive-approximation technique described in Chapter 9.

10-2 ELASTOPLASTIC THERMAL PROBLEM FOR A FINITE PLATE

The first problem considered is that of a thin rectangular plate with a temperature distribution $T(x, y)$ and no external loads. The geometry and the coordinate system used is shown in Figure 10.2.1. Since there are no

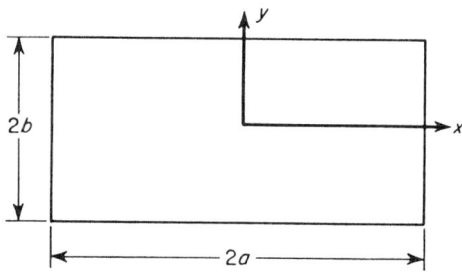

FIGURE 10.2.1 Flat plate and coordinate system.

external loads, the boundary conditions (10.1.19) are used; i.e., ϕ and its normal derivative are zero on the boundaries.

Equation (10.1.16) with the right side zero is the classical biharmonic equation. The elastic thermal stress problem with $g = \Delta g = 0$ has been

Sec. 10-2] Elastoplastic Thermal Problem for a Finite Plate

solved by a variety of methods, including energy, collocation, eigenfunction, and finite-difference methods. In all cases the solution is actually only approximate, although in theory the exact solution can be approached as closely as desired. In solving the plastic flow problem, one or more of these methods can also be used, together with the successive-approximation technique. The simplest and most straightforward approach is to use finite differences, as will now be described.

The plate is divided into a grid of $n \times m$ stations. If symmetry exists about the x and y axes with the origin taken at the center of the plate, only one quadrant of the plate need be considered. At each point of the grid, equation (10.1.16) is written in finite-difference form. For example, at the station designated by i, j in Figure 10.2.2, equation (10.1.16) becomes

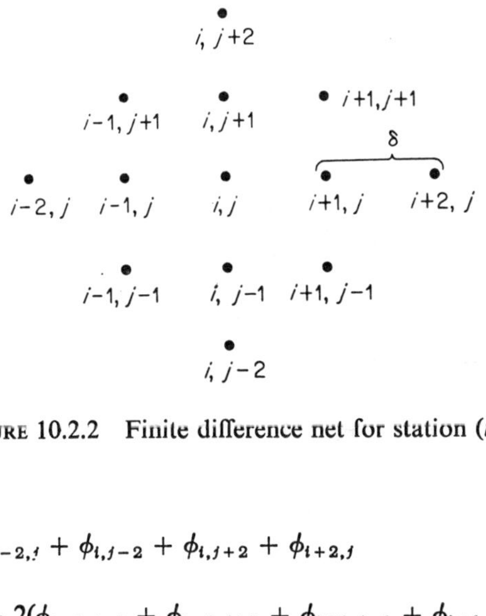

FIGURE 10.2.2 Finite difference net for station (i, j).

$$\frac{1}{\delta^4} [\phi_{i-2,j} + \phi_{i,j-2} + \phi_{i,j+2} + \phi_{i+2,j}$$

$$+ 2(\phi_{i-1,j-1} + \phi_{i-1,j+1} + \phi_{i+1,j-1} + \phi_{i+1,j+1})$$

$$+ 20\phi_{i,j} - 8(\phi_{i-1,j} + \phi_{i,j-1} + \phi_{i,j+1} + \phi_{i+1,j})]$$

$$= -E\nabla^2(\alpha T)_{i,j} - g_{i,j} - \Delta g_{i,j} \qquad (10.2.1)$$

where δ is the grid spacing.

Equations similar to (10.2.1) are written for each of the $n \times m$ stations. There then results $n \times m$ linear equations for the $n \times m$ unknown ϕ's, assuming the right sides are known. These equations can now be solved by any of the numerous methods of solving large sets of simultaneous linear algebraic equations. Once these are solved, the stresses can be computed from the relations

$$\sigma_{x,ij} = \frac{\phi_{i,j+1} + \phi_{i,j-1} - 2\phi_{i,j}}{\delta^2}$$

$$\sigma_{y,ij} = \frac{\phi_{i+1,j} + \phi_{i-1,j} - 2\phi_{i,j}}{\delta^2} \quad (10.2.2)$$

$$\tau_{xy,ij} = \frac{\phi_{i-1,j+1} - \phi_{i-1,j-1} - \phi_{i+1,j+1} + \phi_{i+1,j-1}}{4\delta^2}$$

The strains are computed from (10.1.3) and the plastic strain increments from either (10.1.4) or (10.1.5), together with the stress-strain curve. The function Δg is now changed and the solution obtained again. The process is continued until convergence is obtained. It is to be noted that only the right sides of the set of $n \times m$ simultaneous equations change from iteration to iteration. It should also be noted that although equation (10.2.1) has been written for equal spacing between stations, it is possible to write the finite-difference formulas for unequal spacing.

Once the calculation has converged for a given increment of load (in this case, thermal load) and Δg determined, Δg is added to g, the load is incremented, and a new calculation started to determine the value of the plastic strain increments and stresses due to the new increment of load. It is to be remembered that at any station for which σ_e is less than the yield stress, the plastic strain increments are set equal to zero.

As an example, such a solution was obtained for a square plate with a parabolic temperature distribution given by $T = T_0(y^2 - \frac{1}{3})$, the constant T_0 being raised in increments until large zones of plastic flow occurred. Linear strain hardening was assumed with the strain-hardening parameter m taken to be 0.1. Some of the results are shown in Figures 10.2.3 and 10.2.4. Only one quadrant of the plate is shown, the other three quadrants being identical because of symmetry. In these computations 20 stations were taken in the x direction and 20 in the y direction, resulting in a set of 400 simultaneous equations to solve.

Figure 10.2.3 shows the rate of growth of the regions of incipient flow. The curves are the loci of all the points of incipient plastic flow for a given value of loading parameter $\tau_0 = T_0 E \alpha / \sigma_0$. Plastic flow starts first at the centers of the four sides of the plate and moves rapidly inward. Plastic flow does not start at the center of the plate until it is well developed at the sides. Once plastic flow has started at the center, however, the rate of growth of this zone is greater than at the sides.

Figure 10.2.4 shows the plastic strain trajectories for the maximum load,

Sec. 10-2] Elastoplastic Thermal Problem for a Finite Plate 221

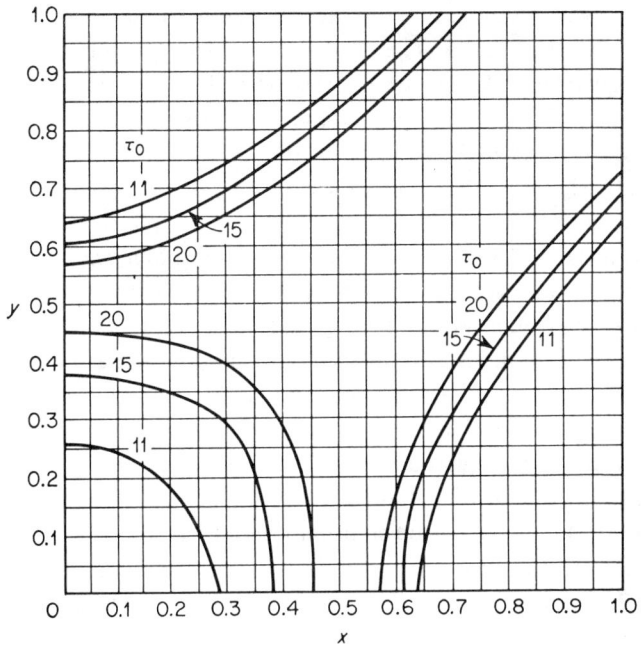

FIGURE 10.2.3 Curves of incipient plastic flow.

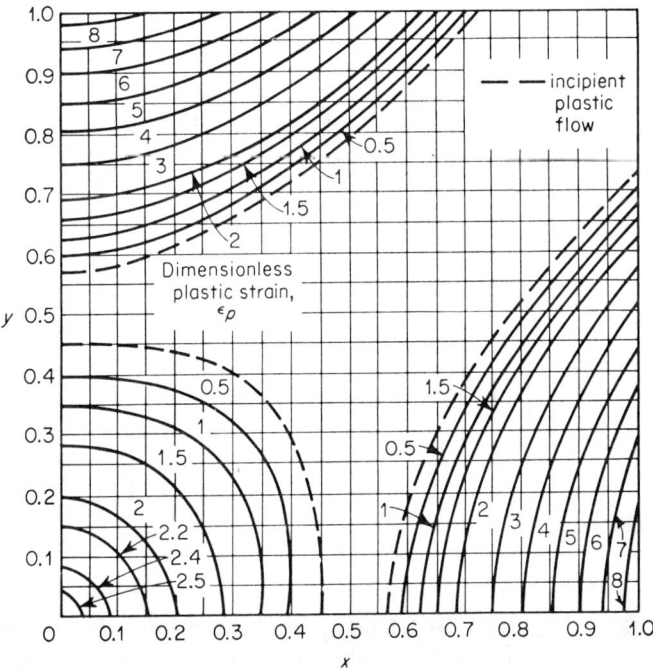

FIGURE 10.2.4 Equivalent plastic strain trajectories.

$\tau_0 = 20$. ϵ_p is the summation of the equivalent plastic strain increment divided by the yield strain. The curves shown are the loci of all points of constant equivalent plastic strain. The dashed curves, labeled curves of incipient plastic flow, which represent the elastoplastic boundary, correspond to the curves labeled $\tau_0 = 20$ in Figure 10.2.3.

Since both Figures 10.2.3 and 10.2.4 show only one quadrant, it is apparent that for the entire plate there will be five regions of plastic flow. There is a region about the center almost circular in shape and four regions identical in shape along the four sides of the plate. There is no plastic flow at the corners of the plate. Maximum plastic flow occurs at the centers of the four sides, where the plastic straining is roughly three times as great as at the center of the plate. Furthermore, the strain gradient is considerably steeper than that near the center.

The accuracy of the method was checked by comparing the results of the elastic solution ($g = 0$) with those obtained by other methods, and by comparing the results of the plastic solution obtained at the center section of a 3×1 plate with the closed-form solution for the infinite plate discussed in Section 9.2. The agreement was excellent, as can be seen in Figure 10.2.5. S and ϵ are defined in Section 9.2; i.e.,

$$S = \frac{\sigma_x}{\sigma_o} \quad \epsilon = \frac{\epsilon_x - \alpha T}{\sigma_o/E}$$

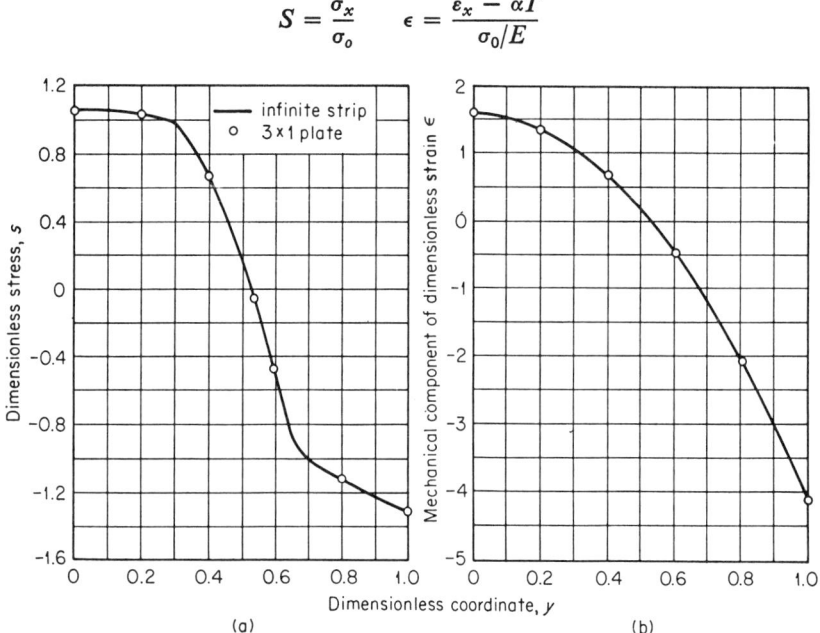

FIGURE 10.2.5 Comparison of finite-difference elastoplastic stress-strain solutions for 3×1 plate with closed-form solution for infinite strip: (a) dimensionless stress; (b) dimensionless strain.

Since the solution obtained was an incremental one using finite increments of load, it was thought worthwhile to run an experiment to determine the effect of increment size $\Delta\tau_0$ on the final solution. It was found that $\Delta\tau_0 = 2.5$ was approximately the largest increment size that produced no appreciable difference in the final stress and strain distribution at $\tau_0 = 20$. Calculations were also performed using the two iterative techniques described in Section 9.1, that is, using equations (10.1.4) to calculate the plastic strain increments and using equations (10.1.5). As expected for very large load increments, the first method (using the Prandtl–Reuss relations) diverged, whereas the second method, using the plastic strain–total strain equations, always converged.

For those cases where both methods converged they gave identical answers. The solution to this problem can be found discussed in greater detail in reference [2].

The above calculations were performed for a very thin plate. For a very thick plate, a condition of plane strain would exist at planes far from the surfaces. To obtain a solution to this problem only minor modifications of previous calculation method are necessary. Thus g and Δg must be computed using equations (10.1.14) instead of (10.1.13). Also it is now necessary to compute σ_z. This can be done as for the case of the cylinder in Section 9.4. From the third of equations (10.1.8),

$$\sigma_z = E\varepsilon_z + \mu(\sigma_x + \sigma_y) + E(\varepsilon_x^P + \varepsilon_y^P + \Delta\varepsilon_x^P + \Delta\varepsilon_y^P) \qquad (10.2.3)$$

Substituting into the relation

$$\int_A \sigma_z \, dA = 0 \qquad (10.2.4)$$

where A is the cross-sectional area of the plate, enables one to determine ε_z, which is a constant, and hence σ_z. The rest of the computations then proceed as for the case of plane stress. The above holds for *generalized* plane strain. For plane strain, $\varepsilon_z = 0$ and σ_z is computed directly from (10.2.3).

10-3 ELASTOPLASTIC PROBLEM OF THE INFINITE PLATE WITH A CRACK

The elastic solution of the infinite plate with a crack was first obtained by Inglis [3] and was used by Griffith [4] in his theory of brittle crack propagation. It was pointed out by Orowan [5] that, for ductile materials, the plastic strain energy was a major factor in the energy balance of the system and could not be ignored in any analysis of crack stability. Since no solution was available for the plastic strain field, various assumptions were introduced to take into account the plastic zone at the tip of a crack, such as Irwin's "equivalent crack length" [6] and Neuber's "plastic particle" [7].

In the present section it will be shown how the previous method can be used to obtain at least approximately the elastoplastic strain field in the vicinity of a crack, for a *strain-hardening material*. Since the solution can be obtained only numerically, the presence of the mathematical singularity at the tip of the crack makes accurate answers very difficult to obtain. The solution to be presented is therefore not intended to provide accurate quantitative results, nor is it necessarily, or even probably, the best method of solving this problem. It is intended primarily to provide qualitative information on the effect of strain hardening and on the differences that might be expected between plane stress and plane strain solutions.

Consider the case of an infinite plate with a central crack 2 units long, with a uniaxial tensile load at infinity perpendicular to the plane of the crack. As before, all the stresses are made dimensionless by dividing by the yield stress and all the strains are divided by the yield strain. The tensile stress at infinity is also divided by the yield stress. Since half the crack length is taken as unity, the x and y coordinates are dimensionless in terms of the half-crack length. We shall attempt a solution using finite differences, as in the case of the rectangular plate of Section 10.2.

We are faced, however, with two problems in applying the finite-difference methods previously described. In the first place, it is obviously impossible to cover an infinite region with a finite-difference grid. Second, the crack tip is a singular point of the stress field and, as pointed out in reference [8], the error in the finite-difference formulation in the vicinity of the singularity spreads to other points. The best procedure in this type of problem is to subtract out the singularity, if possible, giving a new problem with different boundary conditions, but which will be well behaved.

In the present problem we attempt to minimize both the above difficulties by subtracting out the elastic solution from the problem. The elastic solution contains a singularity at the crack tip and also satisfies the boundary conditions. We are thus left with a well-behaved problem with homogeneous boundary conditions. Furthermore, it has been shown experimentally by Dixon [9] that for most materials (mild steel, which has a lower yield point, is one exception), the strain field outside the plastic zone is the same as the elastic strain field; i.e., the elastic solution prevails outside the plastic zone. The assumption is therefore made that the infinite plate can be replaced by a finite rectangle with an edge crack upon whose boundaries the elastic stress field acts. Because of symmetry, only one quadrant of the plate need be considered. We proceed, therefore, in the following manner.

Sec. 10–3] Elastoplastic Problem of the Infinite Plate with a Crack

A rectangular section is chosen as shown in Figure 10.3.1. On the upper, left, and right boundaries of this rectangle, the elastic stress field for the

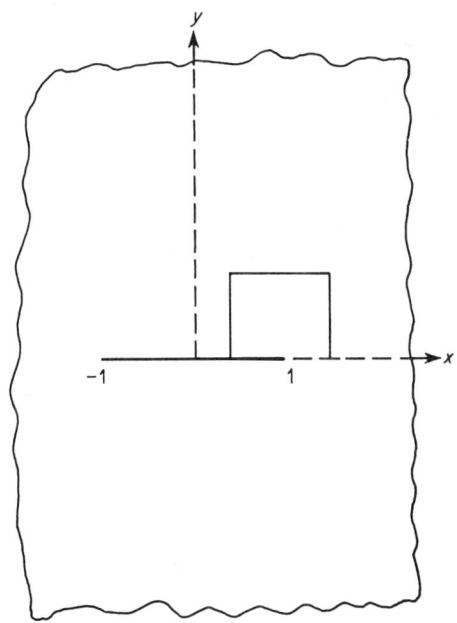

FIGURE 10.3.1   Rectangular boundary for finite-difference grid.

infinite plate acts. The lower boundary is a line of symmetry. The differential equation to be satisfied is equation (10.1.16) without the temperature term; i.e.,

$$\nabla^4 \phi = -g(x, y) - \Delta g(x, y) \tag{10.3.1}$$

Let

$$\phi = \phi_e + \phi_p \tag{10.3.2}$$

where ϕ_e is the elastic solution to the problem. Then ϕ_e satisfies the differential equation

$$\nabla^4 \phi_e = 0 \tag{10.3.3}$$

and the boundary conditions. ϕ_p then satisfies

$$\nabla^4 \phi_p = -g(x, y) - \Delta g(x, y) \tag{10.3.4}$$

and homogeneous boundary conditions. If it is assumed that the elastic solution prevails on the upper, left, and right boundaries of the rectangle shown in Figure 10.3.1, then we are left with the problem of finding a function

ϕ_p, satisfying equation (10.3.4), symmetric about the lower boundary and equal to zero along the crack and on the other three boundaries of the quadrant. This function can be found using finite-difference and successive approximations just as was done for the thermal stress problem of Section 10.2.

The rectangle shown is covered by a grid and the differential equation (10.3.4) is written in finite-difference form for every point of the grid. The solution then proceeds by successive approximations as previously described. The second iterative method employing the plastic strain–total strain equations was used. One additional step is, however, needed. After the function ϕ_p is obtained by solving the set of simultaneous equations, the elastic stress function, ϕ_e, must be added to ϕ_p before computing the total stresses. The rest of the iterative scheme proceeds exactly as before. Both the plane stress and plane strain solutions can be readily obtained, as described in Section 10.2.

The elastic stress function ϕ_e has been obtained in many ways. Using Muskhelishvili's solution [10], it can be written

$$\phi_e = \frac{P}{4} \text{Re} \left[-z\bar{z} + z^2 + 2\bar{z}\sqrt{z^2 - 1} - 2\ln(z + \sqrt{z^2 - 1}) \right] \quad (10.3.5)$$

where Re stands for the "real part of," z is the complex variable $x + iy$, $\bar{z}$ is its conjugate, and P is the stress at infinity divided by the yield stress. For every iteration the value of ϕ_e, as computed from equation (10.3.5), is added to the ϕ_p values at every station before the stresses are computed.

The first computation made was to determine the validity of the assumption that the elastic stress field prevails outside the plastic zone. For this purpose three different-sized rectangles were chosen to enclose the plastic zone, as shown in Figure 10.3.2. For each of these rectangles a plane stress calculation was made by the method described to determine the extent of the plastic zone. Linear strain hardening was assumed with the strain-hardening parameter m taken equal to 0.1, and the load was raised in steps in four increments from a value of 0.2 times the yield stress to 0.5 times the yield stress. The elastoplastic boundaries at a load of 0.5 times the yield stress are shown in the figure. It is seen that there is not too great a difference in the plastic zone size for all three rectangular boundaries used, particularly the last two, indicating that the effect of the plastic zone on the stress field outside the zone dies out rapidly, as indicated by Dixon's experiments [9].

Also shown in Figure 10.3.2 is the elastic yield locus, i.e., the locus of all points which are just at the yield stress as computed elastically. This is commonly assumed to be the boundary of the plastic zone. It can be seen from the figure that this assumption can be appreciably in error. This is further illustrated in Figure 10.3.3, where the elastoplastic boundary is shown

Sec. 10–3] Elastoplastic Problem of the Infinite Plate with a Crack

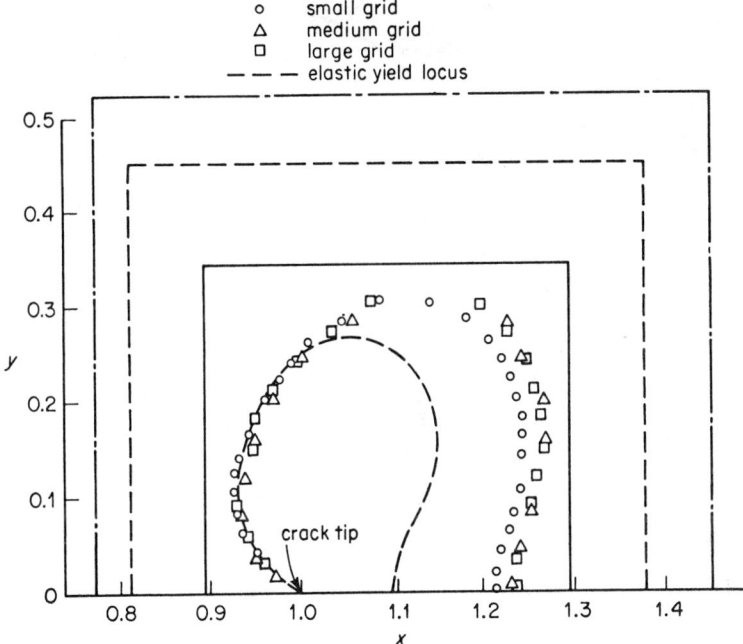

FIGURE 10.3.2 Effect of grid size on plastic zone.

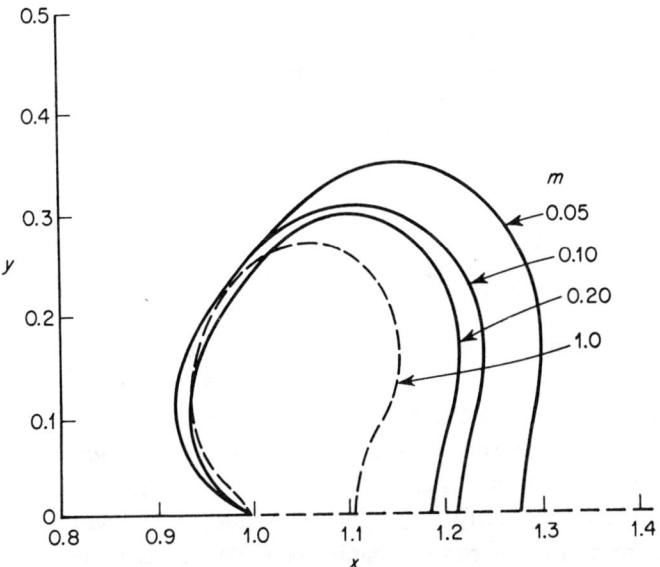

FIGURE 10.3.3 Variation of plastic zone shape with strain-hardening parameter.

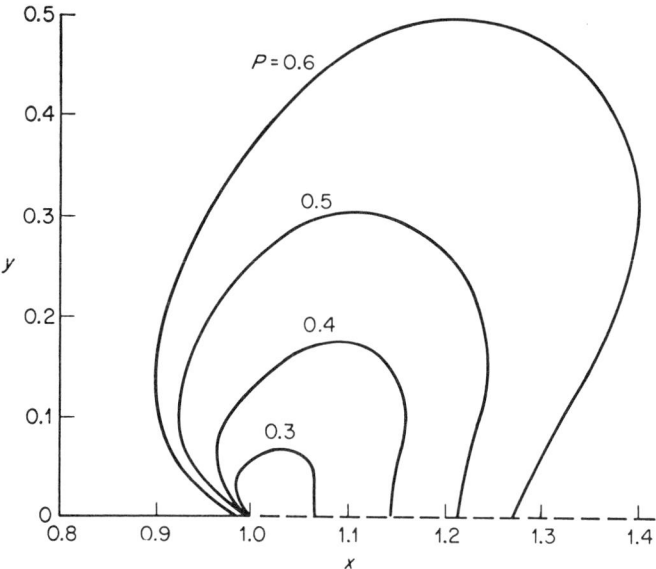

FIGURE 10.3.4 Growth of plastic zone size with load; plane stress.

as a function of the strain-hardening parameter m. Figure 10.3.4 shows the growth of the plastic zone with load for a strain-hardening parameter, $m = 0.1$. It is seen that the shape of the elastoplastic boundaries remain similar to each other as the load increases, again in agreement with the experimental results of reference [9].

All the previous results were for the case of plane stress. The plane strain case provides no additional difficulties. One uses the definition of the functions g and Δg given in equations (10.1.14) and computes σ_z from (10.2.3). Figure 10.3.5 shows the growth of the plastic zone with load for the case of plane strain for linear strain-hardening with m equal to 0.1. Comparison of Figures 10.3.4 and 10.3.5 shows that the areas covered by the plastic zones for plane strain are considerably smaller than those for plane stress for the same load ratios.

As mentioned at the beginning of this section, the above results cannot be looked upon as quantitatively accurate without further verification. This is due to several factors. In the first place, to ensure that the assumption that the elastic solution prevails on the boundary of the rectangle is reasonably correct, the rectangle size was kept at roughly twice the plastic zone size. This meant that a majority of the grid points did not fall within the plastic zone. To keep the number of simultaneous equations to be solved at a reasonable value, the grid spacing was necessarily very coarse with respect to the plastic zone size. Thus, although as many as 800 simultaneous equations

Sec. 10-3] Elastoplastic Problem of the Infinite Plate with a Crack

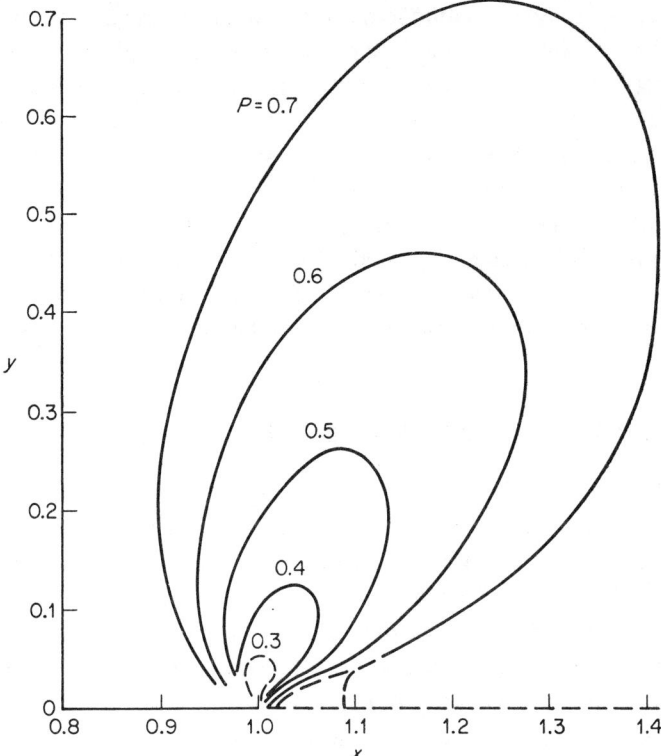

FIGURE 10.3.5 Growth of plastic zone size with load; plane strain.

were solved for one particular case, less than 300 of the grid points were within the plastic zone. It is, of course, possible to improve this situation by using a variable grid spacing, which was not done in the above calculations.

Second, the effect of the singularity on the results of the computation is not at all clear. The elastic solution ϕ_e, which is subtracted as previously described, has a square-root stress singularity. If the plasticity solution has the same order singularity, then the technique used in subtracting the singularity would be effective. However, it is likely that the order of singularity changes as plastic flow progresses. This has been indicated in results obtained by Swedlow [11]. If this is true, then subtracting the elastic singularity may not be so effective and errors due to the singularity may be propagated through the solution. This requires further investigation.

A solution using the successive-approximation technique and finite differences was also obtained in reference [12] for a finite plate with symmetric edge cracks. Again, because of the coarseness of the grid and the stress singularities, the results cannot be expected to be accurate, but do give an indication of the shape and growth of the plastic zone.

A somewhat different technique was employed in reference [11] for solving the problem of a center-cracked symmetrically loaded plate. An equation similar to (10.3.1) was derived in terms of the rate of change (or increment) of the stress function. For the case of plane stress, this equation is quasilinear; i.e., it is linear in $\dot{\phi}$ although nonlinear in ϕ. Assuming the stresses and strains are known at a given time, the increment in ϕ, and consequently in the stresses and strains, can be computed using the stress–strain curve. Both finite-difference and finite-element methods [15] were used in reference [11]. The latter method is said to give better results. The plane strain problem, however, cannot readily be solved by the method of reference [11].

10-4 STRAIN-INVARIANCE PRINCIPLE

A study of the previous examples, as well as other examples involving primarily thermal loading, has led to an interesting observation of practical importance. For thermal stress problems without additional loads, the total strains do not change very much because of the plastic deformation compared to the elastically computed strains. Hence an elastic computation gives approximately the correct values of the total strains for these types of problems. This has been called the *strain-invariance principle* [13, 14].

To compute the plastic strains using strain invariance, the plastic strain–total strain relations are used. If ε_1, ε_2, and ε_3 are known from the elastic solution (principal strains are used for brevity), then ε_{et} is computed from

$$\varepsilon_{et} = \frac{\sqrt{2}}{3} \sqrt{(\varepsilon_1 - \varepsilon_2)^2 + (\varepsilon_2 - \varepsilon_3)^2 + (\varepsilon_3 - \varepsilon_1)^2} \qquad (10.4.1)$$

ε_p can then be obtained from the plot of ε_p versus ε_{et} derived from the stress–strain curve. The plastic strains are computed from

$$\varepsilon_1^P = \frac{\varepsilon_p}{3\varepsilon_{et}} (2\varepsilon_1 - \varepsilon_2 - \varepsilon_3)$$

$$\varepsilon_2^P = \frac{\varepsilon_p}{3\varepsilon_{et}} (2\varepsilon_2 - \varepsilon_3 - \varepsilon_1) \qquad (10.4.2)$$

The stresses are computed from the stress–strain relations

$$\sigma_1 = \lambda(\varepsilon_1 + \varepsilon_2 + \varepsilon_3 - 3\alpha T) + 2G(\varepsilon_1 - \alpha T - \varepsilon_1^P)$$
$$\sigma_2 = \lambda(\varepsilon_1 + \varepsilon_2 + \varepsilon_3 - 3\alpha T) + 2G(\varepsilon_2 - \alpha T - \varepsilon_2^P) \qquad (10.4.3)$$
$$\sigma_3 = \lambda(\varepsilon_1 + \varepsilon_2 + \varepsilon_3 - 3\alpha T) + 2G(\varepsilon_3 - \alpha T - \varepsilon_3^P)$$

This method only applies for a one-step loading problem.

Sec. 10–4] Strain-Invariance Principle

It should be noted that the total strains, being a solution of the elastic problem, satisfy the compatibility relations. The stresses, however, do not satisfy equilibrium equations. If the plastic zone is relatively small compared to the elastic zone, then, since the elastic stresses do satisfy the equilibrium equations, equilibrium will be satisfied approximately on the whole, and the strain-invariance answers would be expected to be reasonably correct. However, if the plastic zone is very large, then equilibrium on the whole will not be satisfied and the strain-invariance answers may be appreciably in error.

The results of strain-invariance answers for the plate problems of Sections 9.2 and 10.2 are shown in Figures 10.4.1 and 10.4.2. Figure 10.4.1 shows a

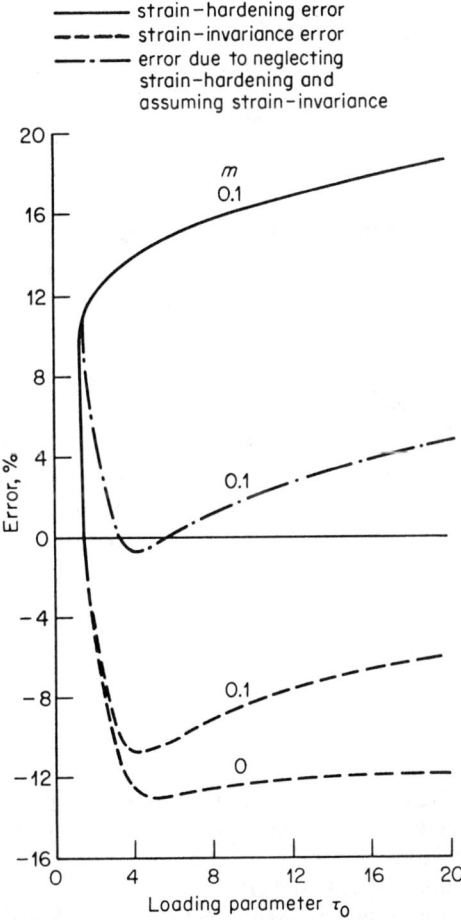

FIGURE 10.4.1 Per cent error in plastic strain using strain invariance; infinite strip.

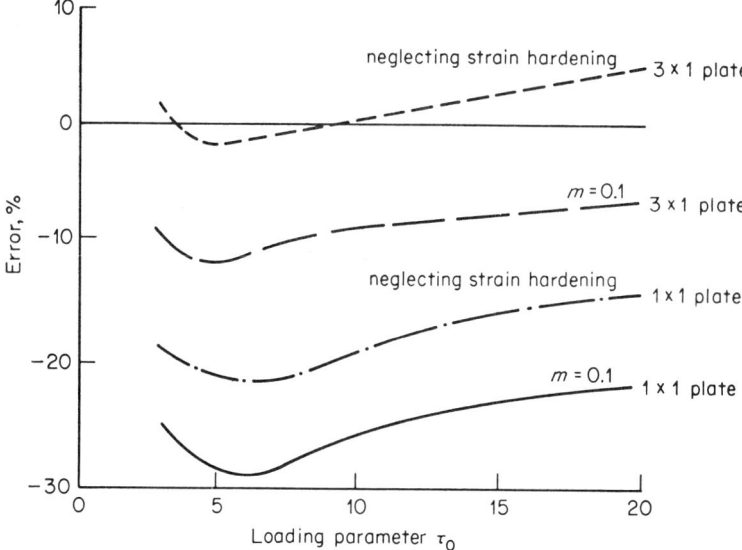

FIGURE 10.4.2 Same as Figure 10.4.1; 1 × 1 and 3 × 1 plates.

plot of the error in the maximum plastic strain assuming strain invariance, as a function of the loading parameter τ_0, for the thin infinite strip of Section 9.2. Also shown is the error in neglecting strain hardening for a strain-hardening parameter $m = 0.1$ (i.e., assuming $m = 0$ instead of $m = 0.1$). It is seen that the error in neglecting strain hardening is on the conservative side (gives larger plastic strains than actually exist); the error in strain invariance is on the nonconservative side. It is interesting to note also, as is shown in the figure, that if one makes both assumptions, i.e., neglects strain hardening and assumes strain invariance, the errors tend to cancel each other. In any case, from an engineering viewpoint a 10 per cent or even a 20 per cent error is not very large, considering our inexact knowledge of material behavior. Figure 10.4.2 shows similar results for the square plate and the 3 × 1 plate of Section 10.2.

Problems

1. Derive equation (10.1.11).
2. Obtain the finite-difference equation corresponding to equation (10.2.1) assuming the grid spacing in the y direction is twice that in the x direction.
3. Determine the expression for ε_z for the case of generalized plane strain by means of equations (10.2.3) and (10.2.4).
4. Show that the stress function given by equation (10.3.5) satisfies the boundary conditions for an infinite plate with a crack 2 units long and loaded at infinity with a tensile stress P normal to the plane of the crack.

References

1. I. S. Sokolinkoff, *Mathematical Theory of Elasticity*, McGraw-Hill, New York, 1956, p. 260.
2. E. Roberts, Jr., and A. Mendelson, Analysis of Plastic Thermal Stresses and Strains in Finite Thin Plate, *NASA TN D-2206*, 1964.
3. C. E. Inglis, Stresses in a Plate Due to the Presence of Cracks and Sharp Corners, *Trans. Inst. Naval Arch.*, **55**, No. 1, 1913, pp. 219–239.
4. A. A. Griffith, The Phenomena of Rupture and Flow in Solids, *Phil. Trans. Roy. Soc. London*, A**221**, 1921, pp. 163–198.
5. E. Orowan, *Fundamentals of Brittle Behavior in Metals, Fatigue and Fracture of Metals*, M.I.T. Press, Boston, Mass., and John Wiley, New York, 1952, pp. 139–169.
6. *Fracture Testing of High Strength Sheet Materials*, A Report of a Special ASTM Committee, *ASTM Bull. 243*, 1960.
7. H. Neuber, Theory of Notch Stresses, *AEC TR-4547*, 1958.
8. L. Fox, *Numerical Solution of Ordinary and Partial Differential Equations*, Pergamon Press, London, 1962, p. 301.
9. J. R. Dixon, The Effect of Local Plastic Deformation on the Stress Distribution Around a Crack, *NEL Rpt. No. 71*, Dec. 1962.
10. N. I. Muskhelishvili, *Some Basic Problems in the Mathematical Theory of Elasticity*, P. Noordhoff, Groningen, 1953.
11. J. L. Swedlow, M. L. Williams, and W. H. Yang, Elasto-Plastic Stresses and Strains in Cracked Plates, *Proceedings of the International Conference on Fracture, Sendai, Japan, Sept. 1965*; also *Galcit SM 65-14*, July 1965.
12. I. S. Tuba, A Method of Elastic–Plastic Plane Stress and Strain Analysis, *J. Strain Anal.*, **1**, No. 2, 1966, pp. 115–120.
13. A. Mendelson and S. S. Manson, Practical Solution of Plastic Deformation Problems in the Elastic–Plastic Range, *NASA TR R-28*, 1959.
14. S. S. Manson, *Thermal Stress and Low Cycle Fatigue*, McGraw-Hill, New York, 1966, p. 196.
15. M. J. Turner, R. W. Clough, H. C. Martin, and L. J. Tupp, Stiffness and Deflection Analysis of Complex Structures, *J. Aerospace, Sci.*, **23**, No. 9, 1965, pp. 805–823.

CHAPTER 11

THE TORSION PROBLEM

In this chapter several problems of torsion will be considered. Although these problems are in general two-dimensional, they are usually simpler than plane stress or plane strain problems. This is due to the fact that the governing differential equation is of the second order rather than the fourth order, as is the case for the plane problems. Furthermore, for bars of circular cross section the loading is radial, so that incremental plasticity theories need not be used at least for this case.

11-1 TORSION OF PRISMATIC BAR. GENERAL RELATIONS

Consider a prismatic bar subject to a twisting couple as shown in Figure 11.1.1. One end of the bar is assumed fixed against rotation but not against warping, and at the other end a couple M with a moment along the z axis is applied. The elastic solution for a beam of arbitrary cross section was first obtained by Saint-Venant [1], using the semiinverse method. In this method some simplifying assumptions concerning the stresses and/or displacements are first made and a solution to the given problem obtained. If this solution satisfies the necessary equations of equilibrium and compatibility and the stress–strain relations, as well as the boundary conditions, then the solution is a correct one.

Sec. 11-1] Torsion of Prismatic Bar. General Relations

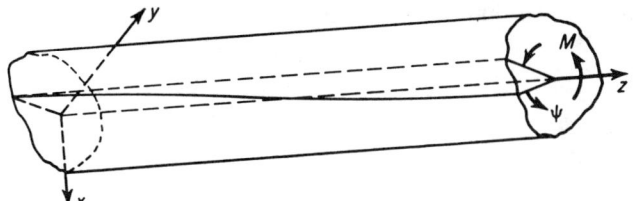

FIGURE 11.1.1 Prismatic bar subject to twisting couple.

For the case of the beam twisted as shown, it is assumed that the angle of twist of a given cross section is directly proportional to its distance from the origin. Thus if the angle of twist per unit length is α, then a section a distance z from the origin will have rotated an amount

$$\psi = \alpha z \qquad (11.1.1)$$

Now consider a point P originally at location (x, y, z). This point will rotate to a position P_1 with coordinates (x_1, y_1, z_1), where

$$x_1 = x + u \qquad y_1 = y + v \qquad z_1 = z + w$$

u, v, and w being the displacements, assumed small. The projection of these points on the xy plane are shown in Figure 11.1.2. If the polar coordinates of the point P are r and θ, then we have

$$x = r \cos \theta \qquad y = r \sin \theta$$
$$x_1 = r \cos (\theta + \psi) \qquad y_1 = r \sin (\theta + \psi) \qquad (11.1.2)$$

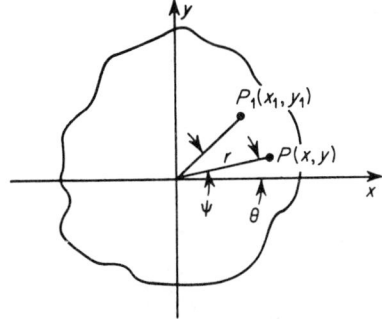

FIGURE 11.1.2 Projection of twisted cross section.

and if it is assumed that ψ is small, so that

$$\cos \psi \cong 1 \qquad \sin \psi \cong \psi$$

then

$$\begin{aligned} u &= x_1 - x = -r\psi \sin \theta = -y\psi \\ v &= y_1 - y = r\psi \cos \theta = x\psi \end{aligned} \tag{11.1.3}$$

or

$$\begin{aligned} u &= -yz\alpha \\ v &= xz\alpha \end{aligned} \tag{11.1.4}$$

and

$$w = w(x, y, \alpha)$$

It is to be noted that equations (11.1.4) are essentially assumptions, since they stem directly from the assumption (11.1.1).

The displacement in the z direction $w(x, y, \alpha)$ is called the *warping function*. For the elastic problem it is directly proportional to the angle of twist per unit length α. For the plasticity problem this is in general no longer true.

Equations (11.1.4) can be substituted into the strain-displacement relations (4.1.12) to obtain the strains. Thus

$$\varepsilon_x = \varepsilon_y = \varepsilon_z = \varepsilon_{xy} = 0$$

$$\varepsilon_{xz} = \frac{1}{2}\left(-y\alpha + \frac{\partial w}{\partial x}\right) \tag{11.1.5}$$

$$\varepsilon_{yz} = \frac{1}{2}\left(x\alpha + \frac{\partial w}{\partial y}\right)$$

Substituting equations (11.1.5) into the compatibility equations (4.7.2), it is seen that the first three and the last of these equations are identically satisfied and the other two become

$$\frac{\partial}{\partial x}\left(-\frac{\partial \varepsilon_{yz}}{\partial x} + \frac{\partial \varepsilon_{xz}}{\partial y}\right) = 0$$

$$-\frac{\partial}{\partial y}\left(-\frac{\partial \varepsilon_{yz}}{\partial x} + \frac{\partial \varepsilon_{xz}}{\partial y}\right) = 0 \tag{11.1.6}$$

or

$$\frac{\partial \varepsilon_{yz}}{\partial x} - \frac{\partial \varepsilon_{xz}}{\partial y} = \text{constant} \tag{11.1.7}$$

Sec. 11-1] Torsion of Prismatic Bar. General Relations

From (11.1.5) it follows that the constant in (11.1.7) must equal α. Therefore, the compatibility equation for this problem becomes

$$\frac{\partial \varepsilon_{yz}}{\partial x} - \frac{\partial \varepsilon_{xz}}{\partial y} = \alpha \tag{11.1.8}$$

It should be emphasized again that the shear strains ε_{yz} and ε_{xz} are the components of the strain tensor as defined by equation (4.1.12) and are therefore equal to one half the engineering shearing strains used by other authors.

Looking now at the equilibrium equations (3.2.2), it is seen by virtue of (11.1.5) and the stress–strain relations that the first two are identically satisfied and the third becomes

$$\frac{\partial \tau_{xz}}{\partial x} + \frac{\partial \tau_{yz}}{\partial y} = 0 \tag{11.1.9}$$

which is the equilibrium equation for the problem.

We now turn to the stress–strain relations. These can be written

$$\varepsilon_{xz} = \frac{1}{2G} \tau_{xz} + \varepsilon_{xz}^P$$

$$\varepsilon_{yz} = \frac{1}{2G} \tau_{yz} + \varepsilon_{yz}^P \tag{11.1.10}$$

where ε_{xz}^P and ε_{yz}^P are the accumulated plastic components of the total shear strains. Substituting (11.1.10) into (11.1.8), the compatibility equation is obtained in terms of the stresses:

$$\frac{\partial \tau_{yz}}{\partial x} - \frac{\partial \tau_{xz}}{\partial y} = 2G\alpha + g(x, y) \tag{11.1.11}$$

where

$$g(x, y) = 2G \left(\frac{\partial \varepsilon_{xz}^P}{\partial y} - \frac{\partial \varepsilon_{yz}^P}{\partial x} \right) \tag{11.1.12}$$

We now introduce a stress function ϕ such that

$$\tau_{xz} = \frac{\partial \phi}{\partial y} \qquad \tau_{yz} = -\frac{\partial \phi}{\partial x} \tag{11.1.13}$$

Then the equilibrium equation (11.1.9) is identically satisfied and the compatibility equation (11.1.11) becomes

$$\nabla^2 \phi \equiv \frac{\partial^2 \phi}{\partial x^2} + \frac{\partial^2 \phi}{\partial y^2} = -2G\alpha - g(x, y) \quad (11.1.14)$$

For the elastic problem $g(x, y)$ is equal to zero.

The boundary conditions for the problem can be obtained directly from equations (3.3.1). The first two of these are identically satisfied and the third reduces to

$$l\tau_{xz} + m\tau_{yz} = 0 \quad (11.1.15)$$

From Figure 11.1.3 it can be seen that

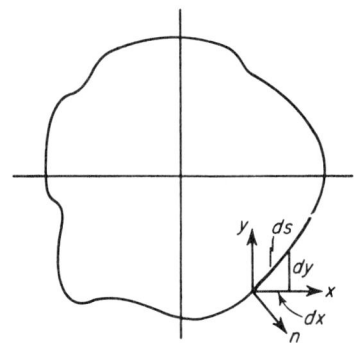

FIGURE 11.1.3 Geometry of boundary.

$$l = \cos(n, x) = \frac{dy}{ds}$$
$$m = \cos(n, y) = -\frac{dx}{ds} \quad (11.1.16)$$

and (11.1.15) becomes, upon combining with (11.1.13),

$$\frac{\partial \phi}{\partial y}\frac{dy}{ds} + \frac{\partial \phi}{\partial x}\frac{dx}{ds} = \frac{d\phi}{ds} = 0 \quad (11.1.17)$$

or ϕ equals a constant along the boundary. In the case of simply connected boundaries, e.g., solid bars, this constant can be chosen arbitrarily, since we are interested only in the derivatives of ϕ. It is, therefore, for convenience chosen to be zero, so that the boundary condition becomes

$$\phi = 0 \quad \text{on the boundary} \quad (11.1.18)$$

Sec. 11-1] Torsion of Prismatic Bar. General Relations

It should be noted that equation (11.1.15) is equivalent to the statement that the resultant shear stress at the boundary is tangent to the boundary. This must always be true if the lateral surface of the bar is force-free. It also follows, since equation (11.1.17) and consequently equation (11.1.15) hold for any line $\phi = $ constant in the cross section, that the resultant shear stress at any point is tangential to the $\phi = $ constant line passing through that point. The lines $\phi = $ constant are called the *stress trajectories*. Furthermore, from (11.1.13) it follows that the resultant shear stress τ at any point is given by

$$\tau = \sqrt{\tau_{xz}^2 + \tau_{yz}^2} = \sqrt{\left(\frac{\partial \phi}{\partial y}\right)^2 + \left(\frac{\partial \phi}{\partial x}\right)^2} \qquad (11.1.19)$$

or

$$\tau = |\text{grad } \phi|$$

The resultant shear stress is thus equal to the gradient of the stress function ϕ.

Let us now calculate the resultant forces and moments acting on any cross section. The force in the x direction is given by

$$Q_x = \iint \tau_{xz} \, dx \, dy = \iint \frac{\partial \phi}{\partial y} \, dx \, dy$$

$$= \int [\phi(x, A) - \phi(x, B)] \, dx = 0 \qquad (11.1.20)$$

where the double integral is taken over the area of the cross section and $\phi(x, A)$ and $\phi(x, B)$ are the values of ϕ at two opposite points of the boundary at a given value of x, and are consequently equal to zero because of the boundary condition (11.1.18). In a similar fashion, it follows that the y component of the resultant stress is zero. Finally, the torque acting on the section is computed from

$$M = \iint (\tau_{yz} x - \tau_{xz} y) \, dx \, dy$$

$$= -\iint \left(\frac{\partial \phi}{\partial x} x + \frac{\partial \phi}{\partial y} y\right) dx \, dy \qquad (11.1.21)$$

Integrating by parts and making use again of the boundary condition (11.1.18) results in

$$M = 2 \iint \phi \, dx \, dy \qquad (11.1.22)$$

If a solution is obtained for equation (11.1.14) subject to the boundary condition (11.1.18), then the stresses can be computed from (11.1.13) and the strains from (11.1.10). (The calculation of the plastic strains will be discussed subsequently.) The equilibrium and compatibility equations, the stress–strain relations, and the boundary conditions will all be satisfied. The assumption (11.1.1), or equivalently (11.1.4), thus leads to the correct solution of the torsion problem. It should be noted, however, that the solution requires the same stress distributions to act on every cross section, including the end sections. However, by Saint-Venant's principle, if the beam is sufficiently long, the solution will be valid for all cross sections far enough away from the ends for any stress distribution acting on the ends, provided the resultant force is zero and the resultant moment is given by (11.1.22).

Equation (11.1.14) is the well-known Poisson's equation, which is encountered so frequently in mathematical physics. Its solution can be obtained by many different techniques. For the elasticity problem, the right side is a constant, and solutions, at least for simple shapes, can readily be obtained. For the plasticity problem the right side is a function of the plastic strains and is therefore unknown until the solution is obtained. (The problem is nonlinear.) For materials with or without strain hardening we can use the method of successive elastic solutions described in Chapter 9. For perfectly plastic materials, a simpler, more specialized approach, to be described subsequently, can sometimes be used. In either case the elasticity solution is a prerequisite to the plasticity solution, and we shall therefore first present some solutions to the elasticity problem.

11-2 ELASTICITY SOLUTIONS

Equation (11.1.14), with $g(x, y)$ equal to zero, can sometimes be solved by guessing a solution, if the boundary of the cross section is of a simple shape. For example, if the bar has an elliptic cross section, the equation of the boundary curve being

$$\frac{x^2}{a^2} + \frac{y^2}{b^2} = 1 \qquad (11.2.1)$$

then choosing for a stress function

$$\phi = C\left(\frac{x^2}{a^2} + \frac{y^2}{b^2} - 1\right) \qquad (11.2.2)$$

Sec. 11-2] Elasticity Solutions 241

the boundary condition (11.1.18), $\phi = 0$, will obviously be satisfied, and substituting (11.2.2) into (11.1.14) results in

$$C = -\frac{a^2 b^2}{a^2 + b^2} G\alpha \tag{11.2.3}$$

Hence

$$\phi = -\frac{a^2 b^2}{a^2 + b^2} G\alpha \left(\frac{x^2}{a^2} + \frac{y^2}{b^2} - 1\right) \tag{11.2.4}$$

is the solution for the elastic case.

The torque required to produce the angle of twist per unit length α is obtained by substituting the solution (11.2.4) into (11.1.22), giving

$$M = \frac{\pi a^3 b^3}{a^2 + b^2} G\alpha = JG\alpha \tag{11.2.5}$$

The constant GJ is called the *torsional rigidity*, or *torsional stiffness*, of the bar, for obvious reasons.

To obtain the shear stresses, the solution (11.2.4) is differentiated to give

$$\tau_{xz} = \frac{\partial \phi}{\partial y} = -2 \frac{a^2}{a^2 + b^2} G\alpha y$$

$$\tau_{yz} = -\frac{\partial \phi}{\partial x} = 2 \frac{b^2}{a^2 + b^2} G\alpha x \tag{11.2.6}$$

and the resultant shear stress is

$$\tau = \sqrt{\tau_{xz}^2 + \tau_{yz}^2} = \frac{2G\alpha}{a^2 + b^2} \sqrt{a^4 y^2 + b^4 x^2} \tag{11.2.7}$$

The maximum stress will occur on the boundary at the point closest to the axis of the bar at $x = 0$, $y = b$. Thus

$$\tau_{max} = \frac{2G\alpha a^2 b}{a^2 + b^2} \tag{11.2.8}$$

To obtain the warping function $w(x, y, \alpha)$, we first compute the strains from (11.1.10):

$$\varepsilon_{xz} = -\frac{a^2}{a^2 + b^2} \alpha y$$

$$\varepsilon_{yz} = \frac{b^2}{a^2 + b^2} \alpha x \tag{11.2.9}$$

and then, from equation (11.1.5),

$$\frac{\partial w}{\partial x} = \frac{b^2 - a^2}{b^2 + a^2} \alpha y$$

$$\frac{\partial w}{\partial y} = \frac{b^2 - a^2}{b^2 + a^2} \alpha x$$

(11.2.10)

Therefore,

$$w = \frac{b^2 - a^2}{b^2 + a^2} \alpha xy + \text{constant}$$

and since at the origin w must vanish, the constant must equal zero, and

$$w(x, y, \alpha) = \frac{b^2 - a^2}{b^2 + a^2} \alpha xy \qquad (11.2.11)$$

The lines of constant w are therefore equilateral hyperbolas.

If the cross section is a circle, then b is equal to a in all the previous formulas, resulting in

$$\phi = -\frac{G\alpha}{2}(x^2 + y^2 - a^2)$$

$$M = \frac{\pi a^4}{2} G\alpha$$

$$\tau_{xz} = -G\alpha y \qquad (11.2.12)$$

$$\tau_{yz} = G\alpha x$$

$$\tau = G\alpha r$$

$$\tau_{\max} = G\alpha a$$

$$w = 0$$

Next consider a bar whose cross section is an equilateral triangle as shown in Figure 11.2.1. The origin of x and y axes is taken at the centroid. Then the equation of the straight lines representing the three sides are

$$x = \frac{a}{\sqrt{3}}$$

$$y = \frac{x}{\sqrt{3}} + \frac{2}{3}a \qquad (11.2.13)$$

$$y = -\frac{x}{\sqrt{3}} - \frac{2}{3}a$$

Sec. 11-2] Elasticity Solutions

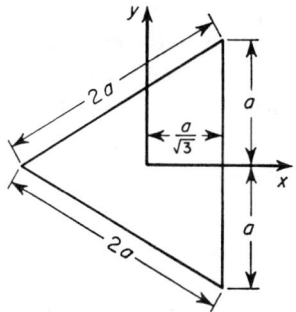

FIGURE 11.2.1 Equilateral triangle.

If we therefore choose as a stress function

$$\phi = C\left(x - \frac{a}{\sqrt{3}}\right)\left(y - \frac{x}{\sqrt{3}} - \frac{2}{3}a\right)\left(y + \frac{x}{\sqrt{3}} + \frac{2}{3}a\right)$$

ϕ will be zero on the boundary. Substituting into (11.1.14) [with $g(x, y) = 0$] gives for C:

$$C = \frac{\sqrt{3}}{2a} G\alpha$$

Hence

$$\phi = \frac{G\alpha}{2a}(\sqrt{3}\,x - a)\left(y - \frac{x}{\sqrt{3}} - \frac{2a}{3}\right)\left(y + \frac{x}{\sqrt{3}} + \frac{2a}{3}\right) \quad (11.2.14)$$

is the solution to the problem. The stresses, the torque, and the warping function can be computed as before.

As a final example, consider a bar of rectangular cross section, as shown in Figure 11.2.2. For this case the solution is not as simple as for the previous

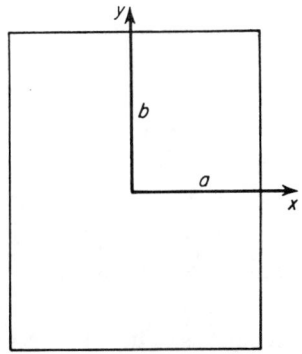

FIGURE 11.2.2 Bar of rectangular cross section.

cases, since it is no longer possible to guess a solution. Equation (11.1.14) must be solved by one of the available methods. For example, the method of separation of variables is used in reference [2], resulting in an infinite-series solution, whereas in reference [3] the Green's function is used, resulting in a double infinite series. Here the finite-difference method will be used, since it is relatively simple and can be directly extended to elastoplastic problems.

We note first that because of symmetry, only one quadrant of the section need be considered. For a square cross section the diagonals are also lines of symmetry and only one octant is used. The quadrant is divided into a grid of $n \times m$ points as shown in Figure 11.2.3. At each of the grid points,

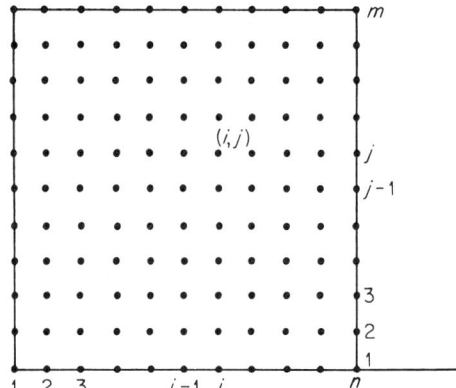

FIGURE 11.2.3 Finite-difference grid for rectangle.

such as the point designated by (i, j), equation (11.1.14) is written in finite-difference form:

$$\phi_{i+1,j} + \phi_{i-1,j} + \phi_{i,j-1} + \phi_{i,j+1} - 4\phi_{i,j} = -2G\alpha h^2 \qquad (11.2.15)$$

where h is the grid spacing assumed constant and is the same in both the x and y directions.

An equation such as (11.2.15) can be written for each of the $n \times m$ grid points, resulting in a set of $n \times m$ simultaneous linear equations for the unknown values of ϕ at each of the points. Once the ϕ's are determined as the solution of this set of equations, the shear stress can be obtained from equations (11.1.13) by numerical differentiation and the torque for a given angle of twist per unit length α from equation (11.1.22) by numerical integration.

Actually the number of equations to be solved is $(m - 1) \times (n - 1)$ rather than $m \times n$, since the boundary conditions require ϕ to be zero at the upper

Sec. 11-3] Membrane Analogy

and right boundaries of the quadrant. For a square section, only the points on the diagonal and to the right of the diagonal (or left) need be considered, resulting in $\tfrac{1}{2}n(n-1)$ equations. Along the lower boundary, because of symmetry, equation (11.2.15) becomes

$$\phi_{i+1,1} + \phi_{i-1,1} + 2\phi_{i,2} - 4\phi_{i,1} = -2G\alpha h^2 \qquad (11.2.16a)$$

and along the left boundary,

$$\phi_{1,j-1} + \phi_{1,j+1} + 2\phi_{2,j} - 4\phi_{1,j} = -2G\alpha h^2 \qquad (11.2.16b)$$

and at the center,

$$4\phi_{2,1} - 4\phi_{1,1} = -2G\alpha h^2 \qquad (11.2.16c)$$

Along the diagonal of a square,

$$2\phi_{i+1,i} + 2\phi_{i,i-1} - 4\phi_{i,i} = -2G\alpha h^2 \qquad (11.2.16d)$$

Such a solution was obtained for a square cross section using an 11 × 11 grid as shown in Figure 11.2.3. The 55 resulting equations were solved by a straightforward Gauss–Seidel process (see reference [9]). The maximum shear stress, which occurs at the centers of the sides, and the torque are compared below with the values from reference [2]:

	Ref. [2]	Finite Difference
$\tau_{max}/G\alpha a$	1.351	1.343
$M/G\alpha a^4$	2.250	2.244
$\phi_{1,1}/G\alpha a^2$	0.589	0.589

It is seen that the solution with this many grid points is sufficiently accurate.

11-3 MEMBRANE ANALOGY

Solutions of the elastic torsion problem can also be obtained experimentally by means of the *membrane analogy* suggested by Prandtl [4]. Consider a membrane such as a soap film having the same shape as the cross section of the bar being twisted. If the edges of this membrane are fixed and a pressure is applied to one side, the membrane will deflect by an amount given by the solution of the following equations (reference [2], p. 269):

$$\frac{d^2 z}{\partial x^2} + \frac{\partial^2 z}{\partial y^2} = -\frac{p}{S} \qquad (11.3.1)$$

where z is the deflection of a point of the membrane, p the applied pressure, and S the constant tension per unit length in the membrane. Comparison of equations (11.3.1) and (11.1.14) [with $g(x, y)$ equal to zero for the elastic problem] shows the analogy immediately. The deflection z of the membrane corresponds to the stress function, and if $2G\alpha$ is equal to p/S, z is equal to ϕ. The maximum slope of the membrane at any point is proportional to the resultant shear stress at the point and the volume under the membrane (between the membrane and the $z = 0$ plane) is proportional to the torque producing a twist of α per unit length. The membrane analogy is useful for determining the stress function for complicated shapes and has been used very successfully. Its extension to plastic torsion for perfectly plastic materials will be discussed in Section 11.4.

11-4 ELASTOPLASTIC TORSION. PERFECT PLASTICITY

If the applied torque is sufficiently large, plastic flow will occur. Since the maximum stress will always occur at the boundary (see Problem 2 and reference [5]), a plastic zone will start at some point on the boundary and spread toward the interior as the torque is increased. Additional plastic zones may subsequently start at other points in the cross section. For the case of the torsion problem, the yield criteria of von Mises and of Tresca both reduce to

$$\tau_{xz}^2 + \tau_{yz}^2 = k^2 \qquad (11.4.1)$$

where k is the yield stress in simple shear. According to the von Mises criterion, k is equal to $\sigma_0/\sqrt{3}$, and according to the Tresca criterion it is equal to $\sigma_0/2$, where σ_0 is the yield stress in simple tension. If the material is perfectly plastic, then equation (11.4.1) must hold everywhere in the plastic region. In terms of the stress function ϕ,

$$\left(\frac{\partial \phi}{\partial x}\right)^2 + \left(\frac{\partial \phi}{\partial y}\right)^2 = k^2 \qquad (11.4.2)$$

Sec. 11-4] Elastoplastic Torsion. Perfect Plasticity

in the plastic region, whereas in the elastic part of the bar

$$\nabla^2 \phi = -2G\alpha \qquad (11.4.3)$$

and $\phi = 0$ on the boundary of the bar. The elastoplastic boundary is unknown and is determined from the conditions that ϕ and its first derivatives (the shear stresses) are continuous across this boundary and that the resultant shear stress is less than or equal to k inside the elastic region. Equation (11.4.2) can also be written

$$|\text{grad } \phi| = k \qquad (11.4.4)$$

inside the plastic region. In other words, the slope of the ϕ surface is a constant, equal to k, in the plastic region, it is not greater than k in the elastic region, and the height and slope of the ϕ surface are continuous across the elastoplastic boundary.

The above conditions on ϕ for a perfectly plastic material suggest the extension of the membrane analogy to a partially plastic bar [6, 7]. A roof of constant slope, proportional to k, is erected with the membrane as its base. As the membrane is pressurized and deflects, it will approach the roof. The region of the membrane corresponding to the region of the bar flowing plastically will be pressed against the roof and will have the same slope as the roof. The rest of the membrane, corresponding to the elastic region, will not be touching the roof and will have a smaller slope.

The *membrane-roof analogy* furnishes a simple physical and intuitive picture of the growth of the plastic zones as the torque is increased. To obtain quantitative results will usually entail a considerable amount of labor. For the case of complete plastic yielding, the solution becomes much simpler. In this case the membrane will be in contact with the whole roof, and it is no longer necessary to use a membrane. Instead, one constructs a roof of the proper slope. This can be done by simply heaping dry sand onto a plate whose shape is similar to the cross section of the bar. Since the torque is equal to twice the volume of sand (see Problem 5), the torque required to produce complete yielding can readily be determined. Thus for a circle of radius a, the volume of the sand hill (in this case, a cone) is

$$V = \tfrac{1}{3}\pi a^2 h$$

where h is the height of the heap. Since the slope of the sand hill corresponds to the shear yield stress k,

$$k = \frac{h}{a}$$

and the torque is given by

$$M_p = \tfrac{2}{3}\pi a^3 k \qquad (11.4.5)$$

Similarly, for a rectangle with dimensions $a \times b$ [8],

$$M_p = \frac{1}{12} a^2(3b - a)k \qquad (11.4.6)$$

Figure 11.4.1 shows the *sand-hill analogies* for the above two cases.

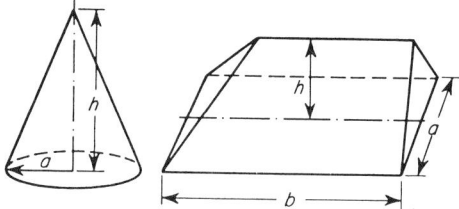

FIGURE 11.4.1 Sand-hill analogies for circular and rectangular cross sections.

11-5 ELASTOPLASTIC TORSION WITH STRAIN HARDENING

The torsion problem for strain-hardening materials has received relatively little attention. In this section it will be shown that the method of successive elastic solutions or successive approximations can readily be adapted to the torsion problem. The perfectly plastic case then becomes a simple limiting case of the more general problem of linear strain hardening. The plastic strain–total strain equations will be used, and for convenience the following dimensionless quantities are introduced.

$$U \equiv \frac{\phi}{2G\varepsilon_0 a} \qquad \beta \equiv \frac{\alpha}{\varepsilon_0}$$

$$\epsilon_x \equiv \frac{\varepsilon_{xz}}{\varepsilon_0} \qquad \epsilon_y \equiv \frac{\varepsilon_{yz}}{\varepsilon_0} \qquad \epsilon_x^P \equiv \frac{\varepsilon_{xz}^P}{\varepsilon_0} \qquad \epsilon_y^P \equiv \frac{\varepsilon_{yz}^P}{\varepsilon_0} \qquad (11.5.1)$$

$$\tau_x \equiv \frac{\tau_{xz}}{2G\varepsilon_0} = \frac{1+\mu}{\sigma_0}\tau_{xz} \qquad \tau_y \equiv \frac{\tau_{yz}}{2G\varepsilon_0} = \frac{1+\mu}{\sigma_0}\tau_{yz}$$

$$\epsilon_p \equiv \frac{\varepsilon_p}{\varepsilon_0} \qquad \epsilon_t \equiv \frac{\varepsilon_{et}}{\varepsilon_0} \qquad M^* \equiv \frac{M}{2G\varepsilon_0 a^3}$$

$$\xi = x/a \qquad \eta = y/a$$

Sec. 11-5] Elastoplastic Torsion with Strain Hardening

where ε_0 and σ_0 are the yield strain and yield stress, respectively, related to each other by $\sigma_0 = E\varepsilon_0$ and a is a characteristic linear dimension of the cross section.

The system of equations to be solved for a simply connected cross section can now be written

$$g(\xi, \eta) = \frac{\partial \epsilon_x^P}{\partial \eta} - \frac{\partial \epsilon_y^P}{\partial \xi} \tag{11.5.2}$$

$$\nabla^2 U = -\beta - g(\xi, \eta)$$
$$U = 0 \quad \text{on boundary} \tag{11.5.3}$$

$$\tau_x = \frac{\partial U}{\partial \eta} \quad \tau_y = -\frac{\partial U}{\partial \xi} \tag{11.5.4}$$

$$\epsilon_x = \tau_x + \epsilon_x^P \quad \epsilon_y = \tau_y + \epsilon_y^P \tag{11.5.5}$$

$$\epsilon_t = \frac{2}{\sqrt{3}} \sqrt{\epsilon_y^2 + \epsilon_x^2} \tag{11.5.6}$$

$$\epsilon_p = f(\epsilon_t) \tag{11.5.7}$$

$$\epsilon_x^P = \frac{\epsilon_p}{\epsilon_t} \epsilon_x \quad \epsilon_y^P = \frac{\epsilon_p}{\epsilon_t} \epsilon_y \tag{11.5.8}$$

The relationship expressed in equation (11.5.7) is obtained from the uniaxial stress–strain curve and, as is evident, relations in terms of plastic strains–total strains are being used.

The successive approximation method proceeds in the usual manner. The plastic strains are assumed to be zero everywhere. Equations (11.5.3) are solved by any available method. The stresses, the total strains, and equivalent total strain are computed by means of equations (11.5.4) through (11.5.7) with the help of the stress–strain curve. If at any point in the cross section the equivalent plastic strain as computed from (11.5.7) is negative, this point is in the elastic region, and the plastic strains at this point are set equal to zero. Otherwise, new approximations to the plastic strains are calculated by means of equations (11.5.8). One then returns to equation (11.5.2) and the process repeated until convergence is obtained. The method will be illustrated for bars of rectangular and circular cross sections. For a circular cross section with linear strain hardening the solution can be obtained in closed form.

11-6 BAR WITH RECTANGULAR CROSS SECTION

The elastic solution for a bar with a rectangular cross section by means of finite differences was presented in Section 11.2. To obtain the elastoplastic solution the function $g(x, y)$ is subtracted from the right side of equation (11.2.15), which in terms of the dimensionless quantities defined in (11.5.1) becomes

$$U_{i+1,j} + U_{i-1,j} + U_{i,j-1} + U_{i,j+1} - 4U_{i,j} = -(\beta + g_{i,j})H^2 \quad (11.6.1)$$

where

$$g_{i,j} = \frac{1}{2H}(\epsilon^P_{x,i,j+1} - \epsilon^P_{x,i,j-1} - \epsilon^P_{y,i+1,j} + \epsilon^P_{y,i-1,j}) \quad (11.6.2)$$

and H is the grid spacing divided by a.

The only difference between equations (11.6.1) and (11.2.15) is in the numbers appearing on the right side, which now depend on the plastic strains and change from iteration to iteration. Equations (11.6.1) can therefore be solved in the same way as equations (11.2.15) for the elastic case. Once the values of U are determined at all the grid points, corresponding values of plastic strains are computed by means of equations (11.5.4) through (11.5.8), as fully described in Section 11.5. The $g_{i,j}$ are then recomputed and equations (11.6.1) solved again, the process being repeated until convergence is obtained.

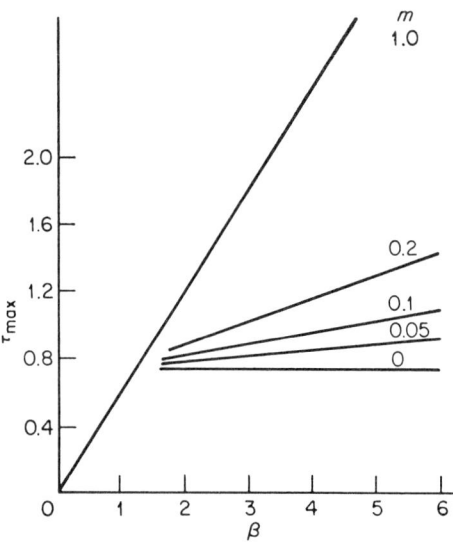

FIGURE 11.6.1 Variation of $\tau_{\max}$ with β.

Sec. 11-6] Bar with Rectangular Cross Section

The results of such a calculation for a square cross section with 11 × 11 grid points as shown in Figure 11.2.3 are shown in Figures 11.6.1 through 11.6.4. In these calculations linear strain hardening was assumed. Equation (11.5.7), relating the equivalent plastic strain to the equivalent total strain, can then be written (see Section 7.9)

$$\epsilon_p = \frac{\epsilon_t - \tfrac{2}{3}(1 + \mu)}{1 + \tfrac{2}{3}(1 + \mu)(m/1 - m)} \qquad (11.6.3)$$

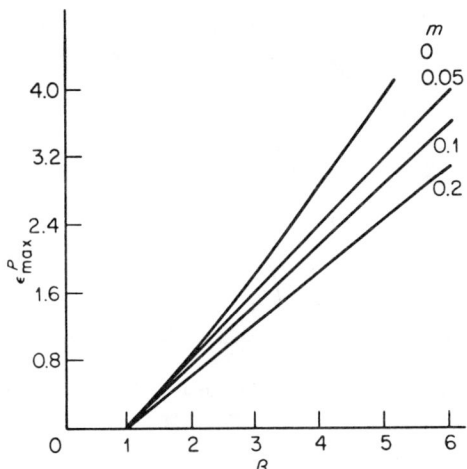

FIGURE 11.6.2 Variation of $\epsilon_{\max}^p$ with β.

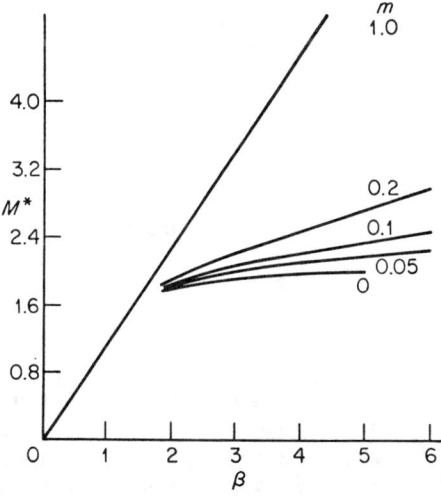

FIGURE 11.6.3 Variation of M^* with β.

where the strain-hardening parameter m is the ratio of the slope of the linear hardening curve to the slope of the elastic curve, as previously defined. For the perfectly plastic case m is equal to 0, and for the elastic case m is equal to 1.

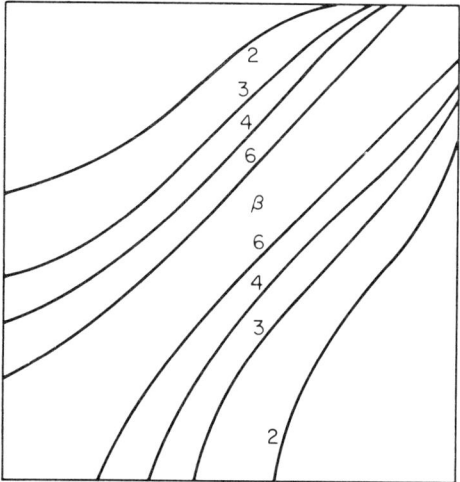

FIGURE 11.6.4 Variation of elastic-plastic boundary with β.

TABLE 11.6.1 Summary of Results for Torsion of a Square

m	β	M^*	τ_{max}	ϵ^P_{max}
0	2	1.786	0.751	0.820
	3	1.918	0.751	1.824
	4	1.955	0.751	2.851
	5	1.977	0.751	3.959
0.05	2	1.813	0.785	0.758
	3	1.997	0.825	1.623
	4	2.094	0.862	2.434
	5	2.166	0.899	3.240
	6	2.228	0.934	4.003
0.10	2	1.838	0.818	0.701
	3	2.073	0.893	1.478
	4	2.223	0.963	2.209
	5	2.347	1.032	2.919
	6	2.465	1.099	3.618
0.20	2	1.890	0.881	0.600
	3	2.220	1.022	1.250
	4	2.471	1.156	1.870
	5	2.717	1.290	2.488
	6	2.966	1.426	3.116

Sec. 11-7] Bar with Circular Cross Section

The figures show the effects of the strain-hardening parameter and the angle of twist on the maximum stress, the maximum plastic strain, the size of the plastic zone, and the torque. The results are also summarized in Table 11.6.1.

Although the calculations are described using deformation theory, a similar calculation, increasing α in steps, gave almost identical results. This is in agreement with similar calculations in reference [11]. As shown in reference [12], incremental and deformation theories give identical results for a perfectly plastic material of any cross section or a strain-hardening material of circular cross section. For strain-hardening materials of noncircular cross sections they will yield different results. It appears, however, that the differences will in general be slight.

11-7 BAR WITH CIRCULAR CROSS SECTION

For a bar with a circular cross section the solution is greatly simplified. In particular, for the case of linear strain hardening, a closed-form solution can be obtained. In polar coordinates the displacements are

$$u_r = 0 \quad u_\theta = \alpha r z \tag{11.7.1}$$

and the only nonzero strain is

$$\varepsilon_{\theta z} = \tfrac{1}{2}\alpha r \tag{11.7.2}$$

The stress–strain relation can therefore be written

$$\frac{1}{2G}\tau_{\theta z} = \tfrac{1}{2}\alpha r - \varepsilon_{\theta z}^P \tag{11.7.3}$$

The von Mises equivalent stress reduces to

$$\sigma_e = \sqrt{3}\,\tau_{\theta z} = \sqrt{3}\,G\alpha r - 2\sqrt{3}\,G\varepsilon_{\theta z}^P \tag{11.7.4}$$

and the equivalent plastic strain is

$$\varepsilon_p = \frac{2}{\sqrt{3}}\varepsilon_{\theta z}^P \tag{11.7.5}$$

Hence

$$\sigma_e = \sqrt{3}\,G\alpha r - 3G\varepsilon_p \tag{11.7.6}$$

Let

$$\frac{\sigma_e}{2G\varepsilon_0} \equiv S_e \qquad \frac{r}{a} \equiv \rho$$

$$\frac{\alpha a}{\varepsilon_0} \equiv \beta \qquad \epsilon_p \equiv \frac{\varepsilon_p}{\varepsilon_0} \qquad \epsilon_\theta^P \equiv \frac{\varepsilon_{\theta z}^P}{\varepsilon_0} \tag{11.7.7}$$

where a is the radius of the bar. Then equation (11.7.6) can be written in dimensionless form as

$$S_e = \frac{\sqrt{3}}{2}\beta\rho - \tfrac{3}{2}\epsilon_p \tag{11.7.8}$$

The stress–strain curve can be written in dimensionless form as

$$S_e = f(\epsilon_p) \tag{11.7.9}$$

and combining with (11.7.8) results in

$$2f(\epsilon_p) = \sqrt{3}\,\beta\rho - 3\epsilon_p \tag{11.7.10}$$

which can be solved iteratively for ϵ_p.

Equation (11.7.10) is valid only in the plastic region. Let this region extend between $\rho = \rho_c$ and $\rho = 1$. To determine the position of the elastoplastic boundary, i.e., ρ_c, let $\varepsilon_p = 0$ when $\sigma_e = \sigma_0$ or when

$$S_e = 1 + \mu$$

Hence from (11.7.8)

$$\rho_c = \frac{2(1+\mu)}{\sqrt{3}\,\beta} \tag{11.7.11}$$

which depends only on Poisson's ratio and the yield strain but not on the stress–strain curve. The value of β at which plastic flow just starts is found from equation (11.7.11) by setting ρ_c equal to 1. Thus the critical value of β will be

$$\beta_c = \frac{2(1+\mu)}{\sqrt{3}} \tag{11.7.12}$$

Sec. 11-7] Bar with Circular Cross Section

and the critical angle of twist per unit length will be

$$\alpha_c = \frac{\varepsilon_0}{a} \beta_c = \frac{2(1+\mu)}{\sqrt{3}} \frac{\varepsilon_0}{a}$$

or

$$\alpha_c G = \frac{1}{\sqrt{3}} \frac{\sigma_0}{a}$$

(11.7.13)

To summarize, the strain-hardening solution is found as follows. The elastoplastic boundary ρ_c is first determined from equation (11.7.11). The stress and strain in the elastic region for $\rho \leq \rho_c$ are then computed from equations (11.7.2) and (11.7.3) with $\varepsilon^P_{\theta z}$ set equal to zero. In the plastic region $\rho > \rho_c$, equation (11.7.10) is solved, usually by an iterative method. $\varepsilon^P_{\theta z}$ can then be computed from equation (11.7.5) and the shear stress from (11.7.3). Once the shear stress is known throughout the section, the torque can be computed by integration.

Let us now consider the case of linear strain hardening. Equation (11.7.9) for the stress–strain curve can be written

$$S_e = (1+\mu)\left(1 + \frac{m}{1-m}\epsilon_p\right)$$

(11.7.14)

Hence equation (11.7.10) becomes

$$2(1+\mu)\left(1 + \frac{m}{1-m}\epsilon_p\right) = \sqrt{3}\,\beta\rho - 3\epsilon_p$$

or

$$\epsilon_p = \frac{\sqrt{3}\,\beta\rho - 2(1+\mu)}{3 + 2(1+\mu)m/(1-m)} \qquad \rho \geq \rho_c$$

(11.7.15)

Note that the critical value of ρ is obtained when the numerator of (11.7.15) vanishes, which results again in equation (11.7.11).

Once the equivalent plastic strain is known from equation (11.7.15), the plastic shear strain and the stress are computed from (11.7.5) and (11.7.3). We have thus obtained a complete solution in closed form. To compute the torque, define $\tau_\theta \equiv \tau_{\theta z}/2G\varepsilon_0$. Then

$$M^* \equiv \frac{M}{2G\varepsilon_0 a^3} = 2\pi \int_0^1 \tau_\theta \rho^2 \, d\rho$$

(11.7.16)

and substituting

$$\tau_\theta = \begin{cases} \frac{1}{2}\beta\rho & \rho \leq \rho_c \\ \frac{1}{2}\beta\rho - \epsilon^P_\theta & \rho \geq \rho_c \end{cases}$$

results in

$$M^* = \frac{\pi\beta}{4} - \sqrt{3}\,\pi[\tfrac{1}{4}A(1 - \rho_c^4) + \tfrac{1}{3}B(1 - \rho_c^3)]$$

where
$$A = \frac{\sqrt{3}\,\beta}{3 + 2(1 + \mu)m/(1 - m)} \quad (11.7.17)$$

$$B = -\frac{2(1 + \mu)}{3 + 2(1 + \mu)m/(1 - m)}$$

Note that for $\rho_c = 1$ (no plastic flow) the torque reduces to the elastic torque as given in equation (11.2.12). For a perfectly plastic material,

$$m = 0 \quad A = \frac{\beta}{\sqrt{3}} \quad B = -\tfrac{2}{3}(1 + \mu)$$

and
$$M^* = \frac{2\pi(1 + \mu)}{3\sqrt{3}}(1 - \tfrac{1}{4}\rho_c^3)$$

or
$$M = \frac{2\pi a^3 \sigma_0}{3\sqrt{3}}\left[1 - \frac{1}{12\sqrt{3}\,a^3}\left(\frac{\sigma_0}{G\alpha}\right)^3\right] \quad (11.7.18)$$

and we recover the classical solution as given in reference [10].

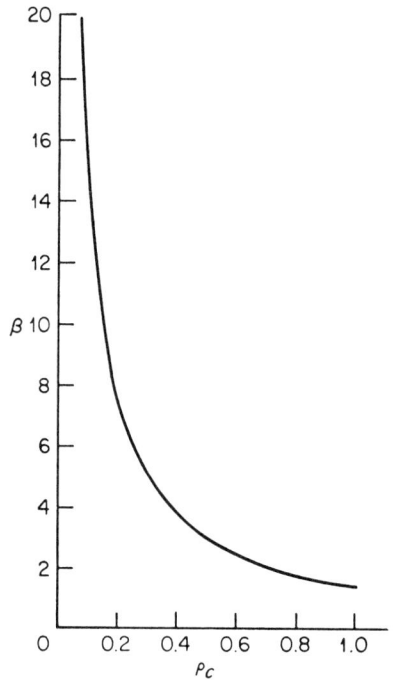

FIGURE 11.7.1 Variation of ρ_c with β.

Sec. 11-7] Bar with Circular Cross Section   257

Results of computations using the above formulas are shown in Figures 11.7.1 through 11.7.3. Figure 11.7.1 shows the elastoplastic boundary as a function of β. Figure 11.7.2 shows the effect of the strain-hardening parameter on the shear stress for $\beta = 5.0$, and Figure 11.7.3 shows the effect of these parameters on the torque.

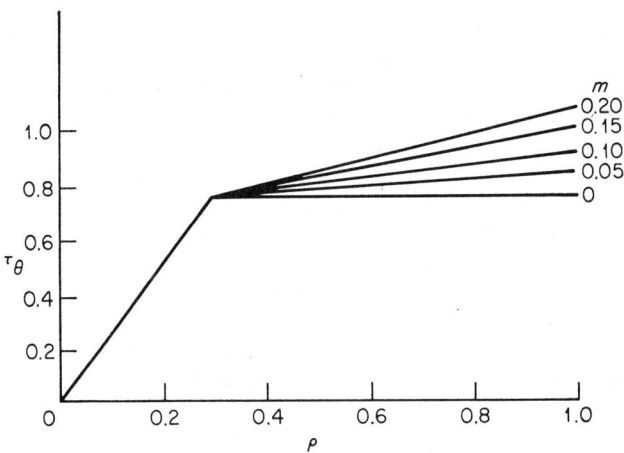

FIGURE 11.7.2   Variation of τ_θ with ρ for various m: $\beta = 5.0$.

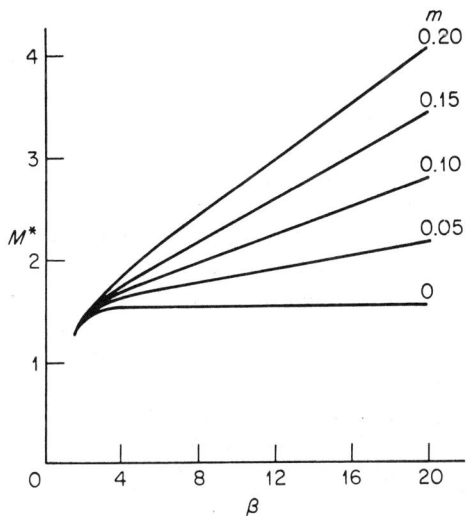

FIGURE 11.7.3   Variation of M^* with β for various m.

Problems

1. Show that the loading is radial for the torsion problem of a bar with a circular cross section, so that total plasticity theories may be used, as long as there is no unloading.
2. Show that the maximum shear stress for a solid bar of elliptic cross section under torsion occurs on the boundary at the point closest to the axis of the bar.
3. Determine the stresses, the torque, and the warping function for the triangular cross section bar of Figure 11.2.1.
4. Show that if the lateral surface of the bar is stress-free, the resultant shear stress must be tangent to the boundary.
5. Show that if p/S is equal to $2G\alpha$, the applied torque acting on a bar is equal to twice the volume between a membrane of the same shape as the cross section of the bar and the $z = 0$ plane.
6. Determine the torque acting on a bar of circular cross section by calculating the volume under a membrane of the same shape.
7. Show that for the torsion problem the yield criteria of von Mises and Tresca both reduce to equation (11.4.1).
8. Using the Saint-Venant assumptions (11.1.4), show that for a bar of circular cross section the radial and tangential displacements become

$$u_r = 0 \qquad u_\theta = \alpha r z$$

and consequently the only nonzero strain is $\varepsilon_{\theta z} = \frac{1}{2}r\alpha$.

9. Calculate the torques required to produce a twist of 0.004 rad/in. in a 2-in.-diameter shaft if the material is perfectly plastic, and if the material strain hardens with $m = 0.1$. Assume $E = 30 \times 10^6$, $\mu = 0.3$, and $\sigma_0 = 30 \times 10^3$.

References

1. B. Saint-Venant, Mémoire sur la torsion des prismes, *Mem. Acad. Sci. Math. Phys.*, **14**, 1856, pp. 233–560.
2. S. Timoshenko and T. N. Goodier, *Theory of Elasticity*, McGraw-Hill, New York, 1951, p. 275.
3. V. Kantorovich and V. I. Krylov, *Approximate Methods of Higher Analysis*, P. Noordhoff, Groningen, 1958, p. 70.
4. L. Prandtl, Zur Torsion von prismatischen Staeben, *Physik. Z.*, **4**, 1903, pp. 758–759.
5. I. S. Sokolnikoff, *Mathematical Theory of Elasticity*, McGraw-Hill, New York, 1956, p. 117.
6. A. Nadai, Der Beginn des Fliessvorganges in einem tortierten Stab, *Z. Angew. Math. Mech.*, **3**, 1923, p. 442–454.
7. A. Nadai, *Theory of Flow and Fracture of Solids*, Vol. 1, McGraw-Hill, New York, 1950.
8. W. Johnson and P. B. Mellor, *Plasticity for Mechanical Engineers*, Van Nostrand, Princeton, N.J., 1962, p. 132.

General References

9. R. S. Varga, *Matrix Iterative Analysis*, Prentice-Hall, Englewood Cliffs, N.J., 1962.
10. W. Prager and P. G. Hodge, Jr., *Theory of Perfectly Plastic Solids*, Wiley, New York, 1951, p. 72.
11. J. H. Huth, A Note on Plastic Torsion, *J. Appl. Mech.*, **22**, 1955, pp. 432–434.
12. W. Prager, An Introduction to the Mathematical Theory of Plasticity, *J. Appl. Phys.*, **18**, 1947, pp. 375–383.

General Reference

Johnson, W. and P. B. Mellor, *Plasticity for Mechanical Engineers*, Van Nostrand, Princeton, N.J., 1962.
Prager, W., and P. G. Hodge, Jr., *Theory of Perfectly Plastic Solids*, Wiley New York, 1951.

CHAPTER 12

THE SLIP-LINE FIELD

12-1 PLANE STRAIN PROBLEM OF A RIGID-PERFECTLY PLASTIC MATERIAL

In the previous chapters it was shown how the successive-approximation method can be applied to a variety of problems, including plane strain and plane stress elastoplastic problems. In these problems the constraints imposed by the elastic parts of the material prevented unrestrained plastic flow. In many metal-forming processes, such as rolling, drawing, forging, etc., large unrestricted plastic flows occur except for very small elastic zones. For such problems it may not be unreasonable to neglect the elastic strains and assume the material to be *rigid-perfectly plastic*, as shown in Figure 2.6.1(b). Any elastic part of the body is then assumed to act as a rigid inclusion and the plastic parts can flow freely at constant equivalent stress.

A great deal of work has been done on solutions of this type of problem under conditions of plain strain, using the theory of *slip lines*. This chapter will be devoted to a brief discussion of this theory. We begin by writing the equations of plane strain for a rigid-plastic body. By plane strain is meant the condition wherein the displacements all occur in parallel planes in the body, say, planes parallel to the xy plane, and all stresses and strains are independent of z; i.e.,

$$\varepsilon_z = \varepsilon_{xz} = \varepsilon_{yz} = \tau_{xz} = \tau_{yz} = 0$$

Sec. 12-1] Plane Strain Problem of a Rigid-Perfectly Plastic Material

$$\varepsilon_x = \varepsilon_x(x, y) \qquad \varepsilon_y = \varepsilon_y(x, y) \qquad \sigma_x = \sigma_x(x, y) \qquad (12.1.1)$$
$$\sigma_y = \sigma_y(x, y) \qquad \sigma_z = \sigma_z(x, y) \qquad \tau_{xy} = \tau_{xy}(x, y)$$

Since $\tau_{xz} = \tau_{yz} = 0$, it follows that the z direction is a principal direction and σ_z is a principal stress.

For a rigid-plastic material, the elastic strains are neglected, so that the total strains and strain increments are equal to the corresponding plastic strains and strain increments, and the Lévy–Mises relations result in [see equation (7.2.5)]

$$\left. \begin{array}{l} d\varepsilon_x = \tfrac{2}{3} d\lambda [\sigma_x - \tfrac{1}{2}(\sigma_y + \sigma_z)] \\ d\varepsilon_y = \tfrac{2}{3} d\lambda [\sigma_y - \tfrac{1}{2}(\sigma_x + \sigma_z)] \\ d\varepsilon_z = \tfrac{2}{3} d\lambda [\sigma_z - \tfrac{1}{2}(\sigma_x + \sigma_y)] \\ d\varepsilon_{xy} = d\lambda \, \tau_{xy} \end{array} \right\} \qquad (12.1.2)$$

and since $d\varepsilon_z = 0$, the last equation gives

$$\sigma_z = \tfrac{1}{2}(\sigma_x + \sigma_y) \qquad (12.1.3)$$

and also the mean stress is

$$\sigma_m = \tfrac{1}{2}(\sigma_x + \sigma_y) = \sigma_z \qquad (12.1.4)$$

The von Mises yield criterion for this case becomes

$$\tfrac{1}{4}(\sigma_x - \sigma_y)^2 + \tau^2 = k^2 \qquad (12.1.5)$$

where k is yield stress in simple shear and τ has been written for brevity instead of τ_{xy}. The equilibrium equations to be satisfied are

$$\frac{\partial \sigma_x}{\partial x} + \frac{\partial \tau}{\partial y} = 0$$
$$\frac{\partial \sigma_y}{\partial y} + \frac{\partial \tau}{\partial x} = 0 \qquad (12.1.6)$$

Equations (12.1.6) and (12.1.5) represent three equations in the three unknowns σ_x, σ_y, and τ. If the boundary conditions are given only in terms of stresses, these equations are sufficient to give the stress distribution without any reference to the stress–strain relations. Such problems are called *statically determinate*. We pointed out a similar situation in discussing the sphere problem in Section 8.3. However, if displacements or velocities are specified over part of the boundary, then the stress–strain relations must be used to relate the stresses to the strains and the problem becomes much more complicated.

The principal stresses in the plastic field can be written in terms of σ_m and k as follows:

$$\sigma_1 = \frac{\sigma_x + \sigma_y}{2} + \sqrt{\left(\frac{\sigma_x - \sigma_y}{2}\right)^2 + \tau^2}$$

$$\sigma_2 = \sigma_z = \tfrac{1}{2}(\sigma_x + \sigma_y) \qquad (12.1.7)$$

$$\sigma_3 = \frac{\sigma_x + \sigma_y}{2} - \sqrt{\left(\frac{\sigma_x - \sigma_y}{2}\right)^2 + \tau^2}$$

or, from (12.1.4) and (12.1.5),

$$\begin{aligned}\sigma_1 &= \sigma_m + k \\ \sigma_2 &= \sigma_m \\ \sigma_3 &= \sigma_m - k\end{aligned} \qquad (12.1.8)$$

The next step is to find the principal directions. We define the *first principal direction* as the direction of the maximum principal stress. Let ϕ be the angle between the first principal direction and the x axis as shown in Figure 12.1.1.

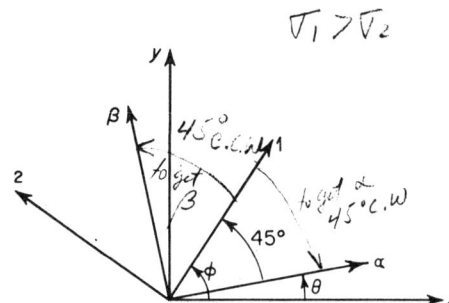

FIGURE 12.1.1 Principal directions and α and β lines.

Then from equation (3.3.6) for the principal directions it follows that

$$\tan 2\phi = \frac{2\tau}{\sigma_x - \sigma_y} \qquad (12.1.9)$$

which gives two values of ϕ differing by 90°. The *second principal direction* is taken 90° counterclockwise from the first.

Sec. 12-1] Plane Strain Problem of a Rigid-Perfectly Plastic Material 263

Having determined the principal stresses and directions, the maximum and minimum shearing stresses and directions can readily be determined. These shearing stresses act on the planes bisecting the principal directions as described in Section 3.4. Their values are given by

$$\tau_{max} = \pm \tfrac{1}{2}(\sigma_1 - \sigma_3) \qquad (12.1.10)$$

The maximum shear directions will be designated by the α and β directions. α, called the *first shear direction*, is taken 45° clockwise from the first principal direction, as shown in Figure 12.1.1, and β, the *second shear direction*, is 90° counterclockwise from the first shear direction or 45° counterclockwise from the first principal direction, as shown.

Let θ be the angle which the first shear direction makes with the x axis (measured counterclockwise). Then

$$\theta = \phi - 45°$$

$$\tan 2\theta = -\frac{1}{\tan 2\phi}$$

and, from (12.1.9),

$$\tan 2\theta = \frac{\sigma_y - \sigma_x}{2\tau} \qquad (12.1.11)$$

It follows therefore that

$$\cos 2\theta = \frac{\tau}{k}$$

$$\sin 2\theta = \frac{\sigma_y - \sigma_x}{2k} \qquad (12.1.12)$$

At every point in the plastic field, the angle which the maximum shear direction makes with the x axis is determined by equations (12.1.11) or (12.1.12). If curves are now drawn in the xy plane such that at every point of each curve the tangent coincides with one of maximum shear directions, then two families of curves called *shear lines*, or *slip lines*, will be obtained. Obviously, since the maximum and minimum shear directions at a point are orthogonal to each other, the two families of slip lines will form an orthogonal set. These two families of curves will be called the α *lines* and β *lines*, respectively.

It should be carefully noted that along an α line α is varying and β is constant, and along a β line β is varying and α is constant. α and β are merely parameters or curvilinear coordinates used to designate the point under consideration, just as x and y designate the point. Thus the point P shown in Figure 12.1.2 can be designated $P(x_1, y_1)$ or $P(\alpha_3, \beta_2)$.

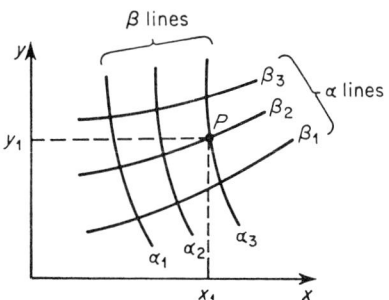

FIGURE 12.1.2 Families of α and β lines.

The normal stresses acting on the maximum shear planes equals the average of the principal stresses, as shown by equation (3.4.5). Thus the stresses acting normal and tangential to the α and β lines are given by

$$\sigma_\alpha = \sigma_\beta = \tfrac{1}{2}(\sigma_1 + \sigma_3) = \sigma_m$$
$$\tau_{\alpha\beta} = \tfrac{1}{2}(\sigma_1 - \sigma_3) = k \qquad (12.1.13)$$

as illustrated in Figure 12.1.3.

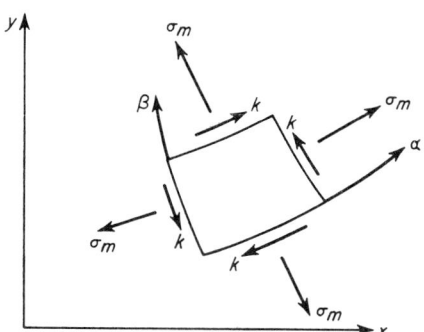

FIGURE 12.1.3 Stresses normal and tangential to α and β lines.

Finally, we can express σ_x, σ_y, and τ in terms of σ_m and θ, as follows:

$$\sigma_x = \sigma_m - k \sin 2\theta$$
$$\sigma_y = \sigma_m + k \sin 2\theta \qquad (12.1.14)$$
$$\tau = k \cos 2\theta$$

All these results can also be obtained by use of Mohr's diagram.

Sec. 12-1] Plane Strain Problem of a Rigid-Perfectly Plastic Material

From (12.1.14) it is seen that the state of stress can be determined in terms of two independent quantities, σ_m and θ. The equilibrium equations can be written in terms of these quantities by substituting (12.1.14) into (12.1.6). Thus

$$\frac{\partial \sigma_m}{\partial x} - 2k \left(\cos 2\theta \frac{\partial \theta}{\partial x} + \sin 2\theta \frac{\partial \theta}{\partial y} \right) = 0$$

$$\frac{\partial \sigma_m}{\partial y} + 2k \left(\cos 2\theta \frac{\partial \theta}{\partial y} - \sin 2\theta \frac{\partial \theta}{\partial x} \right) = 0$$

(12.1.15)

or defining

$$\chi \equiv \frac{\sigma_m}{2k}$$

we can write

$$\frac{\partial \chi}{\partial x} - \cos 2\theta \frac{\partial \theta}{\partial x} - \sin 2\theta \frac{\partial \theta}{\partial y} = 0$$

$$\frac{\partial \chi}{\partial y} - \sin 2\theta \frac{\partial \theta}{\partial x} + \cos 2\theta \frac{\partial \theta}{\partial y} = 0$$

(12.1.16)

Now the choice of the x and y axes is arbitrary. If we choose the x and y axes at a given point to coincide with the α and β directions at this point, then $\theta = 0$ and

$$\frac{\partial}{\partial x} = \frac{\partial}{\partial \alpha} \qquad \frac{\partial}{\partial y} = \frac{\partial}{\partial \beta}$$

and equations (12.1.16) become

$$\frac{\partial \chi}{\partial \alpha} - \frac{\partial \theta}{\partial \alpha} = 0$$

$$\frac{\partial \chi}{\partial \beta} + \frac{\partial \theta}{\partial \beta} = 0$$

(12.1.17)

Equations (12.1.17) are called *compatibility equations* (not to be confused with the strain compatibility equations). Each equation contains derivatives in only one direction. Integrating,

$$\chi - \theta = C_1 \quad \text{along the } \alpha \text{ curve}$$
$$\chi + \theta = C_2 \quad \text{along the } \beta \text{ curve}$$

(12.1.18)

where C_1 and C_2 are constants. These equations were first derived by Hencky in 1923, [14].

From equations (12.1.18) it is apparent that if χ and θ are prescribed on the boundary, it should be possible to proceed along constant α and β lines to determine $\chi = \sigma_m/2k$ and θ. If the displacements or velocities are prescribed over part of the boundary, as is frequently the case, these equations are not sufficient to obtain a solution and the velocity equations following must also be used.

12-2 VELOCITY EQUATIONS

The Lévy-Mises relation (12.1.2) can also be written

$$\frac{d\varepsilon_x - d\varepsilon_y}{d\varepsilon_{xy}} = \frac{\sigma_x - \sigma_y}{\tau_{xy}} \qquad (12.2.1)$$

In addition, the incompressibility condition with $d\varepsilon_z = 0$ is

$$d\varepsilon_x + d\varepsilon_y = 0 \qquad (12.2.2)$$

It is convenient to divide the strains by dt, the increment of time, and write these equations in terms of velocities. Of course, these equations remain homogeneous in t, which acts merely as a scaling parameter. Then

$$\frac{d\varepsilon_x}{dt} = \frac{d}{dt}\frac{\partial u}{\partial x} = \frac{\partial v_x}{\partial x}$$

where v_x is the velocity in the x direction; i.e., $v_x = du/dt$ and similarly $v_y = dv/dt$. Equations (12.2.1) and (12.2.2) become

$$\frac{(\partial v_x/\partial x) - (\partial v_y/\partial y)}{(\partial v_x/\partial y) + (\partial v_y/\partial x)} = \frac{\sigma_x - \sigma_y}{2\tau}$$

$$\frac{\partial v_x}{\partial x} + \frac{\partial v_y}{\partial y} = 0 \qquad (12.2.3)$$

Now since the principal axes of stress and of plastic strain increment coincide (see Section 7.2), it follows that the maximum shear stress lines and maximum shear velocity lines coincide, or that the stress slip lines are the same as the velocity slip lines. Also the strain rates normal to the α and β directions are equal to the mean strain rates [see equation (4.5.4)]. Therefore,

$$\frac{d\varepsilon_\alpha}{dt} = \frac{d\varepsilon_\beta}{dt} = \tfrac{1}{2}(\dot\varepsilon_1 + \dot\varepsilon_2) = \tfrac{1}{2}(\dot\varepsilon_x + \dot\varepsilon_y) = 0 \qquad (12.2.4)$$

Sec. 12-2] Velocity Equations

There are no extensions, only shearing flows in the slip directions—hence their name.

Now consider the velocities in the slip directions. From Figure 12.2.1,

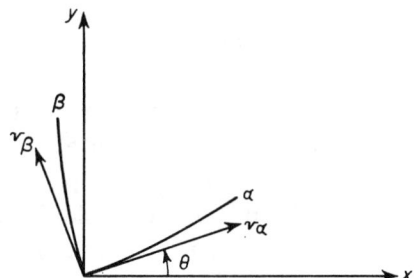

FIGURE 12.2.1 Velocities in α and β directions.

$$v_x = v_\alpha \cos \theta - v_\beta \sin \theta$$
$$v_y = v_\alpha \sin \theta + v_\beta \cos \theta \tag{12.2.5}$$

If $\theta = 0$, the x axis will coincide with the α direction and the condition that the normal strain rates be zero can be written

$$\dot{\varepsilon}_\alpha = \left(\frac{\partial v_x}{\partial x}\right)_{\theta=0} = 0$$

$$\dot{\varepsilon}_\beta = \left(\frac{\partial v_y}{\partial y}\right)_{\theta=0} = 0$$

or, from (12.2.5),

$$\frac{\partial v_\alpha}{\partial x} - v_\beta \frac{\partial \theta}{\partial x} = 0$$
$$v_\alpha \frac{\partial \theta}{\partial y} + \frac{\partial v_\beta}{\partial y} = 0 \tag{12.2.6}$$

or, since the x direction is the same as the α direction and the β direction is the same as the y direction,

$$\frac{\partial v_\alpha}{\partial \alpha} - v_\beta \frac{\partial \theta}{\partial \alpha} = 0$$
$$\frac{\partial v_\beta}{\partial \beta} + v_\alpha \frac{\partial \theta}{\partial \beta} = 0 \tag{12.2.7}$$

If β is kept constant in the first equation and α in the second equation, we can write

$$dv_\alpha - v_\beta \, d\theta = 0 \quad \text{along an } \alpha \text{ line}$$
$$dv_\beta + v_\alpha \, d\theta = 0 \quad \text{along a } \beta \text{ line} \tag{12.2.8}$$

These are the *compatibility equations* for the *velocities* first derived by Geiringer in 1930, [15].

If the problem is statically determined, the slip line field and the stresses can be found from equations (12.1.18) (or their equivalent) and the stress boundary conditions. The velocities can then be computed from (12.2.8) (or their equivalent) using the velocity boundary conditions, since $d\theta$ will now be known from the stress solution. If, however, the problem is not statically determined, which means that the stress boundary conditions are insufficient to obtain a unique slip-line field, then equations (12.1.18) must be solved simultaneously with (12.2.8) using both the stress boundary conditions and the velocity boundary conditions. This is an extremely difficult problem and must usually be done by trial and error. A slip-line field satisfying all stress conditions is assumed. The velocities are then computed and a check made to see if the velocity boundary conditions are satisfied. If not, the slip-line field is modified and the procedure repeated as often as necessary. This is obviously a very laborious process, since the construction of just one slip-line field is a lengthy task.

It is worthwhile to note some of the differences between Hencky's stress equations (12.1.18) and Geiringer's velocity equations (12.2.8).

1. Hencky's equations relate two unknowns, χ and θ, by two equations. Geiringer's equations relate three unknowns, v_x, v_y, and $d\theta$, by two equations.

2. Hencky's equations give the stress state all along a known slip line if the stress state is known at one point on the slip line. Geiringer's equations will not give the velocities along a known slip line, if they are known at one point.

3. Hencky's equations force certain restrictions on the geometry of the slip-line field, as we will shortly see. Geiringer's equations place no restriction on the geometry of the slip-line field, except through the boundary conditions.

12-3 GEOMETRY OF THE SLIP-LINE FIELD

Hencky's equations, as mentioned above, impose some rather severe restrictions on the geometry of the slip-line field, which are a great aid in computation. We will list a number of these but prove only a few to show the method of attack.

Sec. 12-3] Geometry of the Slip-Line Field

1. *Hencky's first theorem* states that the angle between two slip lines of one family at the points where they are cut by a slip line of the other family is constant along their lengths. This is shown in Figure 12.3.1, the angles θ_1 and θ_2 being equal.

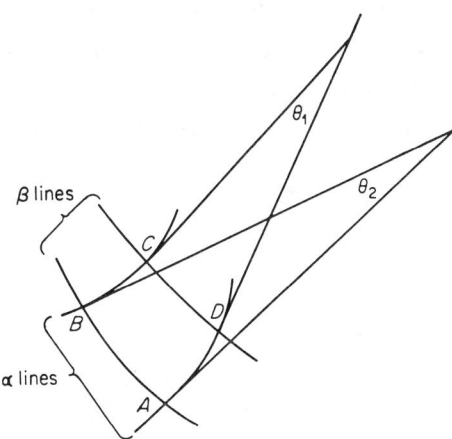

FIGURE 12.3.1 Demonstration of Hencky's first theorem.

2. All α lines (β lines) turn through the same angle in going from one β line (α line) to another.

3. If one α line (β line) is straight between two β lines (α lines), then all α lines (β lines) are straight between these two β lines (α lines). Furthermore, these straight segments have the same length.

4. If both the α and β lines are straight in a certain region, all the stresses in the region are constant. This is called a field of *uniform stress state*.

5. Along a straight shear line the state of stress is constant.

6. If the state of stress is constant along a curve, then either the curve is embedded in a field of constant stress or else the curve is a straight shear line.

7. The radii of curvature of the α lines (β lines) where they intersect a given β line (α line) decrease in direct proportion to the distance traveled in the positive direction of the β line (α line). Therefore, if the plastic zone extends far enough, the radii of curvature eventually become zero, so that neighboring slip lines run together and the solution ends at the envelope of the slip lines. This is *Hencky's second theorem*.

8. As we proceed along a given slip line of one family, the centers of curvature of the slip lines of the other family form an involute of this slip line.

9. The envelope of the slip lines of one family is the locus of the cusps of the slip lines of the other family.

10. The envelope of the slip lines of one family is a limiting line across which the shear lines of the other family cannot be continued.

11. If the radius of curvature of an α line (β line) jumps discontinuously as it crosses a β line (α line), all α lines (β lines) crossing the β line (α line) will suffer the same jump in radius of curvature. This also means that the derivatives of the stresses are discontinuous across the slip line.

There are many similar theorems, but they are not of practical interest.

Hencky's first theorem can easily be proved as follows. Referring to Figure 12.3.1, along the α line AD, the first of Hencky's equations (12.1.18) gives

$$\chi_A - \theta_A = \chi_D - \theta_D$$

and along the β line CD, the second equation gives

$$\chi_D + \theta_D = \chi_C + \theta_C$$

Therefore,

$$\chi_C - \chi_A = 2\theta_D - \theta_A - \theta_C \tag{12.3.1}$$

Also along AB,

$$\chi_A + \theta_A = \chi_B + \theta_B$$

and along BC

$$\chi_C - \theta_C = \chi_B - \theta_B$$

Therefore,

$$\chi_C - \chi_A = \theta_C + \theta_A - 2\theta_B \tag{12.3.2}$$

or, comparing with (12.3.1),

$$\theta_A - \theta_B = \theta_D - \theta_C \tag{12.3.3}$$

which proves Theorem 1.

Theorem 2 is a direct corollary, since from equation (12.3.3),

$$\theta_D - \theta_A = \theta_C - \theta_B \tag{12.3.4}$$

Theorem 3 is also a direct corollary, since if one of the lines is straight, say AD, then $\theta_D - \theta_A = 0$ and therefore $\theta_C - \theta_B = 0$, so the other line is also straight.

Similarly Theorems 4 through 6 follow directly from Hencky's equations (12.1.18). Theorems 7 through 10 are based on the theory of plane curves. Thus let the radii of curvature of the α and β lines at the point A in Figure 12.3.2 be designated by R_α and R_β, respectively. At the point B the radius

Sec. 12-3] Geometry of the Slip-Line Field

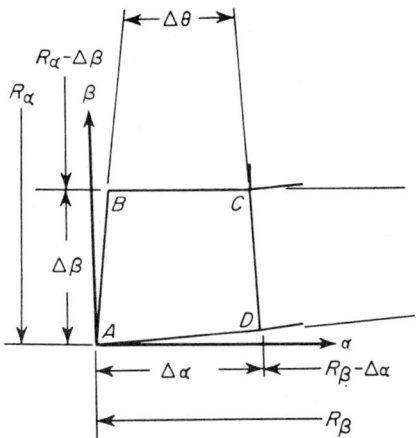

FIGURE 12.3.2 Proof of Hencky's second theorem.

of curvature R_α has decreased by an amount $\Delta\beta$, and at the point D, R_β has decreased by an amount $\Delta\alpha$, to first order of small quantities. In the limit

$$\frac{\partial R_\alpha}{\partial \beta} = -1$$
$$\frac{\partial R_\beta}{\partial \alpha} = -1$$
(12.3.5)

or, more conveniently for computational purposes, since $\Delta\alpha = R_\alpha \Delta\theta$ and $\Delta\beta = -R_\beta \Delta\theta$, equations (12.3.5) can be written

$$\Delta R_\alpha - R_\beta \Delta\theta = 0 \quad \text{along a } \beta \text{ line}$$
$$\Delta R_\beta + R_\alpha \Delta\theta = 0 \quad \text{along an } \alpha \text{ line}$$
(12.3.6)

Equations (12.3.5) are the mathematical statement of Theorem 7, and hence Theorem 7 is proved.

The proofs of the other theorems are given in Prager and Hodge [1] and will not be given here. Figure 12.3.3 illustrates some of these properties. $ABCD$ and $A'B'C'D'$ are neighboring α shear lines. AA', BB', CC', and DD' are infinitesimal arcs of the β lines. The center of curvature of these arcs form an involute $PQRS$ of the slip line $ABCD$. (An involute is the curve obtained by unwinding a flexible string originally lying on the curve, so that the string is always tangent to the curve. The original curve is the evolute of the involute and is the locus of the radii of curvature of the involute.) At the point T where the involute $PQRST$ meets the slip line $ABCDT$, the distance between the

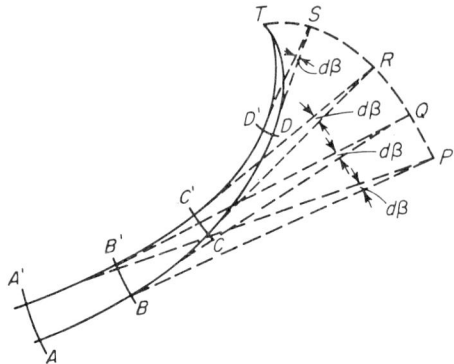

FIGURE 12.3.3 Involute and limiting line. (Reference [1].)

neighboring slip lines becomes zero, and so does the radius of curvature of the β line through T, as stated in Theorem 7. It is a point on the envelope of the α lines and is a cusp of the β line.

12–4 SOME SIMPLE EXAMPLES

State of Uniform Stress

If the stress is constant throughout the field, the slip lines form two sets of orthogonal straight lines. This follows directly from Hencky's equations (12.1.18), for if χ is constant, then θ is constant. This is the converse of Theorem 4 of Section 12.3.

Centered Fan

Consider a slip-line field composed of a set of radial lines originating from a point and a set of concentric circular arcs as shown in Figure 12.4.1. Let

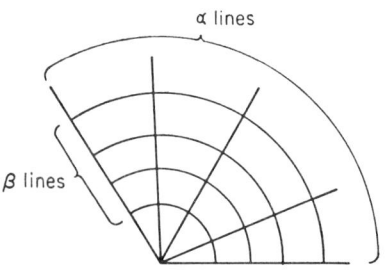

FIGURE 12.4.1 Centered fan.

Sec. 12–4] Some Simple Examples

the straight lines be the α lines and the circular arcs the β lines. Then from Hencky's first equation, since θ is constant along an α line, χ must also be constant along an α line, and, from the second equation, since θ varies linearly with distance along a β line, χ must vary linearly with distance along a β line. Thus the mean stress is constant in the radial direction and varies linearly with the angle measured from the x axis. To find the stress components we then make use of equations (12.1.14). This type of slip-line field is called a *centered fan*. Note that the center of the fan is a singular point of the stress field, since it can have any one of an infinity of values.

Indentation by a Punch

We now consider a problem which combines, or "patches" together, the state of uniform stress and the centered fan—the indentation of a semi-infinite body by a flat rigid punch in the form of an infinite strip. Figure 12.4.2 shows a typical plane [2]. It is assumed that the surface of the punch

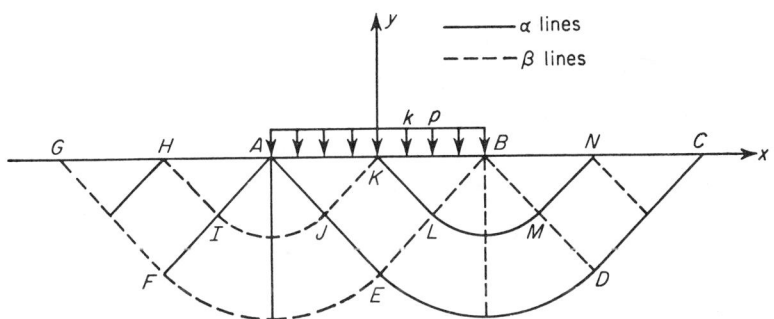

FIGURE 12.4.2 Slip lines under lubricated flat punch.

and body are perfectly lubricated, so that there is no friction between them. It is also assumed that there is a constant pressure over the face of the punch. For the boundary conditions we have that over the segment AB there is a uniform pressure kp, and the rest of the boundary is stress-free. We consider only the case of *incipient* plastic flow, since once plastic flow progresses, the shape of the boundary $GABC$ changes considerably and it is necessary to satisfy the boundary conditions on the deformed boundary.

Assume now that plastic flow occurs over a segment AG of the free boundary as shown. The length of this segment is as yet not known. From the boundary condition on this segment

$$\sigma_y = \tau = 0 \quad \text{on } AG$$

From the yield condition,

$$\left(\frac{\sigma_y - \sigma_x}{2}\right)^2 + \tau^2 = k^2$$

it follows that

$$\sigma_x = \pm 2k$$

Intuitively we would expect σ_x to be compressive and we tentatively assume

$$\sigma_x = -2k \quad \text{on } AG \tag{12.4.1}$$

Since the shear stress is zero, AG is a principal direction and the slip lines must be at $\pm 45°$ with AG. This also follows from the last of equations (12.1.14). The α lines make $45°$ angles with AG and the β lines $135°$ (or $-45°$), as shown in Figure 12.4.2.

Consider the triangular region AGF formed by AG and the slip lines AF and GF. By Theorem 6 of Section 12.3, this is a constant stress region. The slip lines are straight lines with $\theta = \pi/4$. The mean stress χ is a constant and must satisfy equations (12.1.18) throughout this region. Since on the boundary AG

$$\chi = \frac{\sigma_m}{2k} = \frac{\sigma_x}{4k} = -\tfrac{1}{2}$$

it follows that

$$\left.\begin{array}{l}\chi = -\tfrac{1}{2} \\ \theta = \dfrac{\pi}{4}\end{array}\right\} \text{in } AGF \tag{12.4.2}$$

Now consider the boundary AB. Since it has been assumed that there is no friction, $\tau = 0$ along this boundary, so that

$$\left.\begin{array}{l}\sigma_y = -kp \\ \tau = 0\end{array}\right\} \text{along } AB \tag{12.4.3}$$

Therefore,

$$\left(\frac{\sigma_x - \sigma_y}{2}\right)^2 = k^2$$

or

$$\sigma_x = \sigma_y \pm 2k = k(2 - p) \tag{12.4.4}$$

where the plus sign has been chosen (see Problem 7). It follows then, just as for the segment AG, that AB is a principal direction and the slip lines make $\pm 45°$ angles with AB. This time the β lines are at $45°$ with AB and the α lines

Sec. 12-4] Some Simple Examples

are at 135° (or −45°). As before, region AEB is a constant stress region, the slip lines being straight lines with $\theta = \frac{3}{4}\pi$. From the boundary condition,

$$\left. \begin{array}{l} \chi = \dfrac{\sigma_x + \sigma_y}{4k} = \dfrac{1-p}{2} \\ \theta = \tfrac{3}{4}\pi \end{array} \right\} \text{in } AEB \qquad (12.4.5)$$

Now AF and AE are straight α slip lines and it follows from Theorem 3 of Section 12.3 that all the shear lines in between these two are straight, or region FAE is a centered fan. The stresses are then constant along any radial line from A to the arc FE and vary linearly along any arc such as IJ from the value

$$\chi = -\tfrac{1}{2} \text{ along } AF \qquad \text{to} \qquad \chi = \frac{1-p}{2} \text{ along } AE$$

Similar results hold in the regions GBD and BDC.

The pressure p exerted by the punch to produce this state can readily be determined. The line AF is an α line and line $HIJK$ is a β line. The compatibility relation to be satisfied along the β line (Hencky's second equation) is

$$\chi + \theta = \text{constant}$$

Along HI,

$$\chi = -\tfrac{1}{2} \quad \text{and} \quad \theta = \frac{\pi}{4}$$

Along JK,

$$\chi = \tfrac{1}{2}(1 - p) \quad \text{and} \quad \theta = \tfrac{3}{4}\pi$$

Hence

$$-\tfrac{1}{2} + \frac{\pi}{4} = \tfrac{1}{2}(1 - p) + \tfrac{3}{4}\pi$$

or

$$p = 2 + \pi \qquad (12.4.6)$$

The velocity distribution is readily determined from the Geiringer equations (12.2.8). If the punch is moving with a velocity U_0 in the negative y direction, then region ABE moves as rigid body attached to the punch with the same velocity. In region $AEFG$, v_α equals zero and v_β equals $U_0/\sqrt{2}$. Region AEF thus moves out with velocity $U_0/\sqrt{2}$ and region AGF moves in the direction FG with the same velocity.

The above solution was obtained by Prandtl [3]. An alternative solution, given by Hill [4], assumes the rigid-plastic boundary to be $HIJKLMN$ instead of *GFEDC*. An analysis similar to the previous one shows that regions *AHI* and *AJK* are constant state regions and *AIJ* is a centered fan. The stresses in these regions are the same as previously obtained but the velocities are different, the outward flow velocity being twice that of the previous solution.

Actually an infinity of solutions can be obtained between the two limiting solutions discussed above. This illustrates one of the difficulties of the plane strain solution for a rigid-plastic material. More than one solution (or no solution) may be obtained for a given problem, and the "correct" solution may be impossible to ascertain. The only truly satisfactory method is to solve the complete elastoplastic problem using the Prandtl–Reuss relations. This, of course, will in general be extremely difficult. It is often possible, however, to determine the most probable solution, and sometimes a minimum force criterion may be used.

In addition, we note the nonuniqueness of the boundary values due to the quadratic yield conditions. Thus in equation (12.4.1) the negative sign was chosen for σ_x on the basis of intuition. If the plus sign had been chosen, the pressure exerted by the punch, equation (12.4.6), would have come out negative, which is impossible. So we know that the negative choice was correct. However, the correct choice of sign was not really known a priori, and this will often be the case.

12–5 NUMERICAL SOLUTIONS OF BOUNDARY-VALUE PROBLEMS

In the above example of the punch indentation, the slip lines and the solution were obtained completely in closed form from the boundary conditions. In general, however, numerical or graphical methods will be necessary. In this section a brief discussion of the simplest numerical methods will be presented. For this purpose we must distinguish among three types of boundaries, as shown in Figures 12.5.1 through 12.5.3. Figure 12.5.1 shows a case where the boundary curve C_0 is not a slip line. The values of χ and θ are given on this boundary and it is desired to construct the slip-line field. We choose a number of stations on the arc C_0 and try to construct the slip lines passing through these points. The various shear lines are designated 1, 2, 3, etc., and a grid point at the intersection of the *i*th α line with the *j*th β line is designated *i,j*.

Now consider the points (1, 1) and (2, 2) on the curve C_0. At the point (1, 1) draw a straight line with angle θ_{11}, representing the α line through this point. At (2, 2) draw a line with angle $\theta_{22} + \pi/2$. The two lines intersect at

Sec. 12-5] **Numerical Solutions of Boundary-Value Problems** 277

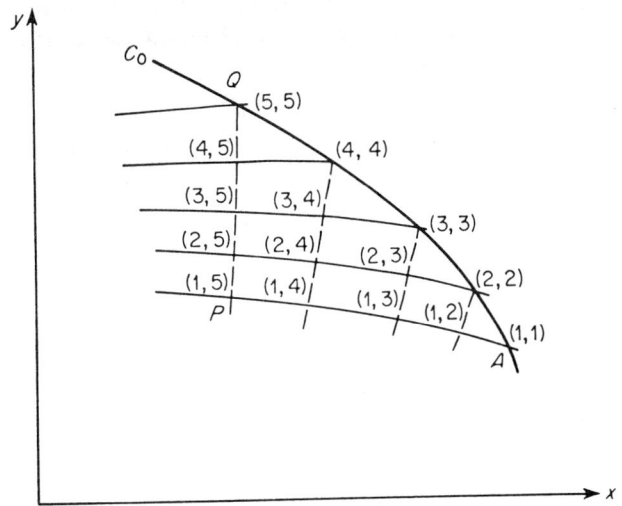

FIGURE 12.5.1 Numerical solution of first boundary-value problem.

(1, 2), which is a rough approximation to the true intersection point of the α and β lines. (If the α and β lines happened to be straight lines, the intersection point would be exact.) The points (2, 3), (3, 4), (4, 5), etc., can be determined approximately the same way. From Hencky's equations we have

$$\chi_{11} - \theta_{11} = \chi_{12} - \theta_{12}$$
$$\chi_{22} + \theta_{22} = \chi_{12} + \theta_{12}$$
(12.5.1)

We can therefore solve for χ_{12} and θ_{12}, giving us

$$\chi_{12} = \frac{\chi_{11} + \chi_{22} + \theta_{22} - \theta_{11}}{2}$$

$$\theta_{12} = \frac{\chi_{22} - \chi_{11} + \theta_{22} + \theta_{11}}{2}$$
(12.5.2)

In the same way we find
$$\chi_{23}, \theta_{23}, \chi_{34}, \theta_{34}, \cdots$$

We can now proceed to find χ and θ at (1, 3), (2, 4), (3, 5), and (4, 6). We thus obtain the slip-line field and the stresses in the entire region bounded by C_0 and the terminal slip lines AP and QP. A little reflection indicates that the solution cannot be carried beyond region APQ without some additional information. This leads us to the following theorem. Given an arc C_0 which is *not* a slip line and all the stresses acting at every point along the arc, then the complete slip-line field and the corresponding stresses can be determined within the region bounded by C_0 and the intersecting terminal slip lines as shown in Figure 12.5.1. Region APQ is called the *region of influence* of the arc C_0.

If the arc C_0 is itself a slip line, then the previous method is obviously inapplicable. If a second slip line, intersecting the first one, as shown in Figure 12.5.2, is also given, a solution can be obtained. For if θ is known

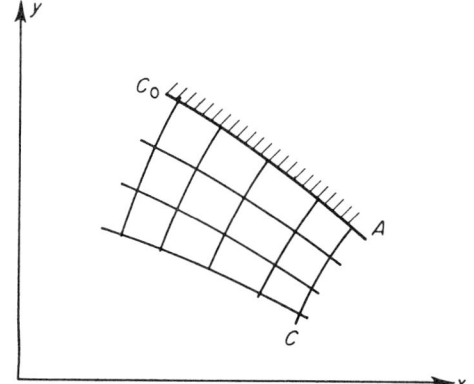

FIGURE 12.5.2 Second boundary-value problem.

along both slip lines, then θ can be determined at the adjoining net points by use of Hencky's first theorem. The complete slip-line field can then be constructed within the quadrilateral shown in Figure 12.5.2. To determine the stresses, it is necessary to know the value of χ at just one point on the boundary slip line, for by use of Hencky's equations (12.1.18) χ can then be computed throughout the region.

Alternatively, if the curve C_0 is a slip line, a solution can be obtained if on a second intersecting curve, not a slip line, either χ or θ is specified. The solution can then be obtained in the region indicated in Figure 12.5.3, using techniques

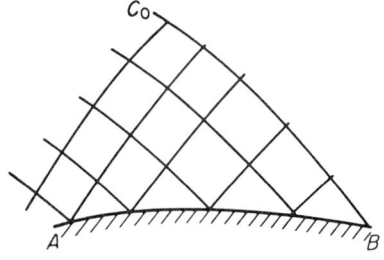

FIGURE 12.5.3 Third boundary-value problem.

Sec. 12-6] Geometric Construction of Slip-Line Fields

similar to those previously outlined. Details of the solution techniques for all three types of boundary-value problems, as well as methods for improving the accuracy, can be found in references [1], [4], and [5].

12-6 GEOMETRIC CONSTRUCTION OF SLIP-LINE FIELDS

A geometric construction for the stress and velocity fields, which is frequently very useful and leads to a better insight into the principles underlying slip-line theory, has been suggested by Prager [6]. For this purpose we make use of two planes, called the *stress plane* and the *physical plane*, as shown in Figure 12.6.1. Consider a point P undergoing plastic flow. The stress vector

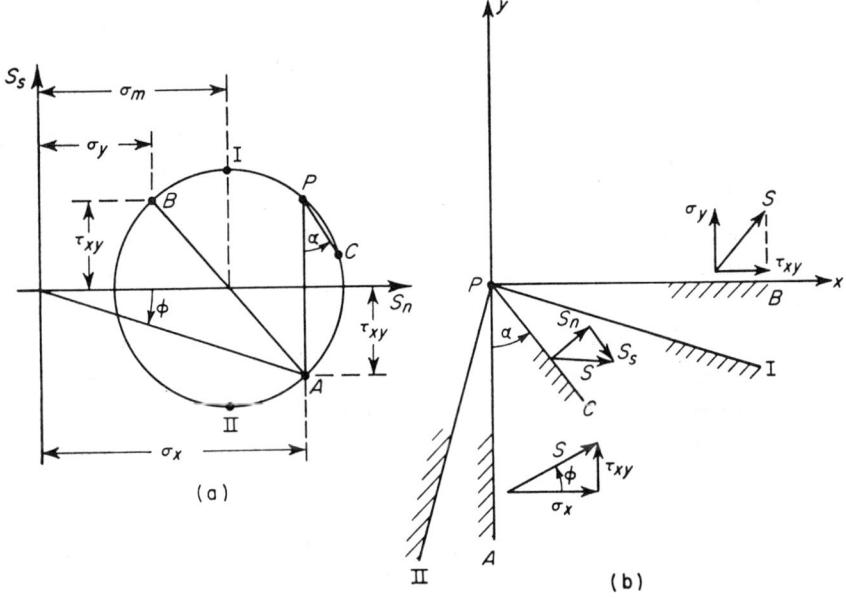

FIGURE 12.6.1 Stress plane (a) and physical plane (b).

acting at the point P will depend on the orientation of the area element through the point P upon which it acts. This is shown in the physical plane of Figure 12.6.1(b). The figure shows the traces of several area elements whose normals lie in the xy plane. These area elements actually contain the point P but are here shown separated for clarity. The shaded side of a given trace represents material, and the stresses shown are those transmitted from the unshaded side to the shaded side. Instead of identifying an area element by the direction of its normal, it is convenient to identify it by the direction of the trace of the element on the xy plane. Thus angles will be measured counterclockwise from the negative y axis instead of the x axis.

On the stress plane, Figure 12.6.1(a), a Mohr circle is plotted for the stress state at the point P of the physical plane. The Mohr's circle is constructed using the following convention. A shear stress which will cause the element to rotate in a clockwise sense is considered as positive, counterclockwise as negative. Thus the stress state $(\sigma_x, -\tau_{xy})$ on plane PA is shown as point A of the stress plane. On plane PB, whose normal stress is σ_y, the shear stress is positive (clockwise) and (σ_y, τ_{xy}) gives the point B in the stress plane. We note that the angle ϕ between the resultant stress S and the normal stress S_n in the stress plane is the same as the angle between the resultant stress and the normal stress in the physical plane, but *it is measured in the opposite direction*; i.e., if ϕ is measured clockwise in the stress plane it is measured counterclockwise in the physical plane as shown.

Once the points A and B are determined, the Mohr's circle can be constructed in the usual way by drawing a circle through A and B whose center lies at the intersection of AB with the S_n axis. The stress vector acting on any plane through P can now be found from the following consideration. If a straight line is drawn in the stress plane through any stress point (e.g., A or B) parallel to the trace in the physical plane upon which the stress acts, it will intersect the Mohr's circle at a point P, and *all such lines will intersect the circle at this same point*. The point P is called the *pole* of the Mohr's circle. In Figure 12.6.1(a) the pole is obtained by drawing a vertical line through A to the point P or a horizontal line through B. For example, to obtain the stress vector on plane OC rotated α degrees counterclockwise from OA, one draws a line through P on the Mohr's circle parallel to OC or α degrees counterclockwise from PA. The point of intersection of this line with the Mohr's circle gives the state of stress on the plane PC.

The top and bottom points of the Mohr's circle, labeled I and II, correspond to the planes upon which the maximum and minimum shearing stresses $\pm k$ act. The directions of the traces of these planes are given by the lines PI and PII. These directions are the first and second shear directions or the α and β directions, respectively, as defined in previous sections.

The geometric construction of the slip-line field can now be obtained making use of the following fact, proved by Prager in reference [6]. As we move along a slip line in the physical plane, *the pole of the Mohr's circle traces out a cycloid in the stress plane*. It does this by rolling without slipping along the top tangent $\tau = k$ if we move along an α line and along the bottom tangent $\tau = -k$ for a β line. Thus assume the stress state is known at the point P in the physical plane of Figure 12.6.2(b). The Mohr's circle and the pole P can be constructed as shown in Figure 12.6.2(a). The directions of

Sec. 12-6] Geometric Construction of Slip-Line Fields

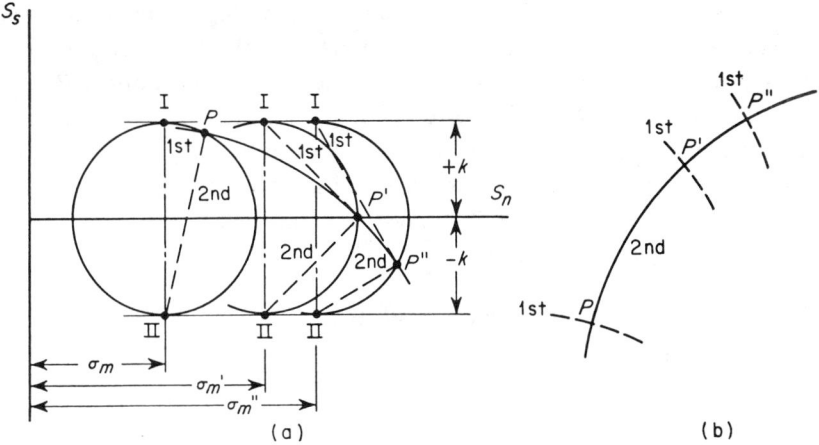

FIGURE 12.6.2 Cycloid trace of pole on stress plane (a) and corresponding slip line in physical plane (b). (Reference [5].)

the first and second slip lines at P are also known from $P\mathrm{I}$ and $P\mathrm{II}$. If we now move along the second shear line (β line) to the point P', the pole P of the Mohr's circle will move to P' as the circle rolls on the tangent line $S_s = -k$. The tangents to the two slip lines at P' are given by the directions $P'\mathrm{I}$ and $P'\mathrm{II}$, as shown by the dashed lines in Figure 12.6.2(a). Alternatively, since $P'\mathrm{II}$ is normal to the cycloid at P' (the point II is the instantaneous center of rotation), the element of the slip line at P' in the physical plane *is normal to the element of the cycloid at P' in the stress plane*. The slip lines at the point P'' can be established the same way. At the same time the stresses σ_m' and σ_m'' are determined from the positions of the center of the circle. It is apparent from the above that cycloids generated in this fashion are the images in the stress plane of the slip lines in the physical plane.

Prager has also shown [6] how to construct the velocity field at the same

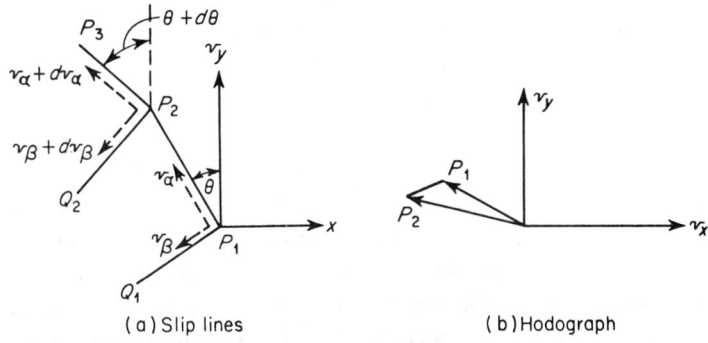

FIGURE 12.6.3 Hodograph construction of velocity field.

time as the slip lines. Let P_1, P_2, and P_3 be neighboring points on the first slip line, as shown in Figure 12.6.3. The line P_1P_2 is in the first shear direction. Let v_α and v_β be the velocity components in the two slip directions at P_1 and let $v_\alpha + dv_\alpha$ and $v_\beta + dv_\beta$ be the corresponding velocities at P_2. The condition that there be no extensions, only shearing flows in the slip directions [see equation (12.2.4)], requires that the projections of the velocity vectors at the points P_1 and P_2 onto the line P_1P_2 must be equal. This yields immediately the Geiringer equation (12.2.8):

$$dv_\alpha - v_\beta \, d\theta = 0$$

The second equation,

$$dv_\beta + v_\alpha \, d\theta = 0$$

is obtained from similar considerations along a β line.

Consider the velocity plane, called the *hodograph*, of Figure 12.6.3(b). The vectors OP_1 and OP_2 represent the velocities of the points P_1 and P_2, respectively. Since, as stated above, the projections of these two vectors onto the line P_1P_2 of Figure 12.6.3(a) must be equal, the vector difference P_1P_2 must be orthogonal to P_1P_2 in the physical plane. In other words, line elements in the velocity plane must be normal to the corresponding line elements of the physical plane. In general, therefore, *corresponding elements of the slip-line field and hodograph are orthogonal, and hence corresponding elements of the stress plane and hodograph are parallel.*

In connection with the above, it should be pointed out that the velocity fields under consideration need not be continuous, and velocity fields with lines of discontinuity are often encountered. Such lines of discontinuity must be shear lines (or the envelope of the shear lines). The velocity components normal to any slip line is continuous across the slip line, but the tangential velocity may have different values on both sides. It follows from the Geiringer equations that the *velocity jump* must be constant along any slip line. This also follows from the fact that the slip lines are the characteristics of the governing differential equations, as will be shown in Section 12.7. In the hodograph, the two sides of a line of velocity discontinuity are mapped into parallel curves, as shown in reference [6].

To illustrate how one begins to construct a slip-line field for given boundary conditions, consider a boundary arc along which the forces are prescribed, as shown in Figure 12.6.4(a). The material adjacent to this arc is known to be in a state of plastic flow, and from the boundary stresses the maximum shear directions and hence the slip-line directions can be determined. A Mohr's circle can be constructed in the stress plane for each of the points A and B of the physical plane. Here a difficulty is immediately encountered, for it is

Sec. 12–6] Geometric Construction of Slip-Line Fields

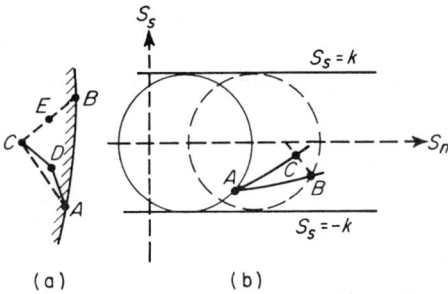

FIGURE 12.6.4 Geometric construction of slip lines.

possible to draw *two circles* for each of the points. For example, in Figure 12.6.5 the stress vector S acting at a point P has been drawn from the origin to the point A of the stress plane. Through the point A, however, two circles can be drawn, each with radius k and with their centers on the S_n axis, as shown. One of these corresponds to a lower mean normal stress than the other and these solutions are therefore called the *weak* and *strong solutions*, respectively. As indicated on the figure, the poles and the shear directions will be different for the two circles. Which of these solutions to choose is usually determined from additional boundary conditions or from looking at the boundary conditions as a whole, rather than at one point. This uncertainty is due to the quadratic nature of the yield criterion, as was indicated for the punch indentation problem of Section 12.5.

Returning now to Figure 12.6.4, the weak solution is adopted. The Mohr's circles for points A and B are shown by the solid and broken-line circles, respectively. As the solid-line circle rolls along the line $\tau = k$, the pole will trace a cycloid which is the image of the *first slip line* through A. As the broken-line circle rolls along $S_s = -k$, the point B will describe a cycloid, shown by the broken line, which will be the image of the *second slip line*

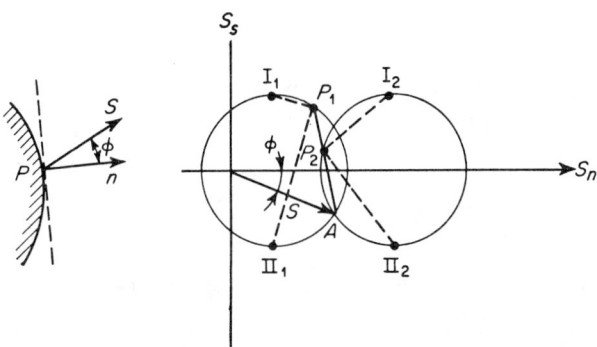

FIGURE 12.6.5 Strong and weak solutions.

through B. The intersection of the two cycloids at C will be the image of the intersection of the two slip lines in the physical plane at C. The directions of the normals to the cycloids at C fix the directions of the tangents to the slip lines at C in the physical plane. The actual location of C in the physical plane is found from the conditions that $AD = CD$ and $BE = CE$. This amounts to assuming that the slip-line arcs AC and BC are circular.

Once the slip lines have been established, the hodograph can be constructed from the known velocities on the boundary using the orthogonality relation between the slip line and hodograph elements.

The above is a brief outline of Prager's geometric construction of slip-line fields. The interested reader is referred to references [6] through [9] for further discussion and examples.

12-7 COMPLETE SOLUTIONS. UPPER AND LOWER BOUNDS

The slip-line solutions discussed heretofore are incomplete in that no attempt has been made to extend these solutions into the rigid regions adjacent to the plastic regions to determine whether an equilibrium stress distribution satisfying the boundary conditions and nowhere exceeding the yield point exists in those regions. Such *partial solutions* are *upper-bound solutions*; i.e., they overestimate the load necessary to produce the plastic flow. This follows from the *upper bound theorem*, derived in reference [10]. We shall discuss the upper and lower bound theorems in greater detail in Chapter 13. For our purpose here, the upper bound theorem can be stated briefly as follows: *A load which produces a kinematically admissible velocity field will be equal or greater than the true load.*

A kinematically admissible velocity field is one which satisfies the velocity boundary conditions, is incompressible, and is continuous with continuous first derivatives, except at certain discontinuity surfaces where the normal velocity must be continuous but the tangential velocity may suffer a jump on crossing the surface. Furthermore, on the boundary we must have

$$\int (Xv_x + Yv_y)dS > 0$$

where X and Y are the boundary forces and dS is an element of boundary area.

Since the slip-line solutions give kinematically admissible velocity fields, they are upper bound solutions. To obtain a *complete solution*, it is necessary to determine if the stress fields in the rigid regions also satisfy equilibrium and the boundary conditions and nowhere exceed the yield stress. If such a *statically admissible* stress field exists in both the plastic and rigid parts, then this solution constitutes a *lower bound*, and since the solution is both an upper bound and a lower bound, it is a *complete solution*, giving the true load for the problem.

The fact that a statically admissible stress field as defined above gives a lower bound is expressed by the *lower bound theorem* [10], which can be stated as follows: *A load which produces a statically admissible stress field will be equal to or less than the true load that will produce plastic flow.*

Lower bound solutions are usually difficult to obtain since it would be necessary to solve the equilibrium equations in the elastic or rigid region. Upper bound solutions can be obtained from slip-line solutions and are of greater value, in as much as they ensure that a certain operation will be performed, since the calculated load will be greater than the required load. Discussions of methods of obtaining upper bounds with examples can be found in references [11] and [5].

12-8 SLIP LINES AS CHARACTERISTICS

The slip lines previously discussed are actually the *characteristics* of the differential equations defining the problem. All the properties of characteristics of hyperbolic equations and all the mathematical methods for solving such equations can therefore be applied directly to the plane strain problem of a rigid–perfectly plastic material. In this section we shall briefly describe the origin and properties of characteristics to give the reader an insight into the mathematical origin of slip-line fields. The reader interested only in the physical description of the previous sections may skip to Chapter 13 without loss of continuity.

Consider the first-order differential equation

$$a \frac{\partial u}{\partial x} + b \frac{\partial u}{\partial y} = c \quad (12.8.1)$$

where a, b, and c are functions of u, x, and y but not of the partial derivatives of u. Such an equation is called *quasilinear*. Introduce the standard notation

$$\frac{\partial u}{\partial x} \equiv p \qquad \frac{\partial u}{\partial y} \equiv q \quad (12.8.2)$$

Then (12.8.1) can be written

$$ap + bq = c \tag{12.8.3}$$

We now pose the following question. Given a curve C in the xy plane along which the function u is specified, does there exist a solution u of equation (12.8.1) satisfying these "initial conditions"? This is the two-dimensional analogue of the ordinary first-order equation, where the initial value is specified at one point. The above problem is called the *Cauchy problem*, and the answer is that it depends on the curve C.

It is obvious that if given the values of u on C, p and q could be determined on C such that (12.8.1) [or (12.8.3)] was satisfied, then u could be computed a small distance away from C by use of a Taylor series expansion in two variables. That is, if the values of x and y on C are designated by x_c and y_c, then

$$u(x, y) = u(x_c, y_c) + (x - x_c)\left(\frac{\partial u}{\partial x}\right)_c + (y - y_c)\left(\frac{\partial u}{\partial y}\right)_c + \cdots \tag{12.8.4}$$

The value of u could thus be obtained at a neighboring curve. By repeating this process, u could be determined over some region, the limits of which, if any, are not yet known. The Cauchy problem, in this case, therefore reduces to the problem of determining p and q on the curve C.

To find p and q we proceed as follows. Let s be the arc length along the curve. Then if the derivatives p and q exist, we must have

$$\frac{du}{ds} = \frac{\partial u}{\partial x}\frac{dx}{ds} + \frac{\partial u}{\partial y}\frac{dy}{ds} = p\frac{dx}{ds} + q\frac{dy}{ds} \tag{12.8.5}$$

Equations (12.8.5) and (12.8.3) give us two equations in the two unknowns p and q. Solving these two equations gives

$$p = \frac{c\,dy - b\,du}{a\,dy - b\,dx}$$
$$q = \frac{a\,du - c\,dx}{a\,dy - b\,dx} \tag{12.8.6}$$

Thus, given a, b, and c, the value of u on the curve C and the shape of the curve, p and q can be computed from (12.8.6) and u in the neighborhood of C obtained from (12.8.4). However, this will not be true for *any* curve C, for if C is such that the denominator in (12.8.6) vanishes; i.e., if

$$\frac{dy}{dx} = \frac{b}{a} \tag{12.8.7}$$

Sec. 12–8] Slip Lines as Characteristics

then there will obviously be no solution unless the numerators also vanish. A curve C whose equation satisfies (12.8.7) is called a *characteristic curve*, and if u is specified along such a curve, there will be no solution to the problem unless u is specified so that

$$\frac{du}{dy} = \frac{c}{b} \quad \text{or} \quad \frac{du}{dx} = \frac{c}{a} \qquad (12.8.8)$$

causing the numerators in (12.8.6) to vanish. In the latter case, i.e., if u is specified on C according to (12.8.8), there will be an infinity of solutions, since (12.8.6) are then indeterminate.

To summarize: In answer to the question posed at the beginning, if u is specified along a curve C, then a solution to (12.8.1) exists provided C is not a characteristic of the differential equation. If C is a characteristic, then either there is no solution, if (12.8.8) is not satisfied, or there are an infinity of solutions, if (12.8.8) is satisfied. Equation (12.8.8) is called the *compatibility equation*.

Let us illustrate these results by a simple example [12]. Consider the differential equation

$$\frac{\partial u}{\partial x} + \frac{\partial u}{\partial y} = 1 \qquad (12.8.9)$$

or
$$p + q = 1$$

Then $a = b = c = 1$ and, from (12.8.7),

$$\frac{dy}{dx} = 1 \qquad (12.8.10)$$

or
$$y = x + A$$

The characteristics are therefore straight lines with slopes of unity, as shown in Figure 12.8.1. For a solution to exist, u must satisfy (12.8.8) or

$$u = x + B = y + (B - A) \qquad (12.8.11)$$

along these lines. Suppose now that u is given on the line segment $0 < x < 1$, $y = 0$. This line segment is the curve C of the previous discussion. Then since C is not a characteristic, we should be able to obtain a unique solution for u. The simplest way to do this is to integrate along the characteristics. Thus considering the characteristic intersecting the x axis at $x = x_i$ as shown,

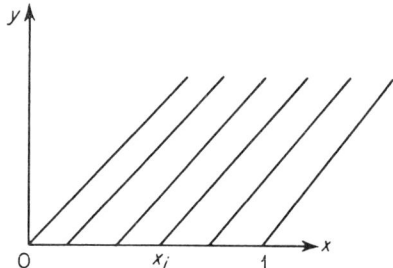

FIGURE 12.8.1   Characteristics for equation (12.8.9).

the value of u specified at this point is u_i, and since along this characteristic u must satisfy (12.8.11), it follows that

$$B - A = u_i \qquad B = u_i - x_i$$

so that

$$u = y + u_i = x + u_i - x_i \qquad (12.8.11a)$$

for all

$$0 < x < 1$$

A solution has thus been obtained for the problem using the initial data and the characteristic and compatibility equations (12.8.10) and (12.8.11). These equations are the equivalent of the original differential equation, which does not have to be used at all. This method of solving a partial differential equation by reducing it to a set of ordinary differential equations for finding the characteristics and for integrating along the characteristics is called *the method of characteristics*. Thus the introduction of characteristics not only serves the purpose of determining whether a unique solution exists but actually provides a method for obtaining the solution.

From the above example it is apparent that the solution is defined and is unique *only in the region bounded by the terminal characteristics at $x = 0$ and $x = 1$*. This region is called the *region of influence*. To determine the solution outside this region requires additional data outside the range $0 < x < 1$. We can state this result as a general theorem.

A partial differential equation having real characteristics has a unique solution within the region bounded by the curve C upon which the initial data is specified, and the two characteristics intersecting the ends of C. If C coincides with a characteristic there is no unique solution.

Let us consider now the case when u is specified along one of the characteristics, say, the line $x = y$. For there to be any solutions, u must be specified on this line such that

$$u = x + u_0 = y + u_0 \qquad (12.8.12)$$

Sec. 12-8] Slip Lines as Characteristics

where u_0 is the value of u at $x = y = 0$. The solution elsewhere, however, is not unique. We can add, for example, a function $K(x - y)$ to u, with K an arbitrary constant and still satisfy the initial conditions along the line $x = y$. Thus

$$u = x + u_0 + K(x - y) \tag{12.8.13}$$

will satisfy the differential equation (12.8.9) and also the initial values on $y = x$. This illustrates the fact that if the initial conditions are specified on characteristic, the solution is not unique. In this case we might say that the solution is not defined uniquely at points not on this line because the terminal characteristics coincide, and the region of influence is just the line itself.

Another very important property of characteristics can now be seen—the possibilities of discontinuities in the solution which are propagated along characteristics. For example, suppose on the initial line $y = 0$ of the previous example we are given

$$\begin{aligned} u &= f_1(x) & 0 < x < x_1 \\ u &= f_2(x) & x_1 < x < 1 \end{aligned} \tag{12.8.14}$$

where $f_1(x_1) \neq f_2(x_1)$; i.e., u is double-valued at this point. Then the solution will be double-valued and hence discontinuous all along the characteristic $y = x - x_1$, passing through this point. The values of u to the left of this line will be determined by $u = f_1(x)$ and to the right by $u = f_2(x)$.

In a similar fashion, discontinuities in the derivatives of u may be propagated along characteristics. For example, if we are given that at $y = 0$,

$$\begin{aligned} u &= 1 & 0 < x < \tfrac{1}{2} \\ u &= x + \tfrac{1}{2} & \tfrac{1}{2} < x < 1 \end{aligned}$$

Then the solution using (12.8.11) is

$$\begin{aligned} u &= y + 1 & \text{to the left of } y = x - \tfrac{1}{2} \\ u &= x + \tfrac{1}{2} & \text{to the right of } y = x - \tfrac{1}{2} \end{aligned} \tag{12.8.15}$$

On the line $y = x - \tfrac{1}{2}$, u is continuous, but the derivatives of u are not, for

$$\begin{aligned} \frac{\partial u}{\partial x} &= 0 & \frac{\partial u}{\partial y} &= 1 & \text{to the left of } y = x - \tfrac{1}{2} \\ \frac{\partial u}{\partial x} &= 1 & \frac{\partial u}{\partial y} &= 0 & \text{to the right of } y = x - \tfrac{1}{2} \end{aligned} \tag{12.8.16}$$

These derivatives are therefore discontinuous all along the characteristic $y = x - \tfrac{1}{2}$.

In the above example the differential equation (12.8.9) was linear. It was therefore possible to determine the characteristics once and for all independently of the solution u. Thus equation (12.8.7) could be integrated first and then equation (12.8.8) solved for u along the characteristics. However, if the equation were quasilinear, i.e., if a and b were functions of u, then equation (12.8.7) could obviously not be solved for the characteristics without first knowing the solution u. In this case it is necessary to solve (12.8.7) and (12.8.8) simultaneously; i.e., u and the characteristics must be determined simultaneously.

Second-Order Equation

Let us now consider the second-order quasilinear partial differential equation

$$A \frac{\partial^2 u}{\partial x^2} + B \frac{\partial^2 u}{\partial x \, \partial y} + C \frac{\partial^2 u}{\partial y^2} = D \qquad (12.8.17)$$

By quasilinear we mean that the coefficients may all be functions of u, $\partial u/\partial x$, $\partial u/\partial y$, as well as of x and y. Using the standard notation,

$$\frac{\partial^2 u}{\partial x^2} \equiv r \qquad \frac{\partial^2 u}{\partial x \, \partial y} \equiv s \qquad \frac{\partial^2 u}{\partial y^2} \equiv t \qquad (12.8.18)$$

equation (12.8.17) is written

$$Ar + Bs + Ct = D \qquad (12.8.19)$$

Now assume that on some initial curve C_0 in the xy plane, the values of u, and its normal derivative are given. Alternatively the partial derivatives p and q are given, for then u and the normal derivative can be computed. The values of u, p, and q along C_0 are called a *strip* of first order and u, p, and q must obviously satisfy the condition

$$du = p \, dx + q \, dy$$

called the *strip condition*.

We now ask ourselves the same question as before. Does there exist a solution u of equation (12.8.17) satisfying these initial conditions? If the second derivatives r, s, and t on the curve can be determined so as to satisfy

Sec. 12–8] Slip Lines as Characteristics

(12.8.17), then as before u can be determined as well as its first derivatives a small distance away from the curve by Taylor series expansions, and in this way the solution can be continued into the region. Let us see then if the second derivatives on the curve can be computed. On the curve

$$du = \frac{\partial u}{\partial x} dx + \frac{\partial u}{\partial y} dy$$

and

$$d\left(\frac{\partial u}{\partial x}\right) = \frac{\partial^2 u}{\partial x^2} dx + \frac{\partial^2 u}{\partial x\, \partial y} dy$$

$$d\left(\frac{\partial u}{\partial y}\right) = \frac{\partial^2 u}{\partial x\, \partial y} dx + \frac{\partial^2 u}{\partial y^2} dy$$

or

$$r\, dx + s\, dy = dp$$
$$s\, dx + t\, dy = dq$$

(12.8.20)

and the differential equation is

$$rA + sB + tC = D$$

A, B, C, D, dq, and dp, being known on the curve, we have three equations to solve for the unknown second derivatives r, s, and t. For example,

$$s = \frac{\begin{vmatrix} dx & dp & 0 \\ 0 & dq & dy \\ A & D & C \end{vmatrix}}{\begin{vmatrix} dx & dy & 0 \\ 0 & dx & dy \\ A & B & C \end{vmatrix}} \quad (12.8.21)$$

If the determinant in the denominator vanishes there will be no solution unless the numerator also vanishes, in which case there are an infinite number of solutions. If the denominator vanishes,

$$A(dy)^2 - B\, dx\, dy + C(dx)^2 = 0$$

(12.8.21a)

or

$$A\left(\frac{dy}{dx}\right)^2 - B\frac{dy}{dx} + C = 0$$

which gives

$$\frac{dy}{dx} = \frac{B \pm \sqrt{B^2 - 4AC}}{2A} \quad (12.8.22)$$

Equation (12.8.22) defines two sets of curves (one using the plus sign, one using the minus sign). These are called *characteristic curves*, and if the data are given along one of these curves, no solution exists unless, as already mentioned, the numerator determinants also vanish. For this case,

$$A\,dp\,dy + C\,dq\,dx - D\,dx\,dy = 0 \qquad (12.8.23)$$

Along the characteristic curves equation (12.8.23) must be satisfied. This is the *compatibility equation*.

Equations (12.8.22) and (12.8.23) replace the partial differential equation (12.8.17) and if the initial curve is not a characteristic, a solution can be obtained by finding the characteristics from (12.8.22) and then integrating along the characteristics using (12.8.23). If the equation is nonlinear, i.e., if A, B, and C are functions of u, p, and q, then (12.8.22) and (12.8.23) must be solved simultaneously, as discussed for the first-order equation. If the data are prescribed along one of the characteristic curves, there is no unique solution.

The above discussion implies that (12.8.22) can be solved obtaining two real families of curves. But this can only be true if $B^2 - 4AC \geq 0$. If $B^2 - 4AC < 0$ there are no real characteristics. There are therefore three types of equations.

1. $B^2 - 4AC > 0$. The equation is called *hyperbolic*. There are two sets of characteristics.

2. $B^2 - 4AC = 0$. The equation is called *parabolic*. There is only one set of characteristics.

3. $B^2 - 4AC < 0$. The equation is called *elliptic*. There are no real characteristics.

The parabolic equation is similar to the hyperbolic with regard to the properties and use of the characteristics. The first-order equations previously discussed are actually parabolic, since they always have one family of characteristics (if the coefficients are real). In fact, since there are no second derivative terms, $B^2 - 4AC = 0$.

The properties of the hyperbolic equation are then essentially the same as previously discussed for the first-order equation. We shall briefly note some of these properties and make a comparison with the elliptic case.

In the elliptic case, if u and its normal derivative is prescribed along any curve C, then a solution can always be found in the neighborhood of C. If the values are prescribed on a closed boundary, which is the usual case, the solution will be determined within the bounded region. It is necessary, however, that u be regular on C, i.e., it possess derivatives of all orders.

Sec. 12-8] Slip Lines as Characteristics

In the hyperbolic (and parabolic) case, if the curve C is not a characteristic, a solution exists. The prescribed values in this case need not be regular and the higher derivatives of the prescribed values may have finite jumps at points on C. These discontinuities are then propagated along the characteristics where they originate. If C is a characteristic curve, there is no solution unless u and its derivatives satisfy a compatibility relation, in which case there are an infinity of solutions. It is also evident that closed boundaries are excluded, for a given characteristic would then intersect the boundary twice, and values of u and its derivatives could not be assigned arbitrarily to both points of intersection.

Boundary-Value Problems

There are generally three types of boundary-value problems for hyperbolic equations, [13].

1. Given u, p, and q on a curve C_0 which is not a characteristic and which intersects each characteristic at most once, u can be determined in a triangular region D_0 bounded by C_0 and a characteristic of each family as shown in Figure 12.8.2(a). The curves $\bar{y}$ constant and $\bar{x}$ constant are the characteristic curves of each family. More specifically, the value of u at each point P of D_0 is determined by the values of u and its derivatives on the portion C_p of C_0 which is bounded by the characteristics through P. The segment C_p is called the *domain of dependence* of the point P.

2. A linear relationship $a(\partial u/\partial n) + bu = c(x, y)$ is prescribed on an arc C_0, and in addition u is prescribed on a characteristic arc C_c passing through one end point of C_0. The solution can again be obtained in the triangular region bounded by C_0, C_c, and the characteristic of the other family intersecting C_c and C_0, as shown in Figure 12.8.2(b).

3. u is prescribed on two intersecting characteristics C_c and C_c', as shown in Figure 12.8.2. Of course, u must satisfy the compatibility relation on both characteristics. The solution is determined in the quadrilateral bounded by the four characteristics as shown.

The reader will recognize the similarity between these three boundary-value problems and the three boundary-value problems for slip-line fields discussed in Section 12.5.

Finally, it follows that by combining these three basic types of boundary conditions, more complicated conditions can be used. For example, in Figure 12.8.2(d), u, p, and q may be given on C_0. But then only one relation of the type $a(\partial u/\partial n) + bu = c(x, y)$ can be prescribed on each of C_0' and C_0''.

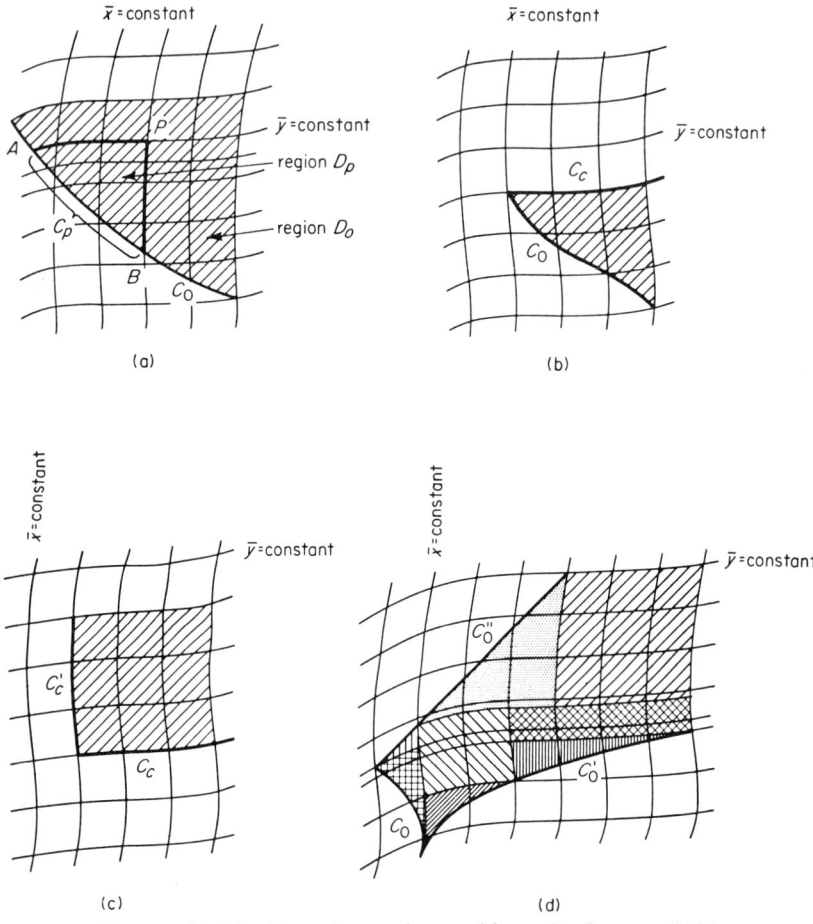

FIGURE 12.8.2 Boundary-value problems (Reference [13].)

The solutions in the various regions shown can then be *patched* together along characteristics so that u is continuous, while $\partial u/\partial x$ and $\partial u/\partial y$ may be discontinuous. Note again that closed boundaries are not admissible.

Numerical Solution

An analytical solution of the hyperbolic equation can be effected in theory, by Reimann's method, which essentially reduces the boundary-value problem of the first kind along an arbitrary curve C_0 [Figure 12.8.2(a)] to a simpler boundary-value problem of the third kind along two intersecting characteristics [Figure 10.8.2(c)]. This is essentially the counterpart of Green's function in solving elliptic equations. In practice, however, only a few very

Sec. 12-8] Slip Lines as Characteristics

simple problems have been solved in this way. The widest use of the method of characteristic is found in conjunction with numerical methods.

Suppose then that u, p, and q are given along a noncharacteristic arc C and we wish to obtain the solution in the region bounded by C and the two terminal characteristics as shown in Figure 12.8.2(a). It is desired to solve equations (12.8.22) and (12.8.23) numerically. If A, B, and C are functions only of x and y, (12.8.22) can be solved separately and the characteristics constructed independently of the initial conditions. Such characteristics are called *fixed*. If A, B, and C are also functions of u, p, and q, then (12.8.22) and (12.8.23) must be solved simultaneously. (Note: If A, B, and C are functions of u, p, and q but *not* of x and y, then it is possible to interchange the roles of the independent and dependent variables and transform the partial differential equation into a linear differential equation with x and y as the dependent variables. This is called a *hodograph transformation*. The solution is then obtained in the plane of u and $\partial u/\partial n$, or p and q, called the *hodograph plane*, instead of the xy plane.)

Equations (12.8.22) can be written

$$dy^+ = f^+ \, dx$$
$$dy^- = f^- \, dx \qquad (12.8.24)$$

where

$$f^+ = \frac{B + \sqrt{B^2 - 4AC}}{2A}$$
$$f^- = \frac{B - \sqrt{B^2 - 4AC}}{2A} \qquad (12.8.25)$$

and (12.8.23) can then be written

$$D \, dy^- - A f^+ \, dp - C \, dq = 0 \quad \text{along an } f^+$$
$$D \, dy^+ - A f^- \, dp - C \, dq = 0 \quad \text{along an } f^- \qquad (12.8.26)$$

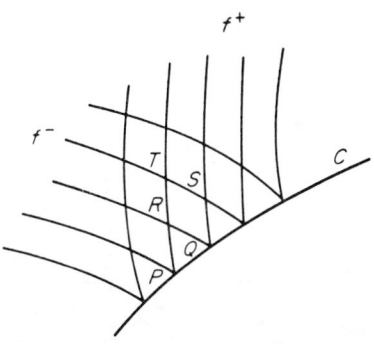

FIGURE 12.8.3 Numerical solution.

Let us then draw the curve C and two families of characteristics as shown in Figure 12.8.3. Consider any two adjacent points P and Q on C and let the f^+ characteristic from P intersect the f^- characteristic from Q at R. If the characteristics are fixed, the position of the point R is known. If the characteristics depend on the solution, the coordinates of R must be determined at the same time that p and q are determined at this point. Equations (12.8.24) and (12.8.26) are now written approximately as follows:

$$y_R - y_P = \tfrac{1}{2}(f_R^+ + f_P^+)(x_R - x_P)$$
$$y_R - y_Q = \tfrac{1}{2}(f_R^- + f_Q^-)(x_R - x_Q)$$

and (12.8.27)

$$\tfrac{1}{2}(D_R + D_P)(y_R - y_P) - \tfrac{1}{2}(A_R f_R^+ + A_P f_P^+)(p_R - p_P)$$
$$- \tfrac{1}{2}(C_R - C_P)(q_R - q_P) = 0$$
$$\tfrac{1}{2}(D_R + D_Q)(y_R - y_Q) - \tfrac{1}{2}(A_R f_R^- + A_Q f_Q^-)(p_R - p_Q)$$
$$- \tfrac{1}{2}(C_R - C_Q)(q_R - q_Q) = 0$$

Also from the condition

$$du = p\,dx + q\,dy$$

$$u_R - u_P = \tfrac{1}{2}(p_R + p_P)(x_R - x_P) + \tfrac{1}{2}(q_R + q_P)(y_R - y_P)$$
or $\quad u_R - u_Q = \tfrac{1}{2}(p_R + p_Q)(x_R - x_Q) + \tfrac{1}{2}(q_R + q_Q)(y_R - y_Q) \quad$ (12.8.28)

Equations (12.8.27) and one of (12.8.28) furnish five equations for the five unknowns x_R, y_R, p_R, q_R, and u_R. These equations are nonlinear and they usually have to be solved by some iterative method. Convergence is usually rapid if the interval PQ is not too large.

In the same way, the solution at other grid points adjacent to the initial curve, such as S in the figure, can be obtained. From R and S we can proceed to T, and so on. The solution is of course defined only in the region bounded by the terminal characteristics, as previously explained.

Two First-Order Equations. Slip Lines

If we have two first-order equations of the form

$$a\frac{\partial u}{\partial x} + b\frac{\partial u}{\partial y} + e\frac{\partial v}{\partial x} + f\frac{\partial v}{\partial y} = d$$
$$A\frac{\partial u}{\partial x} + B\frac{\partial u}{\partial y} + E\frac{dv}{\partial x} + F\frac{\partial v}{\partial y} = D$$

(12.8.29)

Sec. 12-8] Slip Lines as Characteristics

with

$$du = \frac{\partial u}{\partial x} dx + \frac{\partial u}{\partial y} dy$$
$$dv = \frac{\partial v}{\partial x} dx + \frac{\partial v}{\partial y} dy$$
(12.8.30)

then by a similar procedure the equation for the characteristics is

$$(aE - eA)\left(\frac{dy}{dx}\right)^2 + (eB + fA - bE - aF)\frac{dy}{dx} + bF - fB = 0 \quad (12.8.31)$$

and the compatibility relation along the characteristics is

$$\begin{vmatrix} du & dx & dy & 0 \\ dv & 0 & 0 & dx \\ d & a & b & e \\ D & A & B & E \end{vmatrix} = 0 \quad (12.8.32)$$

It will now be shown that the slip lines discussed in Sections 12.1 through 12.7 are characteristics of the governing differential equations and all the properties of characteristics discussed in Section 12.8 apply to them. We start with equations (12.1.16), which are essentially the equilibrium equations with the yield condition included:

$$\frac{\partial \chi}{\partial x} - \cos 2\theta \frac{\partial \theta}{\partial x} - \sin 2\theta \frac{\partial \theta}{\partial y} = 0$$
$$\frac{\partial \chi}{\partial y} - \sin 2\theta \frac{\partial \theta}{\partial x} + \cos 2\theta \frac{\partial \theta}{\partial y} = 0$$
(12.8.33)

These are two simultaneous quasilinear first-order equations, and if they are hyperbolic we should be able to use the method of characteristics. Comparing the coefficients with those in (12.8.29), it is seen that

$$a = 1 \quad b = 0 \quad e = -\cos 2\theta \quad f = -\sin 2\theta \quad d = 0$$
$$A = 0 \quad B = 1 \quad E = -\sin 2\theta \quad F = \cos 2\theta \quad D = 0$$

and equation (12.8.31) for the characteristics becomes

$$\sin 2\theta \left(\frac{dy}{dx}\right)^2 + 2\cos 2\theta \frac{dy}{dx} - \sin 2\theta = 0$$

or

$$\frac{dy}{dx} = \frac{-\cos 2\theta \pm 1}{\sin 2\theta} \quad (12.8.34)$$

or
$$\left(\frac{dy}{dx}\right)^+ = \frac{2\sin^2\theta}{2\sin\theta\cos\theta} = \tan\theta$$
$$\left(\frac{dy}{dx}\right)^- = -\cot\theta = \tan\left(\theta + \frac{\pi}{2}\right) \tag{12.8.35}$$

But $\tan\theta$ and $\tan(\theta + \pi/2)$ are the slopes of the slip lines, so that (12.8.35) shows that the characteristics of the differential equations (12.8.33) are just the slip lines. We can identify the α slip lines with the plus characteristics and the β slip lines with the minus characteristics, as defined by equations (12.8.35).

To determine the compatibility relation that must be satisfied along the slip lines, equation (12.8.32) gives

$$\begin{vmatrix} d\chi & dx & dy & 0 \\ d\theta & 0 & 0 & dx \\ 0 & 1 & 0 & -\cos 2\theta \\ 0 & 0 & 1 & -\sin 2\theta \end{vmatrix} = 0 \tag{12.8.36}$$

or
$$d\chi - \left(\sin 2\theta \frac{dy}{dx} + \cos 2\theta\right)d\theta = 0$$

and, from (12.8.34),
$$d\chi \mp d\theta = 0$$

or
$$\chi - \theta = \text{constant along } \alpha \text{ characteristic}$$
$$\chi + \theta = \text{constant along } \beta \text{ characteristic} \tag{12.8.37}$$

which are Hencky's equations, previously obtained.

Problems

1. Derive equation (12.1.5).
2. Show that the principal stresses for plane strain problems are given by equations (12.1.7) and (12.1.8).
3. Show that the Tresca criterion differs from equation (12.1.5) only by a multiplicative constant.
4. Derive equation (12.1.9).
5. Show that equations (12.1.14) can be obtained by means of equations (12.1.4) and (12.1.12).
6. Prove Theorems 6 and 8 of Section 12.3.
7. Determine the velocity distribution for the condition of Figure 12.4.2 using the Geiringer equations.
8. Obtain the complete solution for the rigid punch indentation of Figure 12.4.2 assuming the rigid plastic boundary to be *HIJKLMN*.

Sec. 12-8] Slip Lines as Characteristics

9. Show that the choice of a plus sign in equation (12.4.1) or a minus sign in equation (12.4.4) would lead to solutions which violate the initial assumptions.
10. Prove that the point of intersection of a line drawn through the pole of the Mohr's circle and the circle will give the stress state acting on a plane whose trace is parallel to this line.

References

1. W. Prager and P. G. Hodge, Jr., *Theory of Perfectly Plastic Solids*, Wiley New York, 1951.
2. P. G. Hodge, Jr., An Introduction to the Mathematical Theory of Perfectly Plastic Solids, *ONR-358*, Feb. 1950.
3. L. Prandtl, Über die Härte Plastischer Koerper, *Goettinger Nachr., Math. Phys. Kl.*, 1920, pp. 74–85.
4. R. Hill, The Plastic Yielding of Notched Bars Under Tension, *Quart. J. Mech. Appl. Math.*, 2, 1949, pp. 40–52.
5. E. G. Thomsen, C. T. Yang, and S. Kobayashi, *Mechanics of Plastic Deformation in Metal Processing*, Macmillan, New York, 1965.
6. W. Prager, A Geometrical Discussion of the Slip-Line Field in Plane Plastic Flow, *Trans. Roy. Inst. Tech. (Stockholm)*, 65, 1953, pp. 1–26.
7. W. Prager, The Theory of Plasticity: A Survey of Recent Achievements, *Proc. Inst. Mech. Eng.*, 169, 1955, pp. 41–57.
8. H. Ford, The Theory of Plasticity in Relation to Engineering Application, *J. Appl. Math. Phys.*, 5, 1954, pp. 1–35.
9. J. M. Alexander, Deformation Modes in Metal Forming Processes, *Proceedings of the Conference on Technical Engineering Manufacture*, Institute of Mechanical Engineers, New York, 1958, Paper No. 42.
10. D. C. Drucker, H. J. Greenberg, and W. Prager, The Safety Factor of an Elastic–Plastic Body in Plane Strain, *J. Appl. Mech.*, 18, 1951, pp. 371–378.
11. W. Johnson and P. B. Mellor, *Plasticity for Mechanical Engineers*, Van Nostrand, Princeton, N.J., 1962.
12. L. Fox, *Numerical Solution of Ordinary and Partial Differential Equations*, Pergamon Press, London, 1962.
13. G. A. Korn and T. M. Korn, *Mathematical Handbook for Scientists and Engineers*, McGraw-Hill, New York, 1961.
14. H. Hencky, Ueber einige statisch bestimmte Faelle des Gleichgewichts in plastischen Koerpern *Z. angew. Math Mech.*, 3, 1923, pp. 245–251.
15. H. Geiringer, Beit zum Vollständigen ebenen Plastizitäts-problem, *Proc 3rd Intern. Congr. Appl. Mech*, 2, 1930, pp. 185–190.

General References

Prager, W., *Introduction to Plasticity*, Addison-Wesley, Reading, Mass., 1959.
Ford, H., *Advanced Mechanics of Materials*, Wiley, New York, 1963.

CHAPTER 13

LIMIT ANALYSIS

13-1 DESIGN OF STRUCTURES

The theory of limit analysis is used primarily in the design of steel structures composed of various elements such as beams, frames, girders, arches, etc. For many years the basis for structural design has been the *allowable stress concept*. The allowable stress was usually taken to be the yield stress of the material, and the design stress was then taken to be some fraction of the allowable stress, depending on the factor of safety used. In some applications the allowable stress was governed by the possibility of buckling or fatigue. The design methods used were always elastic.

It is apparent, however, that the important consideration in an engineering structure is not whether the yield stress is exceeded at some point, but whether the structure will carry the intended loads or perform its intended function, and there is really no reason for assuming that the stress in the structure should never exceed the elastic limit. As a matter of fact, it is fairly evident that in almost all structures, local plastic flow will occur at stress raisers and at points of discontinuity in the geometry, and, furthermore, residual stresses as high as half the yield strength may already be in some elements as they come from the steel mills, before the load is even applied.

The practice is therefore becoming more widespread to design structures into the plastic range, the materials being assumed to behave in an elastic-perfectly plastic manner. In these design procedures no attempt is made to determine the stresses and strains in the structure, but rather what is sought is the load-carrying capacity or limiting load at which the structure will collapse. This type of analysis is called *limit design* or *plastic design*, and the load at collapse is called the *plastic collapse load*.

For the great majority of problems this type of approach makes more sense than a design based on an elastic analysis. Furthermore, since the stress distribution is not sought, it is much simpler. There are other important problems, however, which cannot be resolved on the basis of simple limit analysis. Among these are buckling, fatigue, and fracture.

The first use of plastic design in structures was apparently made by Kazinczy in the design of apartment buildings in 1914 (see reference [1]). Since then many contributions have been made to this theory both in this country and abroad. Plastic design is already part of certain specifications in some countries, and it is being used more and more by engineers in this country.

13-2 SIMPLE TRUSS

To illustrate the concepts behind limit analysis a few simple examples taken from reference [2] will be considered. Let us start with the problem of the simple symmetric truss shown in Figure 13.2.1 and analyze this truss

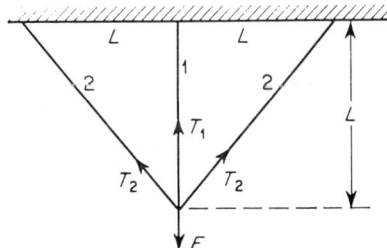

FIGURE 13.2.1 Simple truss.

by the conventional elastic method and by the method of limit analysis. Assume the truss to be made of mild steel with an elastic modulus $E = 30 \times 10^6$ and a yield stress $\sigma_0 = 30,000$. The stress–strain curve is assumed elastic-perfectly plastic. Let $L = 12$ in. and cross-sectional area of all three members, $A = 1$ in.^{2}. The conventional analysis proceeds as follows. Let T_1 be the resultant force in bar 1 and T_2 the resultant forces in each of the bars 2. Then equilibrium of forces requires

$$T_1 + \sqrt{2}\, T_2 = F \qquad (13.2.1)$$

Since this is a redundant structure (having more members than necessary to maintain equilibrium), the equilibrium equation (13.2.1) is obviously not sufficient to obtain the solution and the continuity of displacement must be used. The elongation of rod 1 is

$$\delta_1 = \frac{T_1 L}{AE} \quad (13.2.2)$$

and the elongation of each of rods 2 is (see Figure 13.2.2)

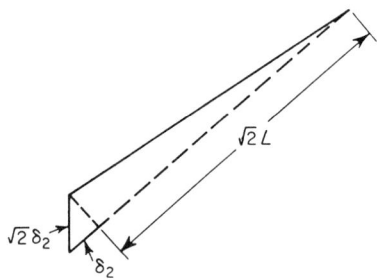

FIGURE 13.2.2 Elongation.

$$\delta_2 = \frac{T_2}{AE} \sqrt{2}\, L \quad (13.2.3)$$

For continuity of displacements,

$$\delta_1 = \sqrt{2}\, \delta_2 \quad (13.2.4)$$

which results in

$$T_1 = 2 T_2 \quad (13.2.5)$$

Substituting into (13.2.1) gives

$$T_2 = \frac{F}{2 + \sqrt{2}}$$

$$T_1 = \frac{2F}{2 + \sqrt{2}} \quad (13.2.6)$$

Thus if $F = 20{,}000$ lb,

$$T_1 = 11{,}720 \text{ lb}$$

$$T_2 = 5{,}860 \text{ lb} \quad (13.2.7)$$

Sec. 13-2] Simple Truss

and the elongations are

$$\delta_2 = \frac{5,860 \times 12 \times \sqrt{2}}{30 \times 10^6} = 0.00331 \text{ in.}$$

$$\delta_1 = \sqrt{2}\,(0.00331) = 0.00468 \text{ in.} \tag{13.2.8}$$

This completes the elastic analysis, the stresses and deformation having been determined. If the question is now asked, How strong is the truss?, the usual answer is to quote a factor of safety giving the ratio of the maximum safe load to the applied load. The conventional practice has been to take the maximum safe load as that load which would just cause the maximum stress to reach the yield stress. In this case, therefore, the factor of safety would be

$$f = \frac{\sigma_0}{T_1/A} = \frac{30,000}{11,720} = 2.56 \tag{13.2.9}$$

and the maximum safe load presumably is

$$F_{\max} = 2.56 \times 20,000 = 51,200 \text{ lb} \tag{13.2.10}$$

However, if the material is ductile, as in the case of mild or structural steel, this safety factor and the corresponding maximum safe load are not too meaningful. The actual load that can be carried by the truss is considerably greater. Assuming the material to be perfectly plastic, if the load is raised to 51,200 lb, the middle bar will just be at the yield. However, bars 2 will only be at $\frac{1}{2}$ their yield, since the stress in these bars is one half the stress in the middle bar, as seen from equation (13.2.5). As the load is increased above 51,200 lb, the stress in the middle bar will remain the same, since it is already at the yield, and the additional load will be carried by the other two bars. Furthermore, since the latter two bars are still elastic, the deformations of the system will still be small. The load can be increased until the outside bars just reach the yield point. From (13.2.1) the load at which this will occur is

$$F = \sigma_0 + \sqrt{2}\,\sigma_0 = (1 + \sqrt{2})\sigma_0 = 72,400 \text{ lb} \tag{13.2.11}$$

This is the plastic collapse load. The deflections just as this load is reached will be

$$\delta_2 = 0.017 \text{ in.}$$

$$\delta_1 = \sqrt{2}\,\delta_2 = 0.024 \text{ in.} \tag{13.2.12}$$

which are still small. If the load is increased beyond 72,400 lb, large deformations will take place and it can be assumed that the structure is no longer usable.

We note that the factor of safety based on the load-carrying capacity is, for the case $F = 20,000$,

$$f_p = \frac{72,400}{20,000} = 3.62$$

rather than 2.56. Moreover, the load-carrying capacity or the plastic collapse load is determined simply from the equilibrium equation (13.2.1) and a knowledge of the yield stress. The detailed elastic analysis previously made was not necessary. This is another great advantage of limit analysis.

Now consider the inverse problem, i.e., the problem of design rather than analysis. Supposing we were given a truss such as shown in Figure 13.2.1, and told to design such a truss to have minimum cross-sectional areas for a working load of 20,000 lb, and a factor of safety of 3, taking all three bars to have equal areas.

According to elastic design, the forces in the bars at a load of $3 \times 20,000 = 60,000$ lb would be, from (13.2.6),

$$T_2 = \frac{60,000}{3.414} = 17,600 \text{ lb}$$

$$T_1 = 35,200 \text{ lb}$$

To keep the stress in the middle bar below the yield stress of 30,000 psi, the area would have to be

$$A = \frac{35,200}{30,000} = 1.17 \text{ in.}^2$$

If plastic analysis were used, however, the stresses in all three bars could be allowed to reach 30,000 psi at a load of 60,000 lb. Therefore, from (13.2.1),

$$A(1 + \sqrt{2}) \times 30,000 = 60,000$$

or

$$A = \frac{2}{2.414} = 0.83 \text{ in.}^2$$

Sec. 13–3] Pure Bending of Beams

Thus, using plastic design, 30 per cent of the material is saved. Furthermore, under the working load of 20,000 lb, the actual stresses in the bars are

$$\sigma_1 = \frac{T_1}{A} = \frac{11{,}720}{0.83} = 14{,}100$$

$$\sigma_2 = \frac{T_2}{A} = 7{,}050$$

which are well within the elastic range.

13–3 PURE BENDING OF BEAMS

The next problem considered is the pure bending of a beam. Consider the beam shown in Figure 13.3.1. The conventions for positive moments and

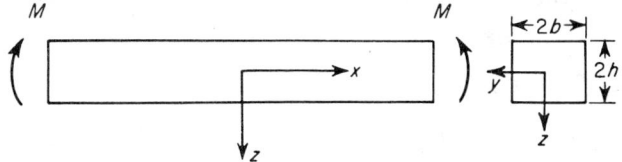

FIGURE 13.3.1   Beam under pure bending.

deflections as shown in this figure will be used throughout. The bending moment is given by

$$M = 2b \int_{-h}^{h} z\sigma_x \, dz \qquad (13.3.1)$$

With the usual assumption of beam theory that plane sections remain plane the strain ε_x is given by

$$\varepsilon_x = kz \qquad (13.3.2)$$

where k is the curvature of the middle surface.

If M is sufficiently small, the stresses will be elastic and

$$\sigma_x = E\varepsilon_x = kEz \qquad (13.3.3)$$

Substituting into (13.3.1) and integrating gives

$$M = 2bkE\tfrac{2}{3}h^3 = \tfrac{4}{3}kbEh^3 \qquad (13.3.4)$$

The maximum stress will occur at the outer fibers and will be tensile at $z = h$ and compressive at $z = -h$. When this maximum stress equals the yield stress σ_0, we have, from (13.3.3),

$$\sigma_0 = kEh \qquad (13.3.5)$$

and substituting into (13.3.4) gives for the moment when the outer fiber just reaches the yield stress

$$M_e = \tfrac{4}{3}bh^2\sigma_0 \qquad (13.3.6)$$

On the basis of elastic analysis this is the maximum allowable moment. However, on the basis of plastic analysis, it is evident that this is not the maximum moment the beam can withstand, since if the moment is increased beyond M_e, the additional load will be transferred to the inner fibers of the beam. The stress distributions for various states of loading are shown in Figure 13.3.2. The first figure (a) shows the distribution when M is less than

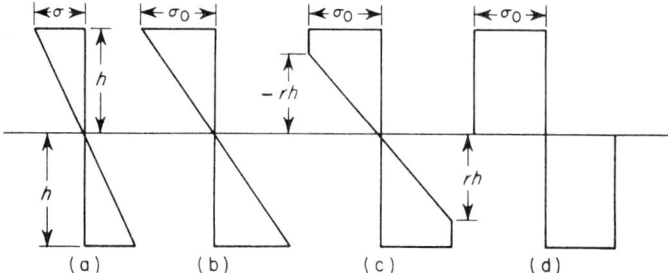

FIGURE 13.3.2 Stress distributions in rectangular beam.

M_e, (b) shows the distribution when M just equals M_e, and as M is increased more of the beam reaches the yield stress but the part between rh and $-rh$ is still elastic, as shown in (c). The moment for the condition shown in (c) is, from (13.3.1),

$$M = 2b \int_{-rh}^{rh} kEz^2 \, dz + 4b \int_{rh}^{h} \sigma_0 z \, dz$$

but, from (13.3.5),

$$kE = \frac{\sigma_0}{rh}$$

Hence

$$M = 4b \left(\int_0^{rh} \frac{\sigma_0 z^2}{rh} \, dz + \int_{rh}^{h} \sigma_0 z \, dz \right) = \tfrac{2}{3}bh^2\sigma_0(3 - r^2) \qquad (13.3.7)$$

Sec. 13-4] Beams and Frames with Concentrated Loads

As M increases, eventually r approaches zero, as shown in (d). At this time the complete beam is flowing plastically and the moment can no longer be increased. The value of the moment at this condition is called the *yield moment, limiting moment,* or *fully plastic moment.* Denoting this moment by M_0, from (13.3.7)

$$M_0 = 2bh^2\sigma_0 \tag{13.3.8}$$

Comparing equations (13.3.8) and (13.3.6), it is seen that the ratio of the fully plastic moment to the maximum elastic moment is

$$\frac{M_0}{M_e} = 1.5 \tag{13.3.9}$$

This ratio is called the *shape factor* of the beam. It will, of course, be different for different-shaped cross sections. For a circular cross section, for

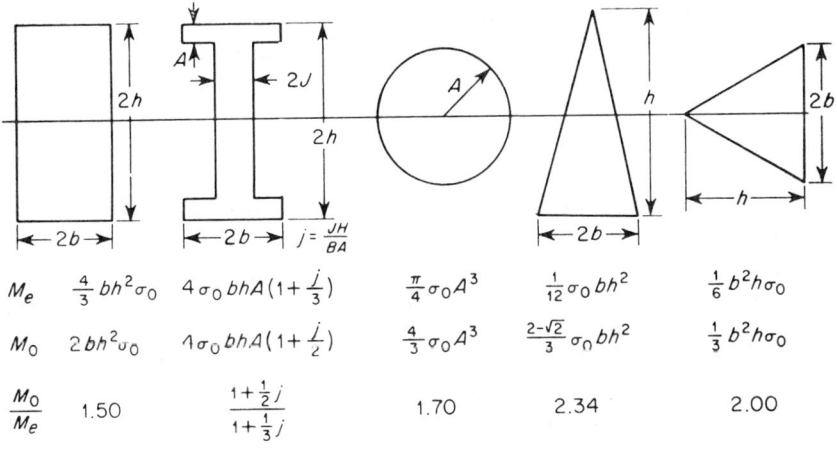

FIGURE 13.3.3 Shape factors for common beam sections.

example, it will be 1.70. Figure 13.3.3 shows the values for several common beam sections.

13-4 BEAMS AND FRAMES WITH CONCENTRATED LOADS

In Section 13.3 the case of pure bending where all sections of the beam behaved the same way was discussed. We now consider concentrated loads where the moment distribution varies along the beam. Our purpose, as before, will be to determine the collapse loads of the structure.

Cantilever Beam with Tip Load

Consider a cantilever beam with a concentrated load at the tip as shown in Figure 13.4.1. The bending moment will obviously vary linearly from zero

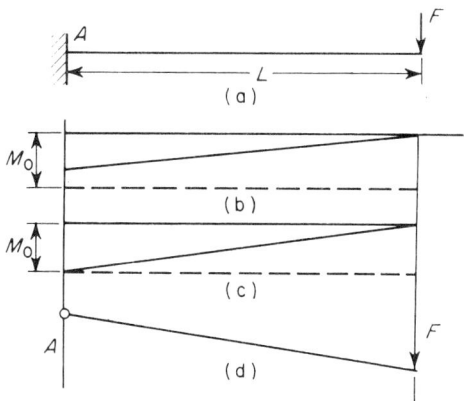

FIGURE 13.4.1 Collapse of cantilever beam.

at the tip to a maximum at the built-in end as shown. If M_0 is the yield moment, then for some load F_0 for which

$$M = LF_0 = M_0 \tag{13.4.1}$$

the section at A will become fully plastic. Since perfect plasticity has been assumed, the beam will now be able to rotate freely about the point A and will thus have collapsed. We say that a *yield hinge* or *plastic hinge* has formed at A. Such a hinge can be thought of as one which is locked in place as long as the moment M is less than M_0, and which becomes free to turn freely when M becomes equal to M_0. The collapse load, from (13.4.1), is obviously

$$F_0 = \frac{M_0}{L} = \frac{2bh^2\sigma_0}{L} \tag{13.4.2}$$

for a rectangular cross-sectional beam.

Indeterminate Beam

The indeterminate beam shown in Figure 13.4.2 is considered next. The end A is built in and the end C is simply supported so it can carry no moment,

Sec. 13-4] Beams and Frames with Concentrated Loads 309

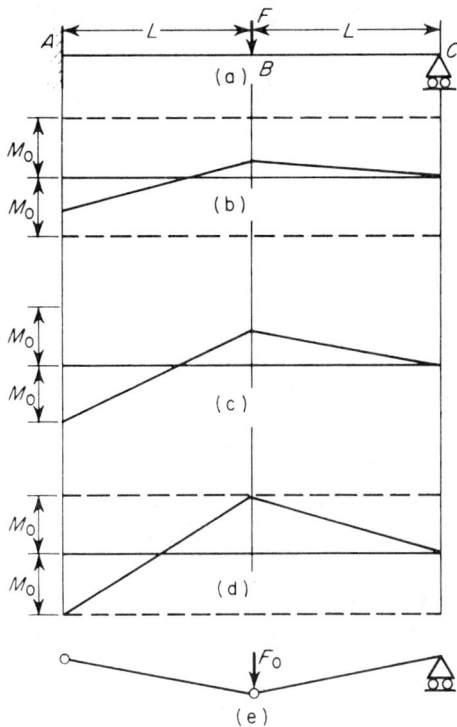

FIGURE 13.4.2 Collapse of simple indeterminate beam.

the load F being applied at the middle. The elastic solution first requires determining the reaction at C, which can be done by first calculating the displacement at C with the support removed and calculating the reaction at C necessary to bring the deflection back to zero with the force F removed. The moment distribution can then be computed. To determine the collapse load, however, none of this is necessary. The moment distribution when everything is elastic is shown in (b), and as the force F is increased, eventually the section at A will become fully plastic, the moment at A being $-M_0$. A yield hinge is thus formed at A, just as in the case of the cantilever previously discussed with the resultant moment distribution as shown in (c). Because of the support at C, however, the beam will not yet collapse. As the load is further increased, the point A cannot carry any more moment, but the rest of the beam can and the moment at B will continue increasing until it reaches the plastic moment M_0, at which time a second yield hinge is formed. The beam will then collapse, as shown in (e).

The calculation of the collapse load for this case is very simple. Let the reaction at C be denoted by R. Then since the moment at A is $-M_0$ and at B it is M_0,

$$M_B = M_0 = RL$$
$$M_A = -M_0 = 2LR - F_0 L$$

Hence

$$F_0 = \frac{3M_0}{L} \qquad R = \frac{M_0}{L} \tag{13.4.3}$$

For a beam of rectangular cross section it follows from (13.3.8) that

$$F_0 = \frac{6bh^2 \sigma_0}{L}$$

At this point it might be well to introduce a term frequently used in limit analysis, the term *mechanism*, or *kinematic mechanism*. By a mechanism is meant an articulated system which can deform without a finite increase in load. Thus in the previous example when the second yield hinge was formed at B, the system became a mechanism as shown in (e). There exist essentially two hinged bars which can deform freely for an infinitesimally applied load. We might say then that one of the objectives of limit analysis is determining when a mechanism will be formed, and in the case when more than one mechanism is possible for a system, to determine which is the collapse mechanism. This last point will be illustrated in the next example.

Simple Frame

The example of a simple frame as shown in Figure 13.4.3 will now be considered. Point 1 is built in and point 5 is pin supported and can therefore

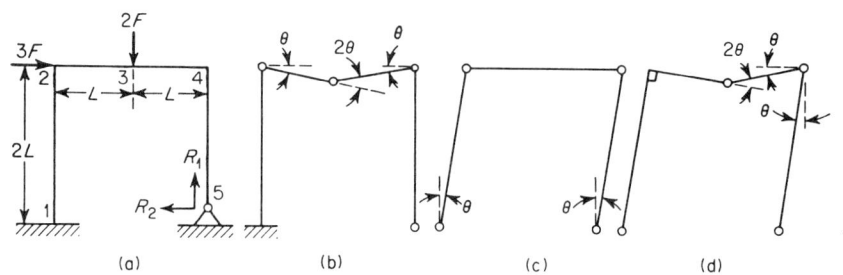

FIGURE 13.4.3 Collapse of simple frame.

Sec. 13-4] Beams and Frames with Concentrated Loads

carry no moments. There are two concentrated forces, a vertical force of amount $2F$ and a horizontal force $3F$ as shown. The critical points where the moments will attain their largest absolute values are obviously the points marked 1, 2, 3, and 4. It is seen that unlike the beam of the previous example, where both critical points had to develop yield hinges, it is not necessary in this case for all four critical points to develop yield hinges. As a matter of fact, there are three distinct possibilities, as shown in (b), (c), and (d) of Figure 13.4.3. Each of these three configurations is a possible mechanism. The question then arises as to which is the correct one. This can be determined as follows.

Let us consider each of these modes of collapse in turn. Let the reactions at point 5 be denoted by R_1 and R_2, and consider the moments at the four critical points.

$$\begin{aligned} M_1 &= 2R_1L - 8FL \\ M_2 &= 2R_1L - 2FL - 2R_2L \\ M_3 &= R_1L - 2R_2L \\ M_4 &= -2R_2L \end{aligned} \tag{13.4.4}$$

Eliminating the two reactions R_1 and R_2 gives

$$\begin{aligned} -M_2 + 2M_3 - M_4 &= 2FL \\ -M_1 + M_2 - M_4 &= 6FL \end{aligned} \tag{13.4.5}$$

If it is now assumed that the frame will collapse by the mechanism shown in (c), then $M_1 = -M_0$, $M_2 = M_0$, and $M_4 = -M_0$. From the second of equations (13.4.5),

$$F = \frac{M_0}{2L}$$

and from the first of (13.4.5),

$$M_3 = \tfrac{1}{2}M_0 \tag{13.4.6}$$

Now consider the mechanism shown in (b). Here

$$M_2 = -M_0 \qquad M_3 = M_0 \qquad M_4 = -M_0$$

Then, from (13.4.5),

$$F = \frac{2M_0}{L} \qquad M_1 = -12M_0 \tag{13.4.7}$$

which is impossible. The second mechanism shown in (b) is therefore not possible.

Finally, consider the third mechanism shown in (d):

$$M_1 = -M_0 \qquad M_3 = M_0 \qquad M_4 = -M_0$$

Then, from (13.4.5),

$$F = \frac{5M_0}{8L} \qquad M_2 = \tfrac{7}{4}M_0 \qquad (13.4.8)$$

But this violates the original assumption for this case—that there is no yield hinge at station 2; i.e., the moment at 2 is less than M_0. This mechanism is therefore also not admissible, leaving the mechanism of (c) as the only possible one. The true collapse load is therefore given by equation (13.4.6).

Several results can now be noted. The collapse loads corresponding to the cases of (d) and (b) are higher than the true collapse load given in (13.4.6). It appears, then, that as the load is slowly increased from zero, the frame will collapse the first chance it gets.

Second, if the load is decreased in these last two cases, so that the maximum moment does not exceed M_0, it would be necessary to divide by 12 and $\tfrac{7}{4}$, respectively, as seen from (13.4.7) and (13.4.8), giving

$$\frac{FL}{M_0} = \frac{1}{6} \quad \text{and} \quad \frac{FL}{M_0} = \frac{5}{14}$$

But these are less than the true collapse load of

$$\frac{FL}{M_0} = \frac{1}{2}$$

It follows then that as the load is slowly increased, the frame will not collapse for any load for which some equilibrium configuration can be found, unless it produces a mechanism.

These two concepts can be formalized into basic theorems of limit analysis to be discussed in Section 13.5.

13-5 THEOREMS OF LIMIT ANALYSIS

The previous problems were simple enough so that the correct collapse load could be obtained in each case without too much difficulty. In more complicated problems, however, it may be difficult or impossible to obtain

Sec. 13-5] Theorems of Limit Analysis

the collapse load exactly, and recourse must then be had to the *upper* and *lower bound theorems*, which provide upper and lower bounds to the true collapse load. These theorems are essentially the same as those presented in Section 12.7 for the load to produce incipient plastic flow in a rigid–perfectly plastic material. They were first presented by Gvozdev [3] and independently proved by Hill [4, 5] for the rigid–perfectly plastic material, by Drucker et al. [6, 7] for the elastic–perfectly plastic material, and in references [8] and [9] for the special cases of beams and frames. We shall present the theorems in their general form. The proofs, which are based primarily on the principle of virtual work, can be found in the cited references and in a particularly elegant form in reference [10].

Lower and Upper Bound Theorems

If an equilibrium distribution of stress can be found which balances the applied load and is everywhere below yield or at yield, the structure will not collapse or will just be at the point of collapse. This gives a lower bound on the limit load, and is called the *lower bound theorem*.

The structure must collapse if there is any compatible pattern of plastic deformation for which the rate at which the external forces do work is equal to or exceeds the rate of internal dissipation. This gives the upper bound on the limit load and is called the *upper bound theorem*.

The lower bound theorem merely says that the structure will withstand the applied load by rearranging the internal stresses to best advantage, if at all possible. It gives lower bounds on, or safe values of, the limit or collapse loading. The maximum lower bound is the limit load.

The upper bound theorem says that if a path of failure exists, the structure will take the path. It gives upper bounds on the limit load. The minimum upper bound is the limit load. Since exact limit loads cannot usually be obtained for practical problems, these two theorems enable one to bracket the answer sufficiently close for engineering purposes.

In addition to the lower and upper bound theorems, the following theorems are sometimes useful [11].

Addition of weightless material cannot result in a lower collapse load.

Increasing the yield strength of the material in any region cannot weaken the body.

Residual or thermal stresses or deflections have no influence on the limit load.

The first two of these are stated in the negative because the converses are not necessarily true; i.e., addition of material or increase in yield strength may, or may not, increase the strength of a body, but it will certainly not decrease it.

The upper and lower bound theorems can be stated in a more precise and elegant fashion by first introducing the following definitions and concepts [2], which are useful in their own right.

Generalized stresses Q_i are used to designate the state of stress in a body. These may be actual stresses, so that $Q_1 = \sigma_x$, $Q_2 = \sigma_y$, etc.; or they may be moments, as in the case of beams, so that $Q_1 = M$; or resultant forces and moments as in the case of shells, where $Q_1 = N_\theta$, $Q_2 = N_\phi$, $Q_3 = M_\theta$, etc. The choice of generalized stresses to be used depends on the particular problem and is a matter of convenience and is not necessarily unique. Once such a choice has been made, however, the corresponding *generalized strains* q_i are defined so that the increment of internal work done is given by

$$dW = Q_i \, dq_i = Q_1 \, dq_1 + Q_2 \, dq_2 + \cdots + Q_n \, dq_n \quad (13.5.1)$$

For example, for the beam problems considered, the generalized stress at a hinge is M, and the corresponding generalized strain is the rotation θ, the internal work at the hinge being equal to $M\theta$.

Instead of using the increment of internal work as given by equation (13.5.1), it is more convenient to introduce the *dissipation function* or *specific power of dissipation*, given by

$$D = Q_i \dot{q}_i = Q_1 \dot{q}_1 + Q_2 \dot{q}_2 + \cdots + Q_n \dot{q}_n \quad (13.5.2)$$

where the dots designate time derivatives. The dissipation function is uniquely determined by the strain rates (even for singular yield loci [10]), and we can consequently write

$$D = D(\dot{q}_1, \dot{q}_2, \ldots, \dot{q}_n) \quad (13.5.3)$$

Consider an arbitrary structure of volume V and surface S subject to certain geometric constraints and upon which a load distribution P_i acts. Let λ be a multiplier and consider the structure under the loads λP_i as λ is slowly increased from zero. The *safety factor f* is defined as the smallest value of λ for which the structure can undergo an increase in deformation without increase in load; i.e., the safety factor is the ratio of the collapse load to the actual load.

We now define a *statically admissible* stress field as one which is in internal equilibrium, is in equilibrium with the external loads λP_i, and nowhere exceeds the yield limit. The multiplier λ^- corresponding to such a statically admissible stress field is called a *statically admissible multiplier*. The lower

Sec. 13-5] Theorems of Limit Analysis

bound theorem can now be stated as follows: *The safety factor is the largest statically admissible multiplier*; i.e., $f \geq \lambda^-$.

A velocity field is called *kinematically admissible* if it satisfies the velocity (or displacement) constraints and if the total external rate of work D_e done by the actual loads on this velocity field is positive.

Let the generalized strain-rate vector associated with a given kinematically admissible velocity field be designated by $\dot{q}_i^*$, where the asterisk is used to indicate that this is not necessarily the true strain-rate vector but one that is kinematically admissible. The internal dissipation function corresponding to $\dot{q}_i^*$ can be determined from equation (13.5.3) and integrated over the complete structure to obtain the total internal dissipation D_i. We now define *a kinematically admissible multiplier* λ^+ as the ratio of the internal to the external energy dissipations:

$$\lambda^+ = \frac{D_i}{D_e} \qquad (13.5.4)$$

The upper bound theorem is then stated as follows: *The safety factor is the smallest kinematically admissible multiplier*; i.e., $f \leq \lambda^+$. Both theorems can be summarized by the relation

$$\lambda^- \leq f \leq \lambda^+ \qquad (13.5.5)$$

The above theorems furnish upper and lower bounds for the load. The question of the uniqueness of the solution must still be considered. The uniqueness of the safety factor follows from equation (13.5.5). It can also be shown [4, 12, 13], using the principle of virtual work, that the stress field at the start of plastic flow is unique except in the rigid regions and at stress points which fall on the straight portions of yield loci (such as the Tresca locus).

As a simple example of the use of these theorems, consider the indeterminate beam shown in Figure 13.5.1. The end A is built in and the end B is

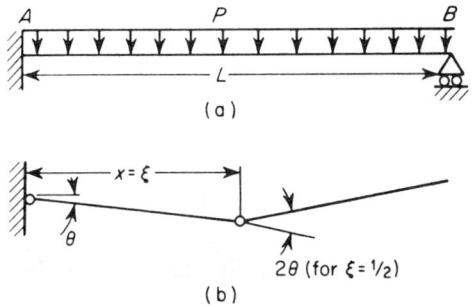

FIGURE 13.5.1 Collapse of beam with distributed load.

simply supported and the distributed load is of intensity P. The beam will collapse when yield hinges are formed at A and at some as yet unknown point designated by $x = \xi$. The correct solution can readily be obtained. Let R be the reaction at the point B. Then the moment at any x is given by

$$M(x) = R(L - x) - \tfrac{1}{2}P(L - x)^2 \qquad (13.5.6)$$

Since yield hinges form at $x = 0$ and $x = \xi$,

$$\begin{aligned} M(0) &= RL - \tfrac{1}{2}PL^2 = -M_0 \\ M(\xi) &= R(L - \xi) - \tfrac{1}{2}P(L - \xi)^2 = M_0 \end{aligned} \qquad (13.5.7)$$

Also, since at $x = \xi$, M must be a maximum,

$$M'(\xi) = -R + P(L - \xi) = 0 \qquad (13.5.8)$$

Solving equations (13.5.7) and (13.5.8) for R, ξ, and P,

$$r = \frac{RL}{M_0} = 2 + 2\sqrt{2} = 4.82843$$

$$p = \frac{PL^2}{M_0} = 6 + 4\sqrt{2} = 11.65685 \qquad (13.5.9)$$

$$\bar{\xi} = \frac{\xi}{L} = 2 - \sqrt{2} = 0.58579$$

Equations (15.5.9) give the exact answer to the problem. However, for more complicated problems the exact solution may become very difficult. An approximate solution can be obtained by finding upper and lower bounds.

We note first that if some arbitrary value ξ were chosen for the hinge, an upper bound would result, for the correct value of ξ will cause the yield limit to be reached at a lower load. Let us then arbitrarily choose the midpoint $\xi = \tfrac{1}{2}$ as the hinge point. Then if the hinge A rotates through an angle θ, ξ will rotate through an angle 2θ and the total internal work will be

$$W_i = 3M_0\theta \qquad (13.5.10)$$

The external load P moves an average distance of $\tfrac{1}{4}L\theta$ and the external work is

$$W_e = PL(\tfrac{1}{4}L\theta) = \tfrac{1}{4}PL^2\theta$$

Sec. 13-5] Theorems of Limit Analysis

Equating the rates of work done, by the upper bound theorem

$$\frac{PL^2}{4} = 3M_0$$

or
$$p^+ = \frac{PL^2}{M_0} = 12 \qquad (13.5.11)$$

Comparing with equation (11.5.9), it is seen that an upper bound has been found which is reasonably close.

A lower bound to P is obtained as follows. Calculating the moment distribution due to the load P [assuming $M(0) = -M_0$] results in

$$m = \frac{M}{M_0} = 5\left(1 - \frac{x}{l}\right) - 6\left(1 - \frac{x}{l}\right)^2 \qquad (13.5.12)$$

Now this moment distribution is not an admissible one (not statically admissible), since it gives values greater than M_0. Thus the maximum value of m occurs at $x/L = \frac{7}{12}$ and is equal to 25/24. If we therefore multiply all the moments and load by 24/25, an admissible moment distribution, which by Theorem 2 is a lower bound, will be obtained. Hence

$$p = 24/25 \times 12 = 11.52 \qquad (13.5.13)$$

is a lower bound for the collapse load.

The collapse load is thus bounded by

$$11.52 \le p \le 12$$

or we can write
$$p = 11.76 \pm 0.24 \qquad (13.5.14)$$

which is sufficiently accurate for engineering purposes.

For greater accuracy one could now take the value $x/L = \frac{7}{12}$, where the previous moment distribution was a maximum, and using this location for the hinge position obtain a new and better upper bound and then obtain as before a new lower bound. The collapse load is then found to be

$$11.65674 \le p \le 11.65714 \qquad (13.5.15)$$

For the above simple example there is little difference in the amount of labor between the exact solution and the method of upper and lower bounds. For more complex problems, however, the method of upper and lower bounds is far superior.

13-6 METHOD OF SUPERPOSITION OF MECHANISMS

In the simple examples previously discussed, it was possible to list all collapse mechanisms and to determine which of these was the correct one. For complicated structures this is no longer possible and more systematic methods are needed to determine the limit loads. One such general method will be briefly discussed in this section. The interested reader is referred to references [2, 14, 15, 16] for more comprehensive treatments.

The *method of superposition of mechanisms* due to Symonds and Neal [17, 18] basically involves combining or superposing various *elementary mechanisms* to obtain the collapse mechanism, the true collapse mechanism being distinguished by the lowest collapse load. The method will be described by means of an illustrative problem taken from reference [17].

Consider the frame shown in Figure 13.6.1, loaded as indicated. Let the fully plastic moment in the legs be M_0 and in the horizontal beams $2M_0$. There are 10 possible sections where plastic hinges can occur, as indicated by the numbers in the figure. The hinge in the beam with the distributed load may occur at some unknown distance x as shown.

The first step is to determine the number of *linearly independent mechanisms* for the structure. It can be shown that this will equal the linearly independent static equations of equilibrium, which in turn is equal to the number of bending moments necessary to specify the bending moment diagram completely minus the degree of redundancy of the structure; i.e.,

$$n = m - r \tag{13.6.1}$$

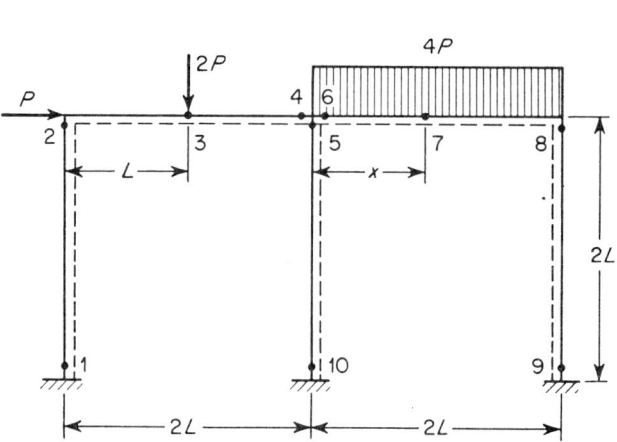

FIGURE 13.6.1 Two-bay frame.

Sec. 13-6] Method of Superposition of Mechanisms

where n is the number of independent equations of equilibrium or number of independent mechanisms, m the number of bending moments necessary to specify the bending moment diagram, and r the number of redundant constraints. For example, the frame of Figure 13.4.3 has four bending moments at the points 1 through 4, it has two redundant constraints, and it therefore has two independent equations of equilibrium, as given by equations (13.4.5). It also has two independent mechanisms, as shown in Figure 13.4.3(b) and (c). The mechanism of Figure 13.4.3(d) can be obtained by combining the other two mechanisms.

For the problem under consideration, Figure 13.6.1, m is equal to 10, as indicated, and the degree of redundancy r is equal to 6. Hence there are four independent mechanisms. For these independent mechanisms we choose the

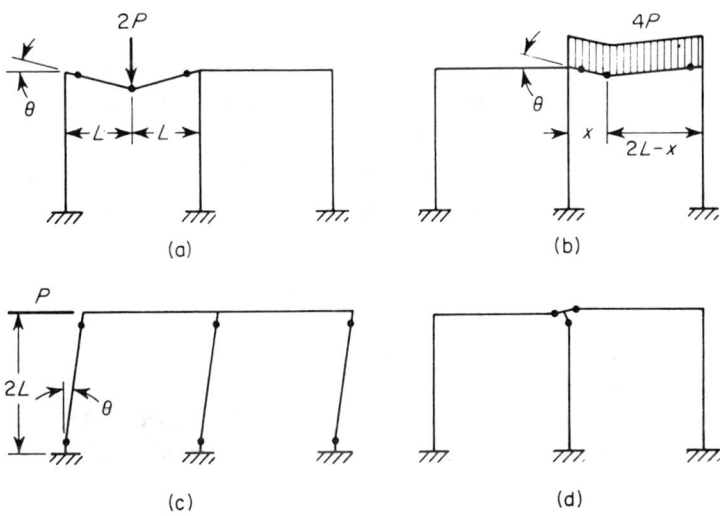

FIGURE 13.6.2 Elementary mechanisms.

basic or *elementary mechanisms* shown in Figure 13.6.2. There are three such elementary mechanisms defined as follows:

1. *Beam mechanisms* as shown in (a) and (b). These are characterized by one hinge at each end and one hinge under each concentrated load and one hinge for a distributed load.

2. *Frame* or *panel mechanisms* as shown in (c). These are characterized by hinges at the joints and supports but not under loads.

3. *Joint mechanisms* as shown in (d). These are *fictitious mechanisms* representing the small rotation of a joint where three or more members are connected.

There is also a fourth type of basic mechanism used in the analysis of gabled frames, called a *portal mechanism*, which will not be used here.

The next step is to compute the collapse load for each of the elementary mechanisms of Figure 13.6.2. For this purpose the *work equation* obtained from the principle of virtual work is used. The principle of virtual work states that if a structure is in equilibrium under a system of external forces, then the external work done by these forces during any virtual displacement is equal to the internal work done by the stresses on the strains due to this virtual displacement, the virtual displacement being any displacement compatible with the constraints. For the case of a frame structure composed of rigid members, this principle takes the form of the following *work equation*:

$$\sum_{i=1}^{k} P_i \delta_i = \sum_{i=1}^{m} M_i \theta_i \qquad (13.6.2)$$

where P_i are the k external loads, δ_i the small displacements of a mechanism motion, M_i the plastic moments, and θ_i the corresponding small rotations. The use of this equation is further simplified by the fact that the product $M_i \theta_i$ is always positive, so that we need not consider the signs of the M_i or the θ_i.

Applying equation (13.6.2) to the mechanisms of Figure 13.6.2 gives for Figure 13.6.2(a),

$$2P(L\theta) = M_0 \theta + 2M_0(2\theta) + 2M_0 \theta$$

or
$$PL = 3.5 M_0 \qquad (13.6.3)$$

In the above equation the hinge at section 2 was assumed to take place in the vertical leg, which is weaker, having a plastic moment of M_0 rather than $2M_0$ as in the beam.

For the mechanism of (b),

$$4P\left(\frac{\theta x}{2}\right) = 2M_0 \theta + 2M_0 \left(\frac{2L\theta}{2L - x}\right) + M_0 \frac{x\theta}{2L - x}$$

$$PL = \frac{(8L - x)L}{2x(2L - x)} M_0 \qquad (13.6.4)$$

For the mechanism of (c),
$$P(2L\theta) = 6M_0 \theta$$

$$PL = 3M_0 \qquad (13.6.5)$$

There is no work equation for the mechanism of Figure 13.6.2(d).

Sec. 13–6] Method of Superposition of Mechanisms

The frame can fail in one of the three mechanisms analyzed or in some combination of them. The correct failure mechanism is the one requiring the smallest load P. Hence in combining the mechanisms we choose only those combinations which are likely to give smaller loads than the mechanisms already available. Examination of the work equation (13.6.2) shows that to reduce the load the work done by the load should be made as large as possible and the internal work of the plastic moments should be made as small as possible. This leads to the following rules for combining mechanisms:

1. Two mechanisms should be combined so as to eliminate at least one common hinge. Only then can the failure load be less than for each of the individual mechanisms.

2. If necessary a joint connecting three or more members may be rotated in order to reduce the internal work at this joint. This corresponds to the elementary mechanism of Figure 13.6.2(d).

3. For a beam with a distributed load, the hinge is first arbitrarily assumed to be at the midpoint of the beam. If the actual collapse mode contains this hinge, its correct position can then be determined. In practice the variation of this hinge position is small and can be neglected.

Let us apply these rules to the present problem. The mechanism of Figure 13.6.2(c) is most likely to be correct, since it corresponds to the smallest value of P. If it is combined with mode (a), the hinge at section 2 is eliminated, so that by rule 1 the resultant value of P should be lower. This combined mode is shown in Figure 13.6.3. The work equation can be obtained for this new mechanism of Figure 13.6.3, but it is usually simpler to modify one of the

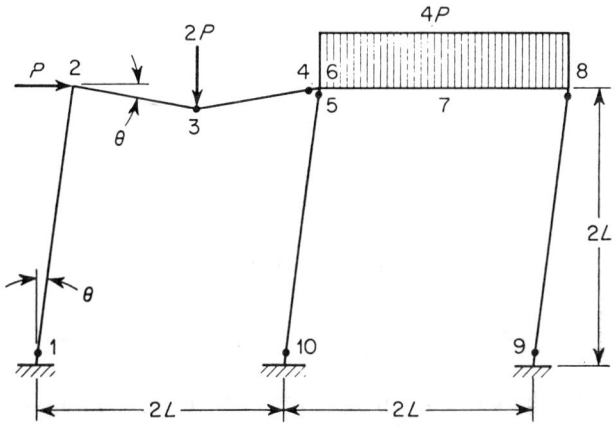

FIGURE 13.6.3 Combined mechanism.

previous work equations, here the one for the mechanism of Figure 13.6.2(c), equation (13.6.5).

$$P(2L\theta) + 2P(L\theta) = 6M_0\theta - M_0\theta + 2M_0(2\theta) + 2M_0\theta$$

or
$$PL = 2.75M_0 \tag{13.6.6}$$

This combined mechanism thus gives a lower collapse load and is therefore more likely than any of the others. To determine whether it is correct, it is necessary to check the other combinations. Neither mechanism (b) or (d) of Figure 13.6.2 combines favorably with Figure 13.6.3, since no hinge is thereby eliminated. However, both of these together will combine with Figure 13.6.3 to eliminate the hinge of section 5 without introducing one at

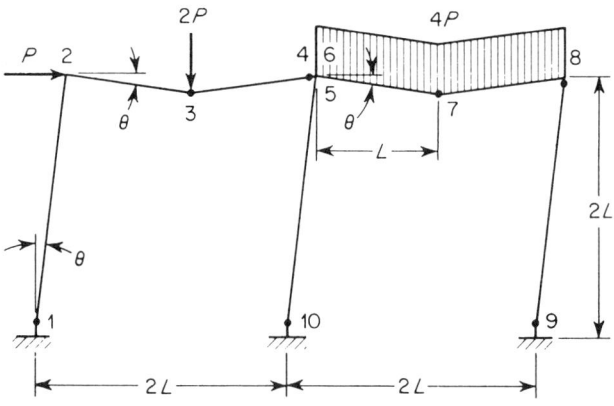

FIGURE 13.6.4   Combined mechanism.

section 6. This new combination is shown in Figure 13.6.4. Whether it is advantageous or not must now be determined.

The work equation for this mechanism is

$$4P(L\theta) + 4P\left(\frac{L\theta}{2}\right) = M_0(5\theta) + 2M_0(6\theta)$$

Hence
$$PL = 2.83M_0 \tag{13.6.7}$$

In the above equation the hinge at section 7 was assumed at the midpoint of the beam according to rule 3. The correct value of x can be determined from the condition that P should be a minimum. This gives a value of x equal to $0.982L$ instead of L. The corresponding value of PL remains the same as given by equation (13.6.7) to three significant figures.

Sec. 13-7] Limit Design

The mechanism yielding the lowest collapse load is the one of Figure 13.6.3, and this is therefore the correct collapse mode, the load being given by equation (13.6.6). However, to verify that this is the correct solution, it is necessary to calculate the remaining moments to make sure that none of them exceeds the yield moment. The four independent equations of equilibrium are

$$2PL = 2M_3 - M_2 - M_4$$
$$2PL = 2M_7 - M_6 - M_8$$
$$2PL = M_2 - M_1 + M_5 - M_{10} + M_9 - M_8$$
$$0 = M_4 + M_5 - M_6$$
(13.6.8)

where for the second equation x was taken to be equal to L. Substituting

$$M_1 = -M_0 \quad M_3 = 2M_0 \quad M_4 = -2M_0$$
$$M_5 = M_0 \quad M_8 = -M_0 \quad M_9 = M_0 \quad M_{10} = -M_0$$

into equations (13.6.8), and solving for the remaining moments gives

$$M_2 = \tfrac{1}{2}M_0 \quad M_6 = -M_0 \quad M_7 = 1.75M_0 \quad (13.6.9)$$

Thus none of the remaining moments exceeds the fully plastic moment for that member, so that the failure mode of figure 13.6.3 is the correct one.

The above example illustrates the use of the method of superposition of mechanisms. Obviously the method requires some skill in combining favorable mechanisms. Such skill can readily be acquired by experience and practice. The method then offers a relatively simple procedure for finding the collapse load of framed structures under proportional loading.

13-7 LIMIT DESIGN

Heretofore the problems discussed have been primarily those of analysis. A given structure was analyzed to determine the maximum safe load. Of equal importance is the problem of design where the loads are specified and the best structure, in some sense, to carry these loads must be determined.

It should be first noted that for some design problems the methods previously discussed can be used directly: For example, if M_0 is the plastic moment in one of the beams of the structure which is considered as a reference beam and $\alpha_i M_0$ are the plastic moments in the other elements of the structure, where α_i corresponds to the ith element and the α_i are all specified, then the design problem becomes one of finding the value of M_0 for collapse to occur under the specified loads. This is called a *restricted design problem* [14]. To find M_0, the upper and lower bound theorems can be used as before, noting that the lower bound theorem now gives an *upper bound* on the required value of M_0 and the upper bound theorem gives a *lower bound* on the value of M_0. Thus in using the method of superposition of mechanisms of Section 13.6, instead of combining mechanisms to obtain the lowest collapse load, we combine the mechanisms to obtain the *largest plastic moment*. The procedures for both types of problems are of course exactly the same.

For the illustrative problem of Section 13.6, as shown in Figure 13.6.1, the highest value of the plastic moment M_0 is obtained for the combined mechanism of Figure 13.6.3 and from equation (13.6.6) is given by

$$M_0 = 0.364PL \tag{13.7.1}$$

The choice of sections to be used in constructing the frame of Figure 13.6.1 is thus determined by equation (13.7.1).

In general, however, the designer seeks not only a safe design but one which will also minimize the cost or the weight. It is to be realized that a minimum weight design does not always result in a minimum cost design. However, because of the many factors entering into the cost of a design, the methods that have been developed for optimum design are primarily based on the minimum-weight criterion. In any case this furnishes a good starting point for the designer.

A design method for minimum weight was first discussed by Heyman [19], followed by Foulkes [20], Livesley [21], and Heyman and Prager [22]. The basic problem is one in *linear programming*, where one seeks to minimize a weight function subject to inequality constraints [23]. However, the number of variables and constraints is so large, except for relatively simple frames, that a straightforward use of linear programming techniques would quickly exhaust the capacity of even a large computer. The method of Heyman and Prager [22], however, does not require as much computer storage and is completely automatic. Furthermore, it can be used for analysis as well as design. A complete description of the method is given in reference [22] and in reference [14].

Problems

1. Consider a truss similar to the one shown in Figure 13.2.1, where the angle between the vertical bar and each of the other two bars is 60°. Find the elastic and plastic safety factors if $F = 20,000$ lb.

2. Calculate the shape factor for a beam of circular cross section.
3. Determine the collapse load for the indeterminate beam of Figure 13.4.2 if the concentrated load F is located at a distance ξL from the built-in end.
4. Obtain the collapse load for a beam built in at both ends with a concentrated load at the center.
5. Find the collapse load for the frame of Figure 13.4.3 if the vertical and horizontal loads are equal.
6. Repeat Problem 5 with the vertical load equal to twice the horizontal load.
7. Determine the collapse load for a uniformly loaded beam built in at both ends.
8. For the problem of Figure 13.6.1, show that if the distributed load is $5P$ instead of $4P$, the other loads remaining unchanged, the frame will fail by the mechanism of Figure 13.6.4 at a load of $2.61 M_0/L$.
9. Repeat Problem 8 assuming the distributed load is $8P$. Show that the failure mechanism is the one of Figure 13.6.2(b) at a load of $1.75 M_0/L$.
10. Derive the equilibrium equations (13.6.8).

References

1. G. Kazinczy, Experiments with Clamped Girders, *Betonszemle*, **2**, Nos. 4, 5, and 6, 1914.
2. P. G. Hodge, Jr., *Plastic Analysis of Structures*, McGraw-Hill, New York, 1959.
3. A. A. Gvozdev. The Determination of the Value of the Collapse Load for Statically Indeterminate Systems Undergoing Plastic Deformation, *Proceedings of the Conference on Plastic Deformations*, Akademiia Nauk SSSR, Moscow, 1938, pp. 19–33. Translated into English by R. M. Haythornthwaite, *Intern. J. Mech. Sci.*, **1**, 1960, pp. 322–355.
4. R. Hill, On the State of Stress in a Plastic-Rigid Body at the Yield Point, *Phil. Mag.*, **42**, 1951, pp. 868–875.
5. R. Hill, A Note on Estimating the Yield-Point Loads in a Plastic-Rigid Body, *Phil. Mag.*, **43**, 1952, pp. 353–355.
6. D. C. Drucker, H. J. Greenberg, and W. Prager, The Safety Factor for an Elastic–Plastic Body in Plane Strain, *J. Appl. Mech.*, **18**, 1951, pp. 371–378.
7. D. C. Drucker, W. Prager, and H. J. Greenberg, Extended Limit Design Theorems for Continuous Media, *Quart. Appl. Math.*, **9**, 1952, pp. 381–389.
8. H. J. Greenberg and W. Prager, Limit Design of Beams and Frames, *Trans. ASCE*, **117**, 1952, p. 447. First published as *Tech. Rept. A18-1*, Brown Univ. Press, Providence, R.I., 1949.
9. M. R. Horne, Fundamental Propositions in the Plastic Theory of Structures, *J. Inst. Civil Engrs. (London)*, **34**, 1949–1950, pp. 174–177.
10. W. Prager, *An Introduction to Plasticity*, Addison–Wesley, Reading, Mass., 1959.
11. D. C. Drucker, Limit Analysis and Design, *Appl. Mech. Rev.*, **7**, No. 10, 1954, pp. 421–423.
12. R. Hill, On the Problem of Uniqueness in the Theory of a Rigid-Plastic Solid, *J. Mech. Phys. Solids*, **4**, 1956, pp. 247–255; **5**, 1957, pp. 1–8, 153–161, 302–307.

13. R. M. Haythornthwaite and R. T. Shield, A Note on the Deformable Region in a Rigid-Plastic Structure, *J. Mech. Phys. Solids*, **6**, 1958, pp. 127–131.
14. C. E. Massonnet and M. A. Save, *Plastic Analysis and Design*, Vol. I, Ginn (Blaisdell), Boston, 1965.
15. B. G. Neal, *The Plastic Methods of Structural Analysis*, Wiley, New York, 1956.
16. J. F. Baker, M. R. Horne, and J. Heyman, *The Steel Skeleton*, Vol. II, Cambridge Univ. Press, New York, 1956.
17. P. S. Symonds and B. G. Neal, Recent Progress in the Plastic Method of Structural Analysis, *J. Franklin Inst.*, **252**, 1951, pp. 383–407, 469–492.
18. B. G. Neal and P. S. Symonds, The Rapid Calculation of the Plastic Collapse Load of a Framed Structure, *Proc. Inst. Civil Engrs. (London)*, **1**, 1952, pp. 58–71.
19. J. Heyman, Plastic Design of Beams and Plane Frames for Minimum Material Consumption, *Quart. Appl. Math.*, **8**, 1951, pp. 373–381.
20. J. Foulkes, The Minimum Weight Design of Structural Frames, *Proc. Roy. Soc. (London)*, Series A223, 1954, pp. 482–494.
21. R. K. Livesley, The Automatic Design of Structural Frames, *Quart. J. Mech. Appl. Math.*, **9**, 1956, p. 257–278.
22. J. Heyman and W. Prager, Automatic Minimum Weight Design of Steel Frames, *J. Franklin Inst.*, **266**, 1958, pp. 339–364.
23. G. B. Dantzig, *Linear Programming and Extensions*, Princeton Univ. Press, Princeton, N.J., 1963.

General References

Baker, J. F., M. R. Horne, and J. Heyman, *The Steel Skeleton*, Vol. II, Cambridge Univ. Press, Cambridge, 1956.

Commentary on Plastic Design in Steel, Progress Reports No. 1 and No. 2 of the Joint WRC–ASCE Committee on Plasticity Related to Design, *Proc. ASCE*, **85**, No. EM3, 1959.

Drucker, D. C., Plastic Design Methods, Advantages and Limitations, *Brown Univ. Div. of Appl. Math. Tech. Rept. No. 24*, 1957.

Hodge, P. G., Jr., *Plastic Analysis of Structures*, McGraw-Hill, New York, 1959.

Massonnet, C. E., and M. A. Save, *Plastic Analysis and Design*, Vol. 1, Ginn (Blaisdell), Boston, 1965.

Neal, B. G., *The Plastic Methods of Structural Analysis*, Wiley, New York, 2nd ed., 1963.

Van den Broek, J. A., Theory of Limit Design, *Trans. ASCE*, **105**, 1940, pp. 638–661.

CHAPTER **14**

CREEP

14-1 BASIC CONCEPTS

When a material stretches under constant load the phenomenon is called *creep*. This was briefly mentioned in Section 2.4, where some typical creep curves were shown. A test which is carried out at constant load is therefore called a *creep test* and the measured strains are called *creep strains*. The slope of the creep curve (strain versus time) at any point is called the *creep rate*.

The shape of the uniaxial creep curve as shown in Figure 14.1.1 has led to a subdivision of the creep curve into three parts.

1. *Primary creep*, where the creep rate is decreasing rapidly.
2. *Secondary* (or steady state) *creep*, where the creep rate is essentially constant.
3. *Tertiary creep*, where the creep rate increases very rapidly.

Since the primary part of the creep curve is usually of short duration and the tertiary part leads quickly to rupture, the greatest interest usually lies in the secondary part, although the primary part cannot always be neglected.

The classical creep experiments were carried out by Andrade [1]. in 1910. Andrade concluded that at constant temperature and constant stress the creep strain can be represented by an equation of the form

$$\varepsilon = (1 + \beta t^{1/3})e^{kt} - 1 \qquad (14.1.1)$$

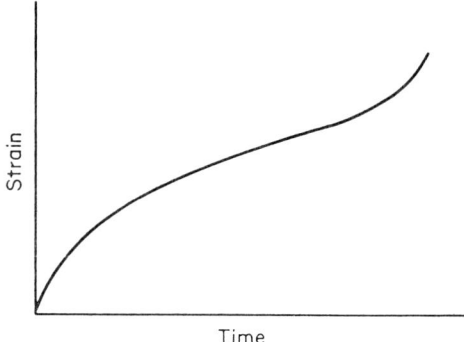

FIGURE 14.1.1 Typical uniaxial creep curve.

Equation (14.1.1) can also be approximated by writing

where
$$\varepsilon = \varepsilon_1 + \varepsilon_2$$
$$\varepsilon_1 = \beta t^{1/3} \qquad (14.1.2)$$
$$\varepsilon_2 = kt$$

ε_1 and ε_2 are now recognizable as the primary and secondary components of the total creep strain, for if we differentiate them with respect to time, we get

$$\dot{\varepsilon}_1 = \tfrac{1}{3}\beta t^{-2/3}$$
$$\dot{\varepsilon}_2 = k \qquad (14.1.3)$$

We see then that $\dot{\varepsilon}_1$ is large for small times but becomes vanishingly small for large times, whereas $\dot{\varepsilon}_2$ is constant with time.

Many experiments on different materials have shown, however, that the exponent $\tfrac{2}{3}$ in equation (14.1.3) is not adequate and can vary between 0.4 and 0.85. Therefore, the relationship generally used is of the form

$$\varepsilon_1 = \beta t^q \qquad (14.1.4)$$

where q and β depend on the material.

The constants β, k, and q will in general depend on the stress and the temperature. The assumption is usually made, however, that q is independent of the stress and that at constant temperature we can write

$$\varepsilon_1 = \beta(\sigma) t^q$$
$$\varepsilon_2 = k(\sigma) t \qquad (14.1.5)$$

Sec. 14-1] Basic Concepts

Equations (14.1.5) imply that the curves for different stress levels are geometrically similar, and this is approximately true.

In particular, the functions $\beta(\sigma)$ and $k(\sigma)$ are frequently taken to be power functions of the stress [2, 3]; i.e.,

$$\beta(\sigma) = B\sigma^m$$
$$k(\sigma) = K\sigma^n \qquad (14.1.6)$$

Such expressions often fit the data fairly well. For example, for a gas turbine alloy steel (Allegheny 418, 12% Cr, 3% W) the data could be fitted fairly well by assuming

$$\varepsilon_1 = 3.36 \times 10^{-30}\sigma^8 t^{1/2}$$
$$\varepsilon_2 = 4.41 \times 10^{-32}\sigma^{6.2} t \qquad (14.1.7)$$

In performing calculations, it is the strain rate at any time which is of importance. To determine the strain rate, we differentiate equations (14.1.5) to get

$$\dot{\varepsilon}_1 = q\beta t^{q-1}$$
$$\dot{\varepsilon}_2 = k(\sigma) \qquad (14.1.8)$$

Alternatively, since from (14.1.5),

$$t = \left(\frac{\varepsilon_1}{\beta}\right)^{1/q} \qquad (14.1.9)$$

we can substitute (14.1.9) into (14.1.8) to eliminate t, giving

$$\dot{\varepsilon}_1 = q\beta^{1/q}\varepsilon_1^{(q-1)/q} = qB^{1/q}\sigma^{m/q}\varepsilon_1^{(q-1)/q}$$
$$\dot{\varepsilon}_2 = k(\sigma) \qquad (14.1.10)$$

Thus the creep rate can be written as a function of stress, temperature, and time, or of stress, temperature, and strain.

Basically, however, expressions of this type pose a fundamental difficulty. For a relation of the form

$$\dot{\varepsilon} = f(\sigma, T, t, \varepsilon) \qquad (14.1.11)$$

implies that there exists a "mechanical equation of state." That is, the creep rate at any time depends on the state of the system at that time and is independent of how or by what path the system got to this state. There is conclusive evidence that this can generally not be true but may be true under certain restricted conditions. This is similar to the problem of using deformation theory in plasticity calculations. However, because there is at present no other theoretical method that is practical, equations such as (14.1.11) are generally used. However, the limitations should be kept in mind.

Creep data are almost always taken at constant stress. The question then arises: How are these data to be used for the case where the stress is varying with time? For example, suppose the stress in a body is constant at a value σ_1 up to a time t_1. The stress then suddenly is changed to σ_2. At what rate will the body begin creeping at the new stress σ_2? To answer this question several different cumulative creep laws have been proposed [11]. Three such laws are illustrated in Figure 14.1.2. These are the *strain hardening, time hardening,* and *life-fraction laws.* The strain-hardening law assumes that in going from one stress level to the next the creep rate depends on the existing strain in the material as indicated in Figure 14.1.2(b). The time-hardening rule assumes that the creep rate depends upon the time from the beginning of the creep

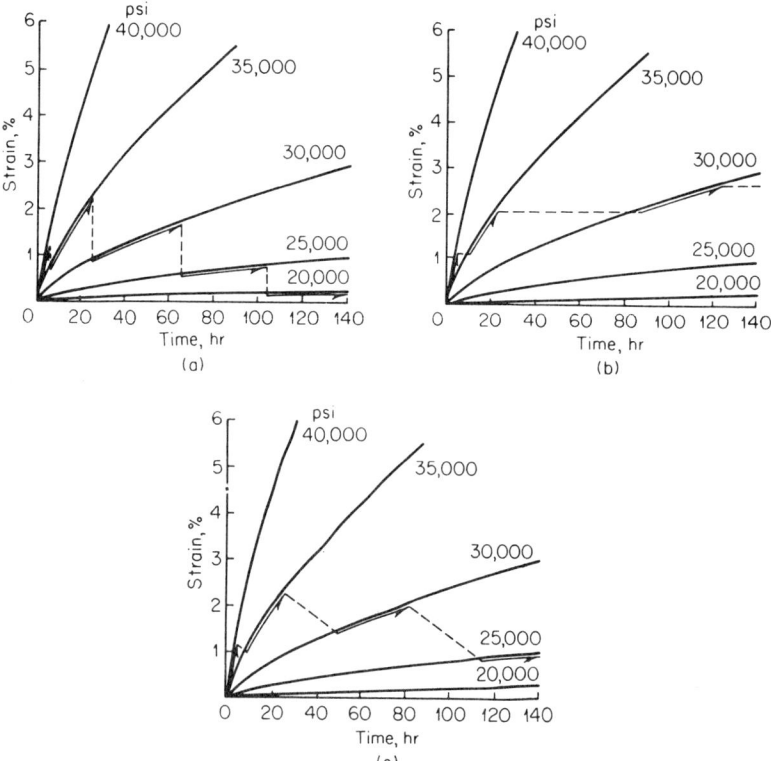

FIGURE 14.1.2 Proposed creep laws: (a) time-hardening rule; (b) strain-hardening rule; (c) life-fraction rule.

Sec. 14-2] Multidimensional Problems

process as shown in Figure 14.1.2(a). The life-fraction law assumes that the creep rate depends upon the fraction of life used up, as shown in Figure 14.1.2(c).

Experimental data seem to lean toward the strain-hardening rule. However, it has also been shown that if the stress is suddenly changed, a transient effect takes place such that the strain rate is higher than that obtained even from the strain-hardening rule. It has therefore been suggested by Rabotnov [4] that a modification be made in the strain-hardening rule, as shown in Figure 14.1.3. Instead of moving from O to A, as in the strain-hardening rule, we

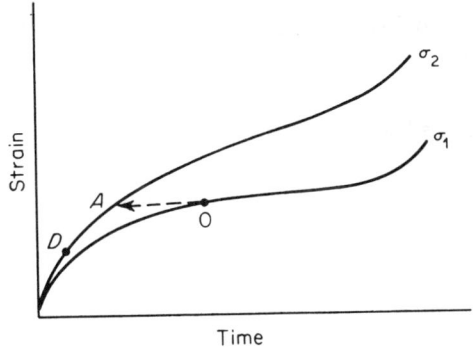

FIGURE 14.1.3 Transient effect on creep rate.

move to a point D which has a higher strain rate. The point D is taken as the point at which the strain is σ_1/σ_2 times the strain at A. The validity of this type of approach requires much more experimental verification.

If one uses the time-hardening rule, then the creep rates at any stress are given by equations (14.1.8). On the other hand, if the strain-hardening rule is used, the creep rates are given by equations (14.1.10). We note that for steady-state or secondary creep, all the rules become the same, since all the stress curves are then parallel straight lines. (This is not evident from Figure 14.1.2, since the primary parts of the curves have been greatly exaggerated).

14-2 MULTIDIMENSIONAL PROBLEMS

As mentioned before, all creep data are of the uniaxial type. How then do we use this data for two-dimensional and three-dimensional problems? The answer is, the same way as for any plasticity problem. Although the physical processes involved in creep are undoubtedly different than in ordinary plastic flow, we assume that the same relations hold for creep strains as for plastic strains. For example, the Prandtl–Reuss relations can be used for computing the creep increments. Thus we assume an equivalent stress defined the same way as in plasticity theory and an equivalent creep strain increment and write

$$\dot{\varepsilon}_x^c = \frac{\dot{\varepsilon}_c}{2\sigma_e}(2\sigma_x - \sigma_y - \sigma_z) \quad \text{etc.} \quad (14.2.1)$$

The computations are then performed as for any plasticity problem. Instead of using the stress–strain curve, however, the creep curves, or equations such as (14.1.8) or (14.1.10) representing these curves, are used. $\dot{\varepsilon}_c$ and σ_e in these equations are the equivalent strain rate and equivalent stress, respectively. The successive-approximation method is very useful in making the calculations.

In general, solutions are obtained incrementally. One starts with a given increment of time and solves the problem by successive approximations. The next increment of time is then taken and the process repeated. The solution can thus be extended to any time. For this purpose it is convenient to write equations (14.1.8) as follows [using equation (14.1.6)]:

$$\Delta\varepsilon_c = qB\sigma_e^m t^{q-1}\,\Delta t$$
or
$$\Delta\varepsilon_c = K\sigma_e^n\,\Delta t \quad (14.2.2)$$

or, for the case of the strain-hardening rule (14.1.10),

$$\Delta\varepsilon_c = qB^{1/q}\sigma_e^{m/q}(\varepsilon_c + \Delta\varepsilon_c)^{(q-1)/q}\,\Delta t \quad (14.2.3)$$

We note that in the case of steady-state creep, the incremental approach is not necessary, since the stress is constant and relations (14.1.8) and (14.1.10) can be integrated directly (if we neglect elastic strains).

With this brief exposition of the creep equations we shall proceed to several examples. The reader interested in more detailed discussions of creep laws and their physical and phenomenological background is referred to the general references at the end of the chapter.

Two examples will be considered. The first is the uniaxial problem of a thin infinite strip with a parabolic temperature distribution across the width. This is exactly the same as the plastic flow problem treated in Section 9.2. Here we will show how the stresses relax with time due to creep. The second problem discussed is that of a rotating disk similar to the problem of Section 9.5. The same method of solution, by successive approximations, will be used here for the creep problems as was used in Chapter 9 for the plastic flow problems. Both these problems are taken from reference [5].

14-3 UNIAXIAL CREEP IN INFINITE STRIP

Consider the simple uniaxial case of a thin infinite plate of width $2c$ as shown in Figure 14.3.1 with a temperature distribution $T(y)$ across the

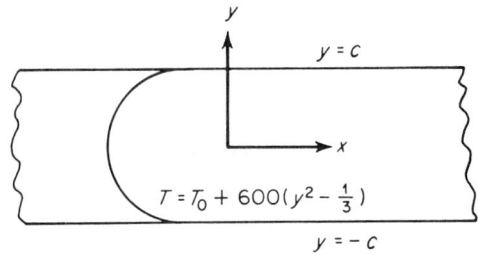

FIGURE 14.3.1 Flat plate with temperature distribution.

width. Under these conditions, the only nonzero stress is

$$\sigma_x = \sigma_x(y) \tag{14.3.1}$$

As in the usual theory of bending, it is assumed that plane sections remain plane. This requires that

$$\varepsilon_x = c_1 + c_2 y \tag{14.3.2}$$

where c_1 and c_2 are constants to be determined. Let ε_x and σ_x be the total strain and stress at the middle of the time interval Δt, ε_x^c the total creep strain up to time t, and $\Delta \varepsilon_x^c$ the additional increment of creep strain during the time interval Δt. Then

$$\varepsilon_x = \frac{\sigma_x}{E} + \alpha T + \varepsilon_x^c + \frac{\Delta \varepsilon_x^c}{2} \tag{14.3.3}$$

from which

$$\sigma_x = E\left(\varepsilon_x - \alpha T - \varepsilon_x^c - \frac{\Delta \varepsilon_x^c}{2}\right) \tag{14.3.4}$$

The boundary conditions require that

$$\int_{-c}^{c} \sigma_x \, dy = 0$$
$$\int_{-c}^{c} \sigma_x y \, dy = 0 \tag{14.3.5}$$

The constants c_1 and c_2 of equation (14.3.2) can be determined by substituting equations (14.3.2) and (14.3.4) into (14.3.5). Substituting these values of c_1 and c_2 back into equation (14.3.2) and assuming E to be constant results in

$$\varepsilon \equiv \varepsilon_x - \alpha T = \frac{1}{2c} \int_{-c}^{c} (\alpha T + \varepsilon_x^c + \tfrac{1}{2}\Delta\varepsilon_x^c) dy$$

$$+ \frac{3y}{2c^3} \int_{-c}^{c} (\alpha T + \varepsilon_x^c + \tfrac{1}{2}\Delta\varepsilon_x^c) y \, dy - \alpha T \quad (14.3.6)$$

As a specific example, let

$$\begin{aligned} T &= T_0 + 600(y^2 - \tfrac{1}{3}) \\ E &= 28 \times 10^6 \\ c &= 1 \\ \alpha &= 9.5 \times 10^{-6} \end{aligned} \quad (14.3.7)$$

Substituting these values into equation (14.3.6) and noting the symmetry of the problem results in

$$\varepsilon = -0.0057(y^2 - \tfrac{1}{3}) + \int_0^1 (\varepsilon_x^c + \tfrac{1}{2}\Delta\varepsilon_x^c) dy \quad (14.3.8)$$

This equation, along with the stress–strain relation (14.3.4) and some relation between stress and creep rate, are all that are needed to solve the plate problem. Let it be assumed that the relation between stress and creep rate is of the form

$$\Delta\varepsilon_x^c = 3 \times 10^{-26} |\sigma_x|^4 \, \Delta t \, \text{sgn} \, \sigma_x \quad (14.3.9)$$

Note that in equation (14.3.9) the sign of $\Delta\varepsilon_x^c$ must be taken the same as the sign of σ_x, as indicated. It will generally be necessary to write the creep law in this form unless the stress exponent is an odd integer. The procedure for obtaining the solution to this problem is now as follows:

1. At the start of the first time interval Δt, ε_x^c is known to be zero and $\Delta\varepsilon_x^c$ is assumed to be zero. Substituting these values into (14.3.8) gives the elastic solution as a first approximation to the total strains.

2. Substitute this first approximation for the strain distribution into the stress–strain relation (14.3.4) and solve for the first approximation to the stress distribution.

3. Substitute this first approximation to the stress distribution into the creep relation (14.3.9) and solve for the second approximation to the incremental creep strains during the first time interval.

Sec. 14–4] Creep in Rotating Disks

4. These incremental strains are substituted into equation (14.3.8) and the iteration proceeds from equation (14.3.8) to (14.3.4) to (14.3.9) back to (14.3.8), etc., until the procedure converges to the correct set of incremental strains and stresses.

5. At the start of the second time increment, the total strains ε_x^c are now known and are equal to the incremental creep strains developed during the previous time increment. In fact, the total creep strains at the beginning of any time interval will always be known and will be equal to the accumulated incremental strains up to that time interval. The procedure for calculating the average stresses and strains for the second or any other time interval is then the same as in steps 1, 2, 3, and 4.

The results of this calculation are presented in Figure 14.3.2. This figure

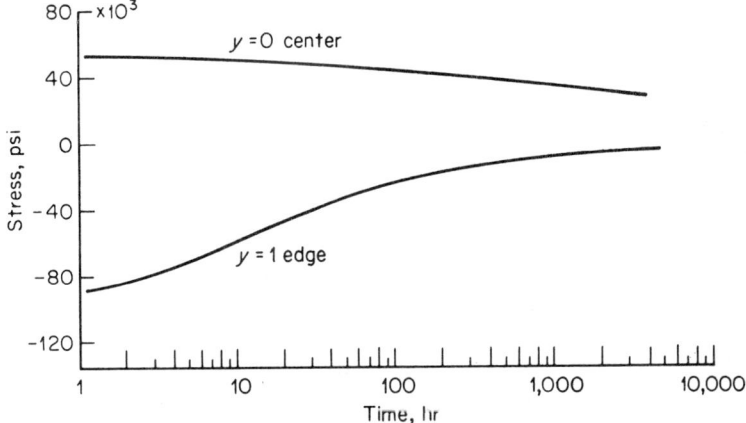

FIGURE 14.3.2 Stress relaxation in plate.

shows a plot of the variation of the stress with time at the center and at the edge of the plate. For this problem as well as the subsequent ones, the unknown integrals were evaluated numerically using Simpson's rule.

14–4 CREEP IN ROTATING DISKS

The simple example of the flat plate of Section 14.3 involves uniaxial stress, making it possible to determine directly the creep strains once the total stresses and strains are known. In most cases of practical interest the stresses are biaxial or triaxial in nature. The general procedure is then the same as discussed in Chapter 9. As an example we consider the case of creep in a rotating disk.

We assume that the von Mises yield criterion and the Prandtl–Reuss stress–strain relations are valid for both plastic flow and creep, as indicated by equation (14.2.1). For problems of thin disks with radial symmetry these relations are

$$\Delta \varepsilon_r^c = \frac{\Delta \varepsilon_c}{2\sigma_e}(2\sigma_r - \sigma_\theta)$$

$$\Delta \varepsilon_\theta^c = \frac{\Delta \varepsilon_c}{2\sigma_e}(2\sigma_\theta - \sigma_r) \qquad (14.4.1)$$

$$\Delta \varepsilon_z^c = -\Delta \varepsilon_r^c - \Delta \varepsilon_\theta^c$$

where as usual

$$\Delta \varepsilon_c = \frac{2}{\sqrt{3}}\sqrt{(\Delta \varepsilon_r^c)^2 + (\Delta \varepsilon_\theta^c)^2 + \Delta \varepsilon_r^c \Delta \varepsilon_\theta^c} \qquad (14.4.2)$$

$$\sigma_e = \sqrt{\sigma_r^2 + \sigma_\theta^2 - \sigma_r \sigma_\theta}$$

The increment in equivalent strain $\Delta \varepsilon_c$ will in general be a function of the equivalent stress σ_e, the total equivalent strain ε_c, the temperature T, the time t, and the strain history of the material. Such a general relationship can be written

$$\Delta \varepsilon_c = f(\sigma_e, \varepsilon_c, T, t) \qquad (14.4.3)$$

For problems of plastic flow without creep, the increment of equivalent strain $\Delta \varepsilon_c$ is a function of the equivalent stress σ_e and the strain history or loading path. For creep problems the strain path becomes time-dependent. For this time dependency of the strain path a number of laws have been proposed such as discussed in Section 14.1. For the examples to be presented we shall use both the time-hardening and strain-hardening rules, although any other law can be used as well.

We proceed essentially as in Section 9.5. The equilibrium equation for the disk is given by

$$\frac{d}{dr}(hr\sigma_r) - h\sigma_\theta + \rho\omega^2 hr^2 = 0 \qquad (14.4.4)$$

where h is the thickness, ρ the density, and ω the rotational speed. The compatibility equation in terms of stresses is derived in Section 9.5 and in the notation used here is

$$\frac{d}{dr}\left(\frac{\sigma_\theta}{E} - \frac{\mu\sigma_r}{E} + \alpha T + \varepsilon_\theta^c + \Delta\varepsilon_\theta^c\right)$$
$$= \frac{1+\mu}{E}\cdot\frac{\sigma_r - \sigma_\theta}{r} + \frac{\varepsilon_r^c - \varepsilon_\theta^c}{r} + \frac{\Delta\varepsilon_r^c - \Delta\varepsilon_\theta^c}{r} \qquad (14.4.5)$$

Sec. 14-4] Creep in Rotating Disks

Equations (14.4.4) and (14.4.5) must now be solved simultaneously for the stresses σ_r and σ_θ subject to the proper boundary conditions for the disk. At any time t the total accumulated creep strains ε_r^c and ε_θ^c up to that time t are known. The increments in creep strain $\Delta\varepsilon_r^c$ and $\Delta\varepsilon_\theta^c$ for the next time interval Δt are not yet known, but a set of values is assumed (such as the incremental strains computed for the previous time interval). Since all the creep strains now have known values, equations (14.4.4) and (14.4.5) form a linear pair of equations which can readily be solved for the stresses σ_r and σ_θ. From the assumed values of $\Delta\varepsilon_r^c$ and $\Delta\varepsilon_\theta^c$, $\Delta\varepsilon_c$ is calculated by the first of equations (14.4.2). This value of $\Delta\varepsilon_c$ corresponds to a particular value of σ_e as given by equation (14.4.3). From these values of $\Delta\varepsilon_c$ and σ_e as well as the stresses computed from equations (14.4.4) and (14.4.5), new values for the creep increments $\Delta\varepsilon_r^c$ and $\Delta\varepsilon_\theta^c$ are now computed from equations (14.4.1). These better approximations to the incremental creep strains are put into equations (14.4.4) and (14.4.5) and the process is repeated until convergence is obtained.

Equations (14.4.4) and (14.4.5) may be solved in a number of ways once the incremental strains are assumed to be known. Two methods have been described in Section 9.5, the finite-difference method and the integral-equation method. The finite-difference method is generally preferable here, since the disk thickness and other dimensions are changing with time. In connection with this, it should be noted that since a disk under creep conditions will grow and change dimensions with time, the values of r and h appearing in equations (14.4.4) and (14.4.5) should be the true values at the time t and not the original values at zero time. Thus if H is the original thickness at the original radial position R, the current values h and r are given approximately by [see equations (9.5.25) and (9.5.26)]

$$h = \frac{H}{(1 + \varepsilon_\theta^c)(1 + \varepsilon_r^c)} \tag{14.4.6}$$

$$r = R(1 + \varepsilon_\theta^c)$$

Furthermore, the strains appearing in the above equations should be natural strains. However, since the strain increments will generally be small and the current dimensions of the disk will always be used, the error in using conventional strains will be negligible.

Several cases of creep in rotating disks will now be considered. The first case treats the same problem as that presented by Wahl et al. [6]. Since reference [6] neglects the transient condition, the same will be done here, in order to compare the results of the present method with those of that reference. This is done by neglecting the elastic strains, since the compatibility relation (14.4.5) then becomes independent of time if the creep rate is given by a function of stress times a function of time. The same problem will then be treated including the elastic strains, to determine the effect of neglecting the transient stress distribution upon the creep strains. The second case will treat a similar disk using more complicated creep laws. The solution to the problem will thus be presented using both the time-hardening and strain-hardening rules.

Case 1(a). Creep in Rotating Disk Neglecting Transient Condition

Consider a constant-thickness constant-temperature disk of 12-in. outside diameter and 2.5-in. inside diameter rotating at 15,000 rpm. Assume steady-state creep with the creep rate as used in reference [6]:

$$\Delta \varepsilon_c = 4.41 \times 10^{-32} \sigma_e^{6.2} \Delta t$$

Neglecting the elastic strains is equivalent to letting E approach infinity in equation (14.4.5), and since the stress distribution now becomes independent of time, we can arbitrarily choose any time interval Δt and compute the stresses. The strains will, of course, be directly proportional to the assumed time interval Δt. One method of solution is therefore as follows. Solve equation (14.4.1) for the stresses. This gives

$$\sigma_r = \frac{2\sigma_e}{3\Delta\varepsilon_c}(2\Delta\varepsilon_r^c + \Delta\varepsilon_\theta^c)$$

$$\sigma_\theta = \frac{2\sigma_e}{3\Delta\varepsilon_c}(2\Delta\varepsilon_\theta^c + \Delta\varepsilon_r^c)$$
(14.4.7)

Letting E approach infinity, integration of equation (14.4.5) and substitution into (14.4.7) results in

$$\sigma_\theta = \frac{4}{3}\frac{\sigma_e}{\Delta\varepsilon_c}\int_a^r \frac{\Delta\varepsilon_r^c - \Delta\varepsilon_\theta^c}{r}\,dr + \frac{4}{3}\frac{\sigma_e}{\Delta\varepsilon_c}C + \frac{2\sigma_e}{3\Delta\varepsilon_c}\Delta\varepsilon_r^c$$
(14.4.8)

where a is the inner radius of the disk.

Integrating (14.4.4) and substituting (14.4.8) into the resultant equation gives

$$\sigma_r = \frac{4}{3r}\int_a^r \frac{\sigma_e}{\Delta\varepsilon_c}\int_a^{r_1} \frac{\Delta\varepsilon_r^c - \Delta\varepsilon_\theta^c}{r}\,dr\,dr_1 + \frac{4C}{3r}\int_a^r \frac{\sigma_e}{\Delta\varepsilon_c}\,dr_1$$

$$+ \frac{2}{3r}\int_a^r \frac{\sigma_e}{\Delta\varepsilon_c}\Delta\varepsilon_r^c\,dr_1 - \frac{\rho\omega^2}{3}\left(\frac{r^3 - a^3}{r}\right)$$
(14.4.9)

Sec. 14-4] Creep in Rotating Disks

where C is determined from the rim loading $\sigma_r(b)$ and is given by

$$C = \frac{b\sigma_r(b) - \frac{4}{3}\int_a^b \frac{\sigma_e}{\Delta\varepsilon_c}\int_a^{r_1}\frac{\Delta\varepsilon_r^c - \Delta\varepsilon_\theta^c}{r}\,dr\,dr_1 - \frac{2}{3}\int_a^b \frac{\sigma_e}{\Delta\varepsilon_c}\Delta\varepsilon_r^c\,dr + \frac{\rho\omega^2}{3}(b^3 - a^3)}{\frac{4}{3}\int_a^b \frac{\sigma_e}{\Delta\varepsilon_c}\,dr}$$

(14.4.10)

and b is the outer radius of the disk.

The solution now proceeds as follows:

1. Assume a value of Δt, say, 1 hour.
2. Assume values for $\Delta\varepsilon_r^c$ and $\Delta\varepsilon_\theta^c$, say, 0.001 everywhere.
3. Calculate

$$\Delta\varepsilon_c = \frac{2}{\sqrt{3}}\sqrt{(\Delta\varepsilon_r^c)^2 + (\Delta\varepsilon_\theta^c)^2 + \Delta\varepsilon_\theta^c\,\Delta\varepsilon_o^c}$$

4. Calculate

$$\sigma_e = \left(\frac{\Delta\varepsilon_c}{4.41 \times 10^{-32}\,\Delta t}\right)^{1/6.2}$$

5. Calculate C from (14.4.10).
6. Calculate σ_r and σ_θ from (14.4.8) and (14.4.9).
7. Calculate new values for $\Delta\varepsilon_r^c$ and $\Delta\varepsilon_\theta^c$ from (14.4.1).
8. Go back to step 3.

The results of carrying out this calculation are shown in Figure 14.4.1 and 14.4.2 and compared with the results of reference [6]. It is seen from the

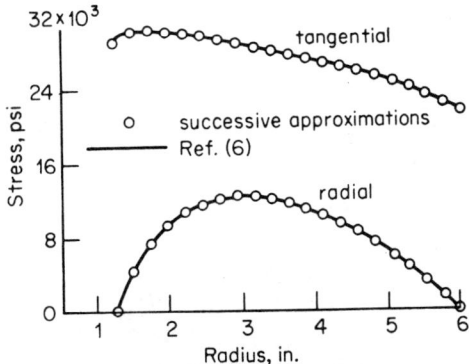

FIGURE 14.4.1 Creep stress in rotating disk neglecting elastic strains.

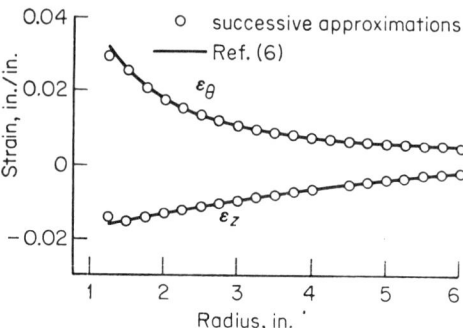

FIGURE 14.4.2 Creep strains in rotating disk neglecting elastic strains.

figures that the agreement is excellent. The strains shown in Figure 14.4.2 were obtained by multiplying the strains computed using a Δt of 1 hour by 180.

Case 1(b). Creep in Rotating Disk Including Transient Conditions

The same disk as in Case 1(a) was considered, with the elastic strains included this time. The modulus of elasticity E was taken to be 18×10^6, corresponding to a disk temperature of 1000°F. The creep rate was the same as previously given for Case 1(a). To start with, a time interval Δt of 0.01 hour was chosen, and as a first approximation the creep increments $\Delta \varepsilon_{rc}^c$ and $\Delta \varepsilon_{\theta c}^c$ were all taken to be constant at 0.00001, the total creep strains ε_r^c and ε_θ^c of course being zero at zero time. The calculation then proceeded as follows:

1. $\Delta \varepsilon_c$ was computed from the relation

$$\Delta \varepsilon_c = \frac{2}{\sqrt{3}} \sqrt{(\Delta \varepsilon_r^c)^2 + (\Delta \varepsilon_\theta^c)^2 + \Delta \varepsilon_r^c \, \Delta \varepsilon_\theta^c}$$

2. Equations (14.4.4) and (14.4.5) were solved for the stresses σ_r and σ_θ by the method described in Section 9.5.

3. σ_e was computed from the relation

$$\sigma_e = \left(\frac{\Delta \varepsilon_c}{4.41 \times 10^{-32} \, \Delta t} \right)^{1/6.2}$$

4. New approximations were obtained for the strain increments $\Delta \varepsilon_r^c$ and $\Delta \varepsilon_\theta^c$ from (14.4.1).

Sec. 14-4] Creep in Rotating Disks

5. Steps 1 through 4 were repeated until there was no change in two successive computations of the strain increments.

The creep strains were thus computed at the end of 0.01 hour. To obtain the creep increments for the next 0.01 hour the same procedure was followed, except that the total creep strains ε_r^c and ε_θ^c were no longer zero but were equal to the accumulated creep strains up to that time. A first approximation to the creep increments $\Delta\varepsilon_r^c$ and $\Delta\varepsilon_\theta^c$ was assumed (usually the values obtained for the previous time interval), and steps 1 through 5 repeated. In this way the incremental and total creep strains were computed up to 180 hours. The time interval was arbitrarily chosen so that the stress during any interval would not drop by more than approximately 1,000 psi. Thus this time interval Δt was rapidly increased as the stress approached steady-state conditions.

The results of this computation are shown in Figures 14.4.3 through 14.4.5. Figure 14.4.3 shows the relaxation of the tangential stress with time at the

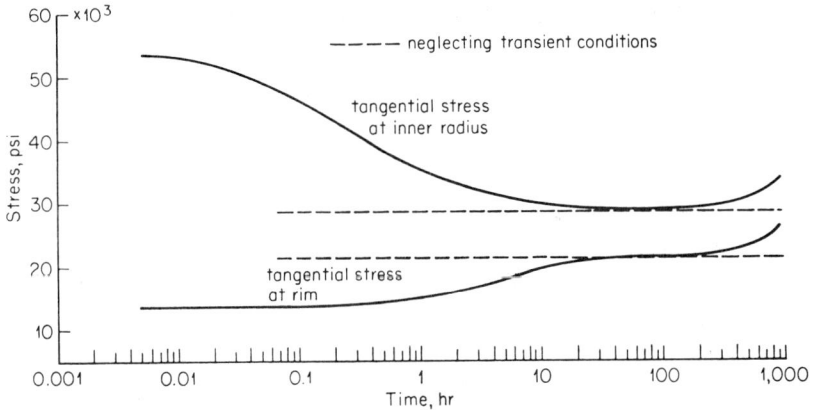

FIGURE 14.4.3 Stress relaxations in rotating disk, case 1(b).

inner and outer radii of the disk. It is seen that steady-state conditions are obtained at about 10 hours and that the tangential stress at the inner surface persists for a long time at a slightly higher value than that computed neglecting the transient stress distribution.

The tangential and axial creep strains at the end of 180 hours are shown plotted in Figure 14.4.4 together with the scatter band of test results given in reference [6]. From Figure 14.4.3 it can be seen that after a long period of stability the stresses start rising. As the disk grows due to creep, the centrifugal loading increases and the thickness at the center decreases, as can be seen

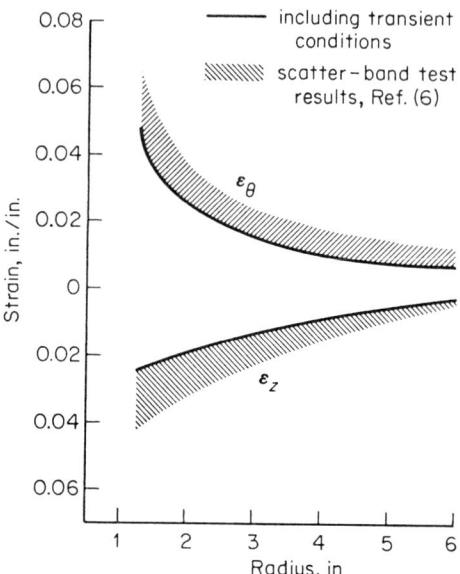

FIGURE 14.4.4 Strain distribution at 180 hours, case 1(b).

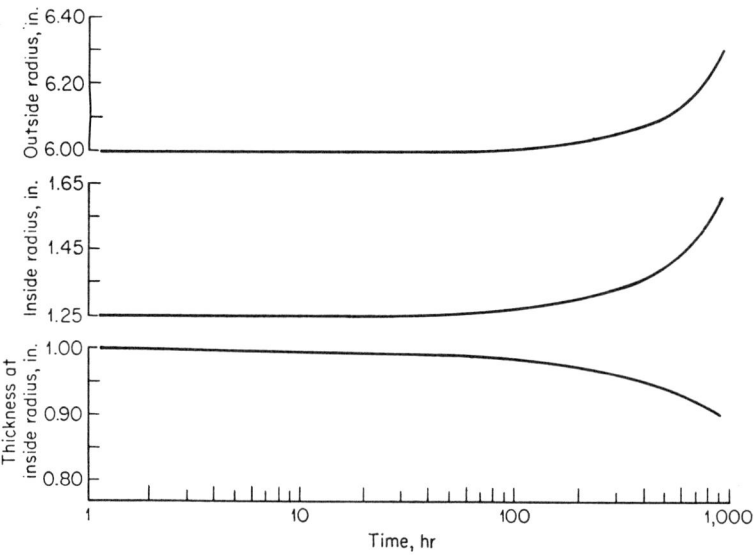

FIGURE 14.4.5 Variation of disk dimensions with time, case 1(b).

from Figure 14.4.5. This increase in centrifugal loading and decrease in cross section will eventually balance and finally surpass the stress relaxation, due to the creep. The stresses will start increasing all over the disk and the disk may start creeping at an accelerating rate. This is illustrated in Figure 14.4.3,

Sec. 14-4] Creep in Rotating Disks

which shows that after a long period of stability the stresses gradually start rising. The disk has now reached an unstable condition which may result in failure.

Case 2. Nonlinear Creep Laws

In the previous examples the creep curves were assumed linear with the creep rate independent of time. It will now be shown that more complicated creep curves and creep laws can be used without any appreciable increase in the complexity of the computations. Any nonlinear creep law can be used; for illustration it will be assumed that the creep data can be represented by

$$\varepsilon_c = B\sigma_e^m t^q \tag{14.4.11}$$

The same disk as in case 1 will now be considered, with the creep data given by equation (4.4.11), using both the time-hardening and the strain-hardening rules previously mentioned.

For the time-hardening rule, since the creep rate depends on the actual time elapsed, the creep rate at constant stress at the middle of the time interval between t and $t + \Delta t$ is obtained directly from (14.4.11):

$$\Delta\varepsilon_c = qB\sigma_e^m \left(t + \frac{\Delta t}{2}\right)^{q-1} \Delta t \tag{14.4.12}$$

where σ_e is the stress at the middle of the time interval and is assumed to be constant during the interval. Solving (14.4.12) for σ_e gives

$$\sigma_e = \left(\frac{\Delta\varepsilon_c}{qB\,\Delta t}\right)^{1/m} \left(t + \frac{\Delta t}{2}\right)^{(1-q)/m} \tag{14.4.13}$$

For the strain-hardening rule, the creep rate depends upon the total accumulated creep rather than the accumulated time. Eliminating the time between equations (14.4.11) and (14.4.12) results in

$$\sigma_e = B^{-1/m} \left(\frac{\Delta\varepsilon_c}{q\,\Delta t}\right)^{q/m} \left(\varepsilon_c + \frac{\Delta\varepsilon_c}{2}\right)^{(1-q)/m} \tag{14.4.14}$$

where ε_c is the total accumulated creep strain up to the beginning of the time interval under consideration and $\Delta\varepsilon_c$ is the strain increment during the time interval. The solution to this problem is now obtained in exactly the same way as for case 1(b), except that in step 3, σ_e is computed by equation (14.4.13) for the time-hardening rule and by equation (14.4.14) for the strain-hardening rule. For the time-hardening rule, track must be kept of the total elapsed time, whereas for the strain-hardening rule, track must be kept of the total accumulated strain.

Part of the results of this calculation for m equal to 6, q equal to $\frac{2}{3}$, and B equal to 1.5×10^{-30} are presented in Figures 14.4.6 and 14.4.7.

The creep curves represented by these constants are shown in Figure 14.4.8. As would be expected, Figure 14.4.6 shows that the stress does not relax quite as rapidly using the strain-hardening rule as with the time-hardening rule, since the creep rates will generally be lower for the strain-hardening rule. This can be used to explain the results of Figure 14.4.7, which

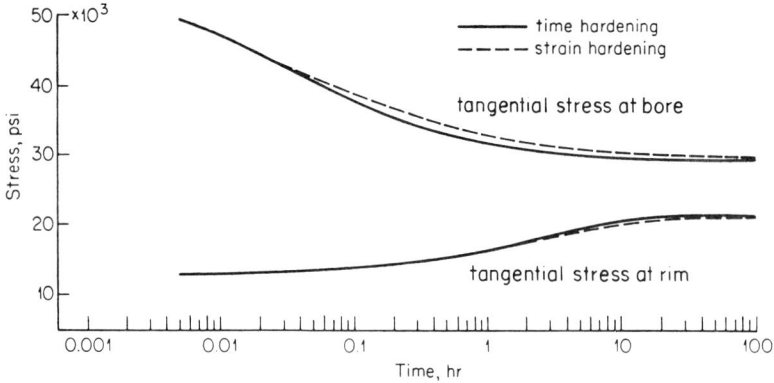

FIGURE 14.4.6 Stress relaxation for strain-hardening and time-hardening rules.

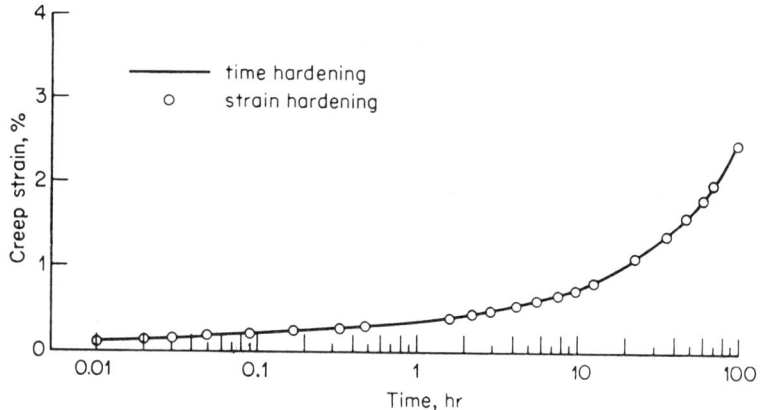

FIGURE 14.47 Creep at bore for strain-hardening and time-hardening rules.

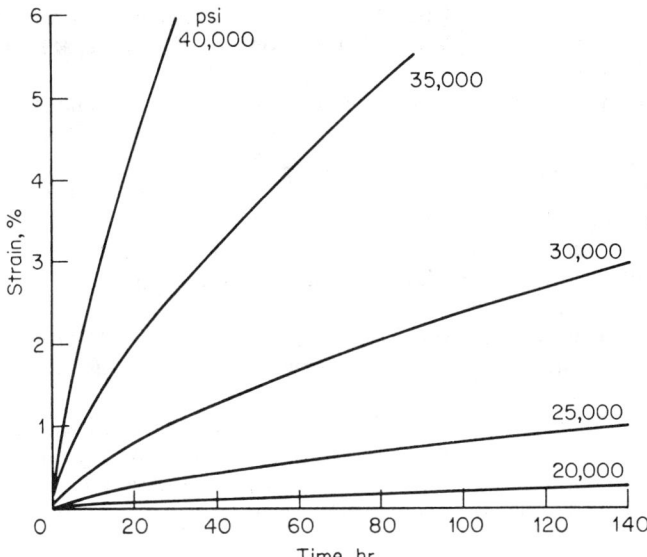

FIGURE 14.4.8 Creep curved used with strain-hardening and time-hardening rules.

shows that the tangential creep strain is essentially the same for both rules. (Although the strain-hardening rule actually gave slightly lower creep strains, the difference is too small to show up on the figure.) Although as seen from Figure 14.1.2, in going to a given stress the creep rate will be smaller for the strain-hardening rule than for the time-hardening rule, this is compensated for in this problem by the fact that the stress does not relax as fast, so that the total creep turns out to be about the same using either rule.

A comprehensive discussion of creep in rotating disks, including the effects of the transient period, can be found in a series of papers by Wahl [6, 7, 8, 9, 10].

References

1. E. N. da C. Andrade, On the Viscous Flow of Metal and Allied Phenomena, *Proc. Roy. Soc. (London)*, **A84**, 1910, p. 1.
2. F. H. Norton, *Creep of Steel at High Temperatures*, McGraw-Hill, New York, 1929, p. 67.
3. R. W. Bailey, The Utilization of Creep Test Data in Engineering Design, *Proc. Inst. Mech. Engrs.*, **131**, London, 1935, pp. 186–205, 260–265.
4. Yu. N. Rabotnov, On the Equation of State of Creep, *Proceedings of the International Conference on Creep*, Vol. 2, New York–London, 1963, p. 117.

5. A. Mendelson, M. H. Hirschberg, and S. S. Manson, A General Approach to the Practical Solution of Creep Problems, *Trans. ASME*, **81D**, 1959, pp. 585–598.
6. A. M. Wahl, G. A. Sankey, M. J. Manjoine, and E. Shoemaker, Creep Tests of Rotating Disks at Elevated Temperature and Comparison with Theory, *J. Appl. Mech.*, **21**, 1954, pp. 222–235.
7. A. M. Wahl, Analysis of Creep in Rotating Disks Based on the Tresca Criterion and Associated Flow Rule, *J. Appl. Mech.*, **23**, 1956, pp. 231–238
8. A. M. Wahl, Stress Distribution in Rotating Disks Subjected to Creep Including Effects of Variable Thickness and Temperature, *J. Appl. Mech.*, **24**, 1957, pp. 299–305.
9. A. M. Wahl, Further Studies of Stress Distribution in Rotating Disks and Cylinders Under Elevated Temperature Creep Conditions, *J. Appl. Mech.*, **25**, 1958, pp. 243–250.
10. A. M. Wahl, Effects of the Transient Period in Evaluating Rotating Disk Tests Under Creep Conditions, *J. Basic Eng.*, **85**, 1963, pp. 66–70.
11. H. R. Voorhees and J. W. Freeman, Notch Sensitivity of Heat-Resistant Alloys at Elevated Temperature, *Wright Air Development Center Tech. Rept. No. 54-175, Part I*, 1954.

General References

Arutyunyan, N. Kh., *Some Problems in the Theory of Creep*, translated by H. E. Nowottny, Pergamon Press, London, 1966.

Finnie, J., and W. R. Heller, *Creep of Engineering Materials*, McGraw-Hill, New York, 1959.

Hult, J., *Creep in Engineering Structures*, Ginn (Blaisdell), Boston, 1966.

Odquist, F. K. G., *Mathematical Theory of Creep and Creep Rupture*, Oxford Univ. Press, London, 1966.

INDEX

Affine transformation, 46
Alexander, J. M., 299
Allen, D. N. de G., 163
Andrade, E. N. daC., 327, 345
Anisotropy, 13, 14
Arutyunyan, N. Kh., 346
Associated flow rule, 115

Bailey, R. W. 345
Baker, J. F., 326
Baldwin, W. M., 22
Bauschinger effect, 13
  theories of, 13, 14
Beam(s)
  indeterminate, 308-10
  pure bending, 305-307
  with tip load, 308
Beltrami's energy theory, 74
Biharmonic operator, 217
Bland, D. R., 119, 133, 134, 163
Boundary-value problems, 276, 293
  numerical solution, 276-78, 294-96
Bridgman, P. W., 16, 22
Brown, W. F., Jr. 22
Budiansky, B., 120, 134

Cauchy problem, 296
Centered fan, 272

Characteristics, 285-98
  fixed, 295
  method of, 288
  region of influence, 288
  slip lines, 296-98
Clough, R. W., 233
Compatibility equations, 60, 165, 265, 287, 292
Compression test, 13
Coulomb, C. A., 71
Coulomb's theory, 73
Creep, 16, 327-45
  of infinite strip, 333-35
  laws, 330, 331
    life-fraction, 330-331
    strain-hardening, 330-331
    time-hardening, 330-331
  multidimensional problems, 331-32
  primary, 327-29
  rotating disks, 335-45
  secondary, 327-29
  tertiary, 327
Creep curves, 16, 328, 330, 345
Cylinder, hollow
  critical pressure, 158
  linear strain hardening, 160-61
  plastic flow, 156-61
  strain-hardening material, 158-59

Cylinder, long solid, 193-97

Dantzig, G. B., 326
Davis, E. A., 212
Deformation
 finite, 51
 pure, 48
Deformation theories, 119-21
Dieter, G. E., 22
Disk, rotating, 197-208
 creep of, 335-45
 finite-difference formulation, 204-208
 integral formulation, 198-203
Displacements, rigid-body, 44
Dissipation function, 314
Distortion energy, 68
Distortion energy theory, 75
Dixon, J. R., 224, 226, 233
Dorn, J. E., 23
Drucker, D. C., 23, 97, 111, 118, 133, 134, 299, 313, 325, 326
Ductility, 12

Effective plastic strain increment, 102
Effective strain, 116
Effective stress, 102, 116
Eichinger, A., 91, 97
Elastic limit, 6
Elastoplastic problems, 2
Elliptic equation, 292
Elongation, percent, 12
Energy
 distortion, 68
 elastic, 67, 68
Equation of state, mechanical, 329
Equilibrium equations, 29, 165
Equivalence of plastic work, 106
Equivalent modified total strain, 124
Equivalent plastic strain increment, 102, 106
Equivalent stress, 102, 106, 107, 108
Essenberg, F., 97

Felgar, R. P., 22
Finite deformation, 51
Finnie, J., 346
Flow rule, 109
 associated, 115
 for Tresca criterion, 108, 109
 for von Mises criterion, 100-104
Flow stress, 6
Flow theories, 119-21

Ford, H., 133, 299
Foulkes, J., 324, 326
Fox, L., 233, 299
Fracture, true strain at, 12
Free thermal expansion, 66

Geiringer, H., 268, 299
Geiringer's equations, 268
Generalized strains, 314
Generalized stresses, 314
Goodier, J. N., 22, 23, 43, 63, 258
Green's function, 185
Greenberg, H. J., 299, 325
Griffith, A. A., 223, 233
Gvozdev, A. A., 313, 325

Haigh-Westergaard stress space, **79-88**
Hardening
 isotropic, 13, 94
 kinematic, 95
Haythornthwaite, R. M., 326
Heller, W. R., 346
Hencky, H., 1, 3, 77, 134, 265, 299
Hencky's equations, 265
Hencky's theorem, 269
Heyman, J., 324, 326
Hill, R., 1, 3, 22, 23, 97, 133, 134, 163, 275, 299, 313, 325
Hirschberg, M. H., 212, 346
Hodge, P. G., Jr., 22, 23, 62, 163, 259, 271, 299, 325, 326
Hodograph, 281-84
 transformation, 295
Hoffman, O., 69, 97, 163
Hooke, R., 64
Hooke's law, 64, 65
Horne, M. R., 325, 326
House, R. N., 3
Hult, J., 346
Huth, J. H., 259
Hydrostatic pressure, effect of, 16
Hyperbolic equation, 292
Hysteresis loop, 6

Ilyushin, A. A., 164, 212
Ince, E. L., 212
Incompressibility condition, 16
Incremental theories, 119-21
Inglis, C. E., 223, 233
Instability point, 6, 11
Internal friction theory, 77
Irwin, G. R., 224
Isotropic hardening, 13, 94

Index

Johnson, W., 22, 23, 134, 163, 258, 259

Kantorovich, V., 258
Kazinczy, G., 301, 325
Kinematically admissible multiplier, 315
Kinematically admissible velocity field, 284, 315
Kinematic hardening, 95
Kinematic mechanism, 310
Kobayashi, S., 299
Koff, W., 97
Koiter, W. T., 134, 156, 163
Korn, G. A., 299
Korn, T. M., 299
Kronecker delta, 27
Krylov, V. I., 258

Lee, E. H., 163
Lévy, M., 1, 3, 100, 133
Lévy-Mises equations, 100, 261
Life-fraction law, 330-31
Limit analysis, 300-26
  of beams, 305-310
    with concentrated loads, 307-310
    in pure bending, 305-307
  design of structures, 300-301
  of frames, 310-12
  of simple truss, 301-305
  theorems, 312-17
    lower bound, 313-15
    upper bound, 313-15
Limit design, 301, 323, 324
  restricted design problem, 324
Livesley, R. K., 324, 326
Loading, definition of, 93
Loading function, 92, 112
Load point, maximum, 6, 8, 9, 11
Lode, W., 88, 89, 97, 109
Lode's
  strain parameter, 109
  stress parameter, 88
Lower bound theorem, 285, 313-15
Lower yield point, 6
Lubahn, J. D., 22
Ludwik, P., 7, 20, 22

Mclean, D., 22
Magnusson, A. W., 22
Manjoine, M. J., 15, 22, 346
Manson, S. S. 134, 212, 233, 346

Marin, J., 11, 22, 97
Martin, H. C., 233
Massonnet, C. E., 326
Maximum load point, 6, 8, 9, 11
Maximum shear theory, 73
Maximum strain energy theory, 74
Maximum strain theory, 72
Maximum stress theory, 71
Mechanism(s), 30
  elementary, 318
    beam, 319
    frame, 319
    joint, 319
    panel, 319
    portal, 320
  fictitious, 319
  linearly independent, 318
  rules for combining, 321
Melan, E., 119, 133
Mellor, P. B., 22, 23, 134, 163, 258, 259
Membrane analogy, 245
Membrane-roof analogy, 247
Mendelson, A., 134, 212, 233, 346
Metal forming processes, 2, 12, 15, 260
Mikhlin, S. G., 212
Minimum weight design, 324
Mises, R. von, 1, 3, 77, 100, 133
Mises, von, criterion, 75, 115, 129
  flow rule, 100-104
  stress-strain relation, 100-104
Mises, von, ellipse, 76
Mohr, O., 43
Mohr's circle(s), 37-39, 280
  for plastic strain increments, 101
  pole of, 280
  for stress, 101
Mohr's diagram, *see* Mohr's circles
Mohr's theory, 77
Multiplier
  kinematically admissible, 315
  statically admissible 314, 315
Muskhelishvili, N. I., 226, 233

Nadai, A., 258
Naghdi, P. M., 97, 134
Navier equations, 68
Neal, B. G., 318, 326
Necking, 6, 9, 12
Neuber, H., 224, 233
Neutral loading, 93
Norton, F. H., 345
Novozhilov, V. V., 63

Octahedral planes, 35
Odquist, F. K. G., 346
Offset yield strength, 6
Optimum design, 324
Orowan, E., 223, 233
Osgood, W. R., 20, 23

Parabolic equation, 292
Pearson, C. E., 43
Pearson, K., 133
Perfectly plastic material, 17, 115, 116
Pi plane, 81-87
Picard's method, 164
Plane elastoplastic problem, 213-30
Plane strain, 137, 213-18, 260
Plane stress, 137, 213-18
   of finite plate, 218-23
   of plate with crack, 223-30
Plastic collapse load, 301
Plastic deformation, 6
Plastic design, 301, 323, 324
Plastic hinge, 308
Plastic moment, 307
Plastic potential, 119
Plastic strain charts, 191-93
Plastic stress-strain relations, 98-134
   general derivation of, 110-119
   Lévy-Mises, 100, 261
   Prandtl-Reuss, 100-104
Plastic work, 104
   equivalence of, 106
Plate, infinite, see Strip, infinite
Plate, infinite, with hole, 208-11
Polar coordinates, 137
Power of dissipation, specific, 314
Prager, W., 19, 22, 62, 109, 133, 259, 271, 279, 280, 281, 284, 299, 324, 325, 326
Prandtl, L., 1, 3, 100, 133, 245, 258, 275, 299
Prandtl-Reuss equations, 100-104, 114, 115, 125, 167, 214
   experimental verification of, 109, 110
Principal directions
   of slip line field, 262
   of strain tensor, 53-55
   of stress tensor, 30-34
Principal planes, 32, 53
Proof strength, 6
Proportional limit, 5
Proportional loading, 99, 120
Punch indentation, 273-76

Pure deformation, 48
Pure shear, 41-42

Quinney, H., 90, 91, 97, 98, 109, 110

Rabatnov, Yu. N., 331, 345
Radial loading, 99, 120
Ramberg, W., 20, 23
Rankine theory, 71
Reduction in area, 10, 12
Reuss, E., 100, 133
Rigid body, 44
Rigid body displacements, 44
Rigid-perfectly plastic material, 17, 260
Roberts, E. Jr., 233
Ros, M., 91, 97
Rotating disk, see Disk, rotating
Rotation Tensor, 48

Sachs, G., 69, 97, 163
Safety factor, 303, 304, 314, 315
Saint-Venant, B. de, 1, 3, 100, 111, 133, 234, 258
Saint-Venant's principle, 240
Saint-Venant theory, 72
Sand-hill analogy, 247-48
Sankey, G. A., 346
Save, M. A., 326
Schwartzbart, H., 22
Shaffer, B. W., 3
Shape factor, 307
Shear lines, see Slip lines
Shell, thin circular, 183-93
Sherby, O. D., 23
Shoemaker, E., 346
Simple shear, 41-42
Singular yield conditions, 123
Slip lines, 263
   compatibility equations, 265
   geometry of, 268-72
   geometric construction, 279-84
      hodograph, 281
      physical plane, 279
      stress plane for, 279
      strong solution, 283
      weak solution in, 283
   velocity equations, 268
Slip line theory, 260-99
   complete solutions, 284
   lower bound theorem, 285
   principal directions, 262
   principal stresses, 262
   shear directions, 263

Index

upper bound theorem, 284
Sokolnikoff, I. S., 43, 62, 63, 69, 233, 258
Sopwith, D. G., 163
Spero, S. W., 212
Sphere
  compatibility equation for, 136
  critical pressure, 140
  equilibrium equation, 135
  internal pressure, 138-48
  with linear strainhardening, 153-56
  plastic flow in, 138-56
  Prandtl-Reuss relations, 136
  residual stresses, 145-47
  shakedown pressure, 147
  strain-hardening material, 150-56
  stress-strain relation, 136
  thermal loading, 148-50
Spherical coordinates, 135-37
Statically admissible multiplier, 314, 315
Statically admissible stress field, 285, 314
Statically determinate problem, 142, 261
Steele, M. C., 163
Stockton, F. D., 97
Strain, 44-63
  compatibility of, 59-61
  conventional, 4
  deviator tensor, 58
    modified, 124
  diametral, 9
  effective, 116
  engineering, 4
  equivalent modified total, 124
  generalized, 314
  invariants, 53, 55
  logarithmic, 8
  maximum shear, 55-57
  modified total, 124, 167
  natural, 8
  octahedral shear, 55, 57
    plastic increment, 101
  physical interpretation, 48-50
  principal, 53-55
  shear, 50
    maximum, 55-57
    octahedral, 55, 57
    octahedral plastic increment, 101
  true, 8, 10, 13
Strain Energy, *see* Energy
Strain hardening, 6
  hypothesis, 107
  linear, 17, 20
Strain-hardening exponent, 11
Strain-hardening hypothesis, 107
Strain-hardening law, 330, 331
Strain-hardening parameter, 153
Strain-invariance principle, 230-32
Strain rate, effect of, 15
Strain Tensor, 44-62
  Eulerian, 52
  Lagrangian, 62
  see also, Strain
Strength coefficient, 11
Stress
  boundary conditions, 32
  convention, 28, 29
  deviator, 39-41
    invariants of, 40, 41
  effective, 102, 116
  equivalent, 102, 106, 107, 108, 116
  generalized, 314
  invariants, 30, 33, 34
  maximum shear, 34, 35
  nominal, 4
  normal, 28
  octahedral shear, 35, 36, 101
  principal, 30-33
  principal directions, 32
  shear, 28
    maximum, 34, 35
    octahedral, 35, 36, 101
  spherical, 39
  true, 7, 8, 13
  unit, 28
Stress function, 216, 237
Stress-strain curve, 4-11
  conventional, 5-7
  dynamic models, 16, 17
  empirical equations, 20, 21
  idealizations, 16, 17
  kinematic models, 18-20
  strain-rate, effect of, 15
  true, 7-11
Stress-strain relations
  complete, 127-132
  elastic, 64-69
  general, 166
  incremental, 119, 120
  plastic, 98-134
    general, 114, 118
    general derivation of, 110-119
    Lévy-Mises, 100, 261
    Prandtl-Reuss, 100-104, 114, 115

Stress-strain relations (*continued*)
   for Tresca criterion, 108, 109, 115
   plastic strain-total strain, 123-127
   total, 119, 121
Stress tensor, 24-43, *see also*, Stress
Stress trajectories, 239
Strip, first order, 290
Strip, infinite thin, 172-82
   linear strainhardening, 178-82
   nonlinear strainhardening, 175-78
Strip condition, 290
Successive approximations, method of
   circular shell, 183-93
   convergence of, 169-71
   creep problems, 332-45
   finite plate, 218-23
   general description, 164-71
   infinite plate with hole, 208-11
   infinite thin strip, 172-82
   plate with crack, 223-30
   rotating disk, 197-208
   solid cylinder, 193-97
   sphere, 150-53
   torsion of bar, 249-55
      circular cross section, 253-55
      rectangular cross section, 250-53
   tube, 156-59
Successive elastic solutions, method of (*see*, Successive approximations)
Superposition of mechanisms, method of, 318-23
Swedlow, J. L., 229, 233
Symonds, P. S., 318, 326

Taylor, G. I., 90, 91, 97, 98, 109, 110
Temperature
   effect on metal properties, 15
   transition, 15
Tensile strength, 6
Tensile test, 4-7
Tensor, 24-27
   notation, 24-27
   relative displacement, 47
   rotation, 48
   skew-symmetric, 26
   substitution, 27
   summation convention, 25, 26
   symmetric, 26
Thomsen, E. G., 299
Time-hardening law, 330, 331
Timoshenko, S., 43, 63, 258

Todhunter, I., 133
Torsion of prismatic bar, 234-57
   elasticity solution, 240-45
      circular cross section, 242
      elliptic cross section, 240-42
      equilateral triangle, 242, 243
      membrane analogy, 245
      rectangular cross section 243-45
   general relations, 234-40
   plasticity solution, 246-58
      circular section, 253-57
      membrane-roof analogy, 247
      perfect plasticity, 246-48
      rectangular section, 250-53
      sand-hill analogy, 247-48
      strain hardening, 248-58
Torsional rigidity, 241
Total theories, 119-21
Transition temperature, 15
Tresca, H., 1, 3
Tresca criterion, 73
   flow rule, 108, 109
   kinematic model, 95, 96
   stress-strain relations, 108, 109, 131
Trozera, T. A., 23
Truss, simple, 301-305
Tuba, I. S., 163, 210, 212, 233
Tube (*see* Cylinder, hollow)
Tupp, L. J., 233
Tupper, S. J., 163
Turner, M. J., 233

Ultimate strength, 6
Uniform stress state, 272
Uniqueness condition, 111
Unloading, 6, 93
   definition of, 93
Upper bound theorem, 284, 313-15

Van den Broek, J. A., 326
Varga, R. S., 259

Wahl, A. M., 133, 337, 346
Wang, C. T., 212
Warping function, 236
Westergaard, H. M., 1, 3, 97
White, G. N., 163
Williams, M. L., 233
Work equation, 320
Work hardening, 6
   definition of, 111
   measures of, 104-107
   hypothesis, 106, 107

Index

Work-hardening hypothesis, 106, 107
Work increment, 26

Yang, C. T., 299
Yang, W. H., 233
Yield
  cylinder, 82, 86
  effect of hydrostatic pressure, 16
  function, 79, 92
  hinge, 308
  locus, 82, 83, 87
  moment, 307
  point, 5, 6
    lower, 6
    offset, 6
    upper, 6
  surface, 79, 92
    convexity, 120-23
    corners, 120, 123
    singular points, 120, 123
    subsequent, 92-96
Yield criterion, 70-79
  Beltrami's energy, 74
  Coulomb's, 73
  distortion energy, 75
  internal friction, 77
  maximum shear, 73
  maximum strain, 72
  maximum strain energy, 74
  maximum stress, 71
  von Mises, 75
  Mohr's, 77
  Rankine's, 71
  Saint-Venant's, 72
  Tresca's, 73